ZERO-TRUST LEARNING

Applications in Modern Network Security

ZERO-TRUST LEARNING

Applications in Modern Network Security

Edited by

Chetna Kaushal, PhD

Amandeep Kaur, PhD

Mohit Angurala, PhD

Aryan Chaudhary, PhD

AAP | APPLE ACADEMIC PRESS

First edition published 2026

Apple Academic Press Inc.
1265 Goldenrod Circle, NE,
Palm Bay, FL 32905 USA

760 Laurentian Drive, Unit 19,
Burlington, ON L7N 0A4, Canada

CRC Press
2385 NW Executive Center Drive,
Suite 320, Boca Raton FL 33431

4 Park Square, Milton Park,
Abingdon, Oxon, OX14 4RN, UK

Apple Academic Press exclusively co-publishes with CRC Press, an imprint of Taylor & Francis Group, LLC

Library and Archives Canada Cataloguing in Publication

Title: Zero-trust learning : applications in modern network security / Chetna Kaushal, PhD, Amandeep Kaur, PhD, Mohit Angurala, PhD, Aryan Chaudhary, PhD, editors.
Names: Kaushal, Chetna, editor.
Description: First edition. | Includes bibliographical references and index.
Identifiers: Canadiana (print) 20250180650 | Canadiana (ebook) 20250180677 | ISBN 9781779643568 (hardcover) | ISBN 9781779643575 (ebook)
Subjects: LCSH: Computer networks—Security measures. | LCSH: Computer security.
Classification: LCC TK5105.59 .Z47 2025 | DDC 005.8—dc23

Library of Congress Cataloging-in-Publication Data

CIP data on file with US Library of Congress

ISBN: 978-1-77964-356-8 (hbk)
ISBN: 978-1-77964-357-5 (ebk)

About the Editors

Chetna Kaushal, PhD
Assistant Professor, Chitkara University Institute of Engineering and Technology, Chitkara University, Punjab, India

Dr. Chetna Kaushal, PhD, is an Assistant Professor at Chitkara University, Punjab, India, with 12 years of experience in research, training, and academics. Her expertise spans machine learning, soft computing, pattern recognition, image processing, and artificial intelligence. She has published 60 research papers in reputed international and national journals, books, and conferences, and has filed and published over 50 patents. Recognized for her contributions, she received the Patent Award in Computer Science and Engineering in 2023 and the Outstanding Innovator in Education Award at the National Symposium on Innovations in Teaching and Learning Practices in Higher Education. Currently supervising two PhD scholars and serving as a reviewer for prestigious journals, Dr. Kaushal remains dedicated to advancing human health and well-being through pioneering research. She has completed her PhD in Computer Science and Engineering from Chitkara University, Punjab, an MTech from DAV University, Punjab, and a BTech in Information Technology from Punjab Technical University.

Amandeep Kaur, PhD
Professor, Chitkara University Institute of Engineering and Technology, Chitkara University, Punjab, India

Amandeep Kaur, PhD, is currently working as a Professor at Chitkara University Institute of Engineering and Technology, Chitkara University, Punjab, India. With more than 24 years of experience, she has about 100 publications in renowned international journals and fully refereed international conferences. She has filed and published more than 100 patents. Dr. Kaur has supervised more than 35 PhD and postgraduate students. Her main areas of research are medical informatics, machine learning, IoT, artificial intelligence, and cloud computing. Dr. Kaur has won excellence awards in the category of "Filing Patent" in the years 2021, 2022, and 2023, and a best

PhD supervision award in the year 2023. She is among the most prominent names among Indian researchers and is included in Stanford University's top 2% most influential scientists list. Dr. Kaur earned her doctorate degree from I. K. Gujral Punjab Technical University, Jalandhar, India. Both her MTech (CSE) and BTech (CSE) degrees were earned with distinction. She qualified for UGC-NET in Computer Science in her first attempt.

Mohit Angurala, PhD
Assistant Professor, Guru Nanak Dev University College, Pathankot, Punjab, India

Mohit Angurala, PhD, is currently employed as an Assistant Professor at Guru Nanak Dev College, Pathankot, Punjab, India. Dr. Angurala has previously served as an Associate Professor at Chandigarh University, as a Senior Assistant Professor at VIT Chennai, and as an Assistant Professor and Head at Khalsa College of Engineering and Technology in Amritsar, Punjab, India. He is currently pursuing his postdoctoral research in the field of security in wireless sensor networks at the Singapore Institute of Technology, Singapore. He completed his PhD in Computer Science and Engineering at I. K. Gujral Punjab Technical University, Kapurthala, Punjab, in 2021. Dr. Angurala earned his Bachelor of Technology in Information Technology from Beant College of Engineering and Technology in Gurdaspur, Punjab, in 2011, and a Master of Technology in Computer Science and Engineering from Punjab Institute of Technology, PTU Main Campus, Kapurthala, Punjab, in 2014. Dr. Angurala has also obtained an MBA degree from Guru Nanak Dev University, Amritsar, Punjab, India.

Aryan Chaudhary, PhD
Chief Scientific Advisor, Bio Tech Sphere Research, India

Aryan Chaudhary, PhD, is the Chief Scientific Advisor at BioTech Sphere Research, India. Aryan continues to make groundbreaking contributions to the industry. Having served as the Research Head at Nijji HealthCare Pvt Ltd, he has demonstrated his expertise in leveraging revolutionary technologies such as artificial intelligence, deep learning, IoT, cognitive technology, and blockchain to revolutionize the healthcare landscape. His relentless pursuit of excellence and innovation has earned him recognition as a thought leader in the industry. His dedication to advancing healthcare is evident through his vast body of work. He has authored several influential academic papers on public health and digital health, published in prestigious international journals. His research primarily focuses on integrating IoT and sensor technology for efficient data collection through one-time and ambulatory monitoring. As a testament to his expertise and leadership, he is not only a keynote speaker at numerous international and national conferences but also serves as the series editor of a CRC book series and is the editor of several books on biomedical science. His commitment to the advancement of scientific knowledge extends further, as he acts as a guest editor for special issues in renowned journals. Recognized for his significant contributions, he has received prestigious accolades, including the "Most Inspiring Young Leader in Healthtech Space 2022" by Business Connect and the title of best project leader at Global Education and Corporate Leadership. Moreover, he holds senior memberships in various international science associations, reflecting his influence and impact in the field. Adding to his accomplishments, Aryan Chaudhary is currently serving as a guest editor for a special issue in the highly regarded journal, *EAI Endorsed Transactions on AI and Robotics,* and he has joined the editorial board of *Biomedical Science and Clinical Research* (BSCR). Additionally, he is a respected professional member of the Association for Computing Machinery (ACM).

Contents

Contributors

Madhav Aggarval
Chitkara University, Punjab, India

Mohit Angurala
Department of Computer Science, Guru Nanak Dev University College,
Pathankot (Constituent College of Guru Nanak Dev University, Amritsar), Punjab, India

C. V. Suresh Babu
Professor, Department of Information Technology, Hindustan Institute of Technology and Science,
Chennai, Tamil Nadu, India

Samruddhi Bhabad
Department of Chemical Engineering, Bapatla Engineering College, Bapatla, Andhra Pradesh, India

Harpreet Kaur Channi
Department of Electrical Engineering, Guru Nanak Dev Engineering College, Ludhiana, Punjab, India

B. Rupa Devi
Associate Professor, Department of CSE, Annamacharya Institute of Technology and Sciences, Tirupati,
Andhra Pradesh, India

Poonam Dhiman
Government PG College, Ambala Cantt, Haryana, India

Pragati Duggal
Chitkara University, Punjab, India

Maitri Dulla
MMCOE, Pune, Maharashtra, India

Kinjal Gandhi
CSE, ITM(SLS) Baroda University, Gujarat, India

Siddhant Gautam
Graphic Era University, Dehradun, Uttarakhand, India

Deepika Ghai
School of Electronics and Electrical Engineering, Lovely Professional University, Punjab, India

Renato R. Maaliw III
College of Engineering Southern Luzon State University Lucban, Quezon, Philippines

Neha Jain
Department of Computer Engineering, MMCOE, Pune, Maharashtra, India

D. Jesline
Assistant Professor, Department of Computing Technologies, Faculty of Engineering and Technology,
SRM Institute of Science and Technology, Kattankulathur, Chennai, Tamil Nadu, India

S. Arul Jothi
Research Analyst, Kanakapushparagam Tech, Erode, Tamil Nadu, India

B. Kanisha
Associate Professor, Department of Computing Technologies, Faculty of Engineering and Technology, SRM Institute of Science and Technology, Kattankulathur, Chennai, Tamil Nadu, India

Christian Kaunert
Professor, International Security, Dublin City University, Ireland

Amandeep Kaur
Chitkara University Institute of Engineering and Technology, Chitkara University, Punjab, India

Gunseerat Kaur
School of Computer Science and Engineering, Lovely Professional University, Punjab, India

Mandeep Kaur
Department of Computer Science and Engineering, Chitkara University Institute of Engineering and Technology, Chitkara University, Punjab, India

Tejinder Kaur
Chitkara University, Punjab, India

Chetna Kaushal
Chitkara University Institute of Engineering and Technology, Chitkara University, Punjab, India

Ismail Keshta
Computer Science and Information Systems Department, College of Applied Sciences, AlMaarefa University, Riyadh, Saudi Arabia

Raghavendra Kulkarni
Assistant Professor, Department of Computer Science, School of Science, Gandhi Institute of Technology & Management (GITAM) – Deemed to be University, Rudraram, Patancheru Mandal, Hyderabad, Telangana, India

Pramod Kumar
Assistant Professor, Computer Science and Technology, Ganga Institute of Technology and Management, Kablana, Jhajjar, Haryana, India

T. Ravi Kumar
Associate Professor, Department of CSE, Aditya Institute of Technology and Management, Tekkali, Andhra Pradesh, India

Monica Luthra
Department of Computer Science, Chandigarh University, Chandigarh, Punjab, India

Madhuri
School of Computer Science and Engineering, Lovely Professional University, Punjab, India

Mahalakshmi
Assistant Professor, Department of Computer Science,
College of Computer Science and Information Technology, Jazan University, Jazan, Saudi Arabia

Shubhangi Mishra
Apex Institute of Technology (CSE), Chandigarh University, Mohali, Punjab, India

S. Muralimohan
Professor, Department of ECE, Mother Theresa Institute of Engineering and Technology, Palamaner, Andhra Pradesh, India

Nitish Ojha
Amity University, Uttar Pradesh, India

J. Vishnu Priyan
Scholar, Department of Information Technology, Hindustan Institute of Technology and Science, Chennai, Tamil Nadu, India

Priyanka
Department of Computer Science, Chandigarh University, Chandigarh, Punjab, India

M. Ramkumar Raja
Associate Professor, Department of Electrical Engineering, College of Engineering, King Khalid University, Abha, Saudi Arabia

Betty Elezebeth Samuel
Lecturer, Department of Computer Science, College of Computer Science and Information Technology, Jazan University, Jazan, Kingdom of Saudi Arabia

Ramandeep Sandhu
School of Computer Science and Engineering, Lovely Professional University, Punjab, India

Arijeet Chandra Sen
Joint Director, Government of India, MTech Cyber Security, BITS Pilani, Delhi, India

Lakshay Sharma
Chitkara University, Punjab, India

Neha Sharma
Apex Institute of Technology (CSE), Chandigarh University, Mohali, Punjab, India

Yashika Sharma
Apex Institute of Technology (CSE), Chandigarh University, Mohali, Punjab, India

V. Sindhu
Assistant Professor, Department of Computer Science, CHRIST University, Bangalore, Karnataka, India

Bhupinder Singh
Professor, Sharda School of Law, Sharda University, Greater Noida, Uttar Pradesh, India

Vikram Singh
Professor, Computer Science and Engineering, Chaudhary Devi Lal University, Sirsa, Haryana, India

U. Sivaji
Associate Professor, Institute of Aeronautical Engineering, Department of IT, Dundigal, Hyderabad, Telangana, India

Mukesh Soni
Department of CSE, University Centre for Research & Development Chandigarh University, Mohali, Punjab, India

Manoj Ram Tammina
Sr. Software Developer, Innovation, Bread Financial, Columbus, Ohio, USA

R. Tharun
Assistant Professor, Department of CE, Mother Theresa Institute of Engineering and Technology, Palamaner, Andhra Pradesh, India

Sanjay Tyagi
Assistant Professor, Department of CSA, Kurukshetra University, Kurukshetra, Haryana, India

Niraj Upadhayaya
Professor (CSE), SRM University, Amaravati, Andhra Pradesh, India

Abhishek Vaish
Indian Institute of Information Technology, Allahabad, Uttar Pradesh, India

M. Vinay
Associate Professor, CHRIST University, Bangalore, Karnataka, India

Krish Wason
Chitkara University, Punjab, India

Preface

In an age defined by the convergence of technology and security, we stand at the threshold of a new era—a fusion of zero-trust principles with the transformative potential of artificial intelligence (AI). As we embark on this journey to explore the role of zero trust in AI, we enter uncharted territory that promises to reshape our approaches to network security, education, and technological advancement.

Our exploration begins with a foundational overview of zero-trust principles, setting the stage for a fundamental shift in security thinking. In the following sections, we examine the significance of zero-trust in transforming network security, highlighting evolving paradigms that adapt to safeguard our digital landscapes.

Zero-trust security offers transformative advantages, revolutionizing our approach to threat protection, vulnerability reduction, and challenge response. With a focus on practicality, we undertake comprehensive studies of various zero-trust frameworks, shedding light on options available for organizations seeking to strengthen their defenses.

Looking to the future, we explore the AI- zero-trust alliance—an innovative partnership that redefines cybersecurity by leveraging the power of AI. Our journey takes an exciting turn as we examine the role of AI within zero-trust learning environments, where AI is reshaping education to be both dynamic and secure. In the following sections, we discuss the application of AI and zero-trust principles to enhance educational security, empowering educators and safeguarding learners in our rapidly evolving digital landscape.

Further along, we delve into malicious software classification, introducing RGB visualization techniques that, combined with zero-trust security, bolster defenses against cyber threats. In subsequent sections, we explore the synergy between zero-trust principles and anonymous identity authentication in strengthening network security. This innovative approach ensures that the right individuals access appropriate resources while preserving user privacy.

Our path leads us to an evolutionary approach for cybersecurity situational assessment within zero-trust networks, where we explore dynamic, real-time threat monitoring and adaptive security measures. In the following sections, we examine the powerful combination of AI and zero-trust in enhancing security. This partnership enables organizations to proactively and effectively address security challenges.

Chapter 12 delves into machine learning-based zero-trust frameworks, where data-driven insights and predictive capabilities transform network security. In the next section, we tackle the critical issue of lateral movement within networks, discussing strategies and technologies designed to mitigate this risk and reinforce zero-trust defenses.

We emphasize the importance of understanding zero-trust principles as a core foundation for organizations seeking to establish strong security postures. Our journey challenges the traditional concept of perimeter security, exploring why it is insufficient in addressing evolving threats. We highlight key observations and open challenges that call for innovative solutions.

Chapter 16 explores the essential topic of key selection within zero-trust network security, enhancing fault tolerance and resilience, especially against side-channel attacks. In the following section, we envision the future evolution of zero-trust principles and their integration with cloud computing security. We bring emerging threats, regulatory concerns, and innovative solutions incorporating blockchain consensus algorithms into focus.

In the subsequent sections, we extend the application of zero-trust principles to the healthcare sector, enhancing security in an environment where data privacy and patient safety are paramount. In the final sections, we introduce a novel solution that combines zero-trust principles with blockchain and homomorphic encryption, ensuring both privacy and security in data transactions.

Our closing chapters explore the future of AI-powered security within zero-trust learning environments, envisioning enhanced safety and adaptability that signal a promising horizon for education and network security.

As we navigate these sections, we invite readers to immerse themselves in the exploration of zero trust's transformative role in AI. This book aims to envision a future where security transcends traditional boundaries, evolving into a dynamic, adaptable, and resilient ecosystem. It is an invitation to engage, question, and innovate—paving the way toward a more secure and technologically empowered world.

—Editors

CHAPTER 1

Zero-Trust Fundamentals

SHUBHANGI MISHRA,[1] MOHIT ANGURALA,[2,*] NEHA SHARMA,[1] and YASHIKA SHARMA[1]

[1]*Apex Institute of Technology (CSE), Chandigarh University, Mohali, Punjab, India*

[2]*Department of Computer Science, Guru Nanak Dev University College, Pathankot (Constituent College of Guru Nanak Dev University, Amritsar), Punjab, India*

*Corresponding author

ABSTRACT

This chapter discusses an in-depth analysis of the foundations of the Zero Trust (ZT) security architecture, providing insights into its main tenets and elements. Traditional security methods have proven insufficient in protecting sensitive data and critical assets as organizations increasingly face complex cyber threats and evolving attack vectors. The ZT security model has emerged as a paradigm shift in cybersecurity in response to these challenges. The guiding principle of ZT is "never trust, always verify." Unlike traditional security models that rely on perimeter defenses and assume trust within the network, ZT promotes continuous verification and authentication of users, devices, and systems, regardless of their location. This chapter provides a comprehensive discussion of the core principles of ZT, including strict access rules, least privilege access, micro-segmentation, ongoing monitoring, and data encryption security.

Zero-Trust Learning: Applications in Modern Network Security. Chetna Kaushal, Amandeep Kaur, Mohit Angurala, Aryan Chaudhary (Eds.)

1.1 INTRODUCTION

1.1.1 DEFINITION OF ZERO TRUST

The cybersecurity paradigm known as Zero Trust (ZT) is based on the phrase "never trust, always verify." Zero Trust operates under the presumption that threats might come from both internal and external sources, in contrast to standard security approaches that depend on implicit trust within the network perimeter [1]. Due to this paradigm shift, a security strategy that places a strong emphasis on stringent access controls and ongoing monitoring to protect important assets and data has emerged.

1.1.2 KEY PRINCIPLES OF ZERO TRUST

The core values of ZT include a number of important concepts. First, the idea of implicit trust is completely abandoned. Based on their position within the network, users, devices, and systems are not given the benefit of the doubt. Second, reliable authentication and verification [2] processes are essential. Any organization requesting resources must go through strict verification procedures to prove its authenticity. This applies to users, gadgets, and any other connecting entity.

1. **Enforcing Security Through Segmentation:** The ZT model's foundation is micro-segmentation. Networks are split up into smaller, isolated portions to prevent potential threats from moving laterally. Only the entities that require particular resources can access each segment, reducing exposure and risk. This strategy ensures that a breach in one place does not instantly provide unauthorized access to the whole network.
2. **Embracing Least Privilege:** ZT relies heavily on the least privilege concept. The bare minimum of access is provided to users and devices so they can complete their assigned jobs. Due to the restricted scope of unauthorized access, this strategy minimizes the possible effects of a security breach. Over-assigning rights, which can result in unforeseen vulnerabilities, is avoided by the approach.
3. **Continual Monitoring and Adaptation:** Continuous monitoring is a key component of ZT's threat detection system. Anomalies and potential breaches are quick to spot when network traffic [3], user behavior, and device activity are all monitored in real-time.

Organizations are able to quickly respond to new threats and reduce risks before they become serious thanks to this proactive approach.

4. **Stricter Access Control and Multi-Factor Authentication:** The ZT paradigm employs strict and context-sensitive access restrictions. Entities requesting access are required to abide by dynamic policies that take into account user identity, device security status, and the present situation. Multi-factor authentication (MFA), which mandates that users produce multiple forms of authentication before gaining access, is also in place. The likelihood of unauthorized access is greatly decreased by this extra protection layer.

5. **Emphasis on Encryption and Auditability:** Data protection is highly valued by ZT. To protect the confidentiality and integrity of all data, encryption is used both during storage and transmission. The approach also supports meticulous activity audits and logging. This audit trail aids in monitoring user behavior and offers information about possible irregularities or security violations.

The ZT model is an innovative strategy for cybersecurity [4] that questions conventional wisdom and takes into account the changing threat landscape. Organizations seek to strengthen their security defenses, reduce risks, and protect sensitive assets and data from potential threats coming from both internal and external sources by implementing a zero-trust approach.

1.2 HISTORICAL CONTEXT AND EVOLUTION OF SECURITY MODELS

1.2.1 PHYSICAL SECURITY AND EARLY COMPUTING (PRE-1950S)

The main issue in the early years of computing, which predate the 1950s, was physical security. Large, secure rooms were used to store computers, and access to them was strictly regulated. The main goal of security measures was to stop unauthorized individuals [5] from physically accessing these early computing systems.

1.2.2 MAINFRAMES AND ACCESS CONTROLS (1950S–1970S)

Access controls became a key area of security concern in the 1950s with the introduction of mainframe computers. The user authentication techniques used in security models during this time were often straightforward passwords. Only those with proper authorization were permitted admittance into the computer area, which had restricted access. The relatively basic security measures were intended to counter insider attacks.

1.2.3 DISCRETIONARY ACCESS CONTROL (1970S–1980S)

The term discretionary access control (DAC) sprang to prominence in the 1970s and 1980s. Users had more control with this paradigm since they could regulate access permissions for their own resources. However, because users may give others excessive powers, this also presents security issues.

1.2.4 MANDATORY ACCESS CONTROL (1980S–1990S)

The security community created mandatory access control (MAC) approaches to solve the DAC's limitations. In order to prevent users from changing permissions, MAC built a centralized policy structure where administrators specified access rules. MAC systems were difficult to manage even if they worked well in secure settings.

1.2.5 ROLE-BASED ACCESS CONTROL (1990S–2000S)

Role-based access control (RBAC) began to gain popularity in the 1990s. By classifying people into roles and granting these roles access rights, RBAC made access management simpler. By lowering the possibility of overprivileged accounts, this strategy decreased administrative burden and improved security.

1.2.6 NETWORK SECURITY AND FIREWALLS (1990S–2000S)

As networking technology developed, the 1990s and 2000s saw a change in emphasis towards network security. For defending networks against external attacks, firewalls, and intrusion detection systems (IDS) became essential technologies. It was at this point that the focus shifted from protecting individual computers to protecting entire network infrastructures.

1.2.7 WEB AND APPLICATION SECURITY (2000S–2010S)

Security risks have grown as a result of the spread of the internet and web applications, and now encompass web-based dangers like cross-site scripting (XSS) and SQL injection. Web application firewalls came into existence during this time, and secure coding practices were encouraged.

1.2.8 CLOUD AND MOBILE SECURITY (2010S–2020S)

The 2010s saw the emergence of new mobile and cloud security issues. Concepts like Identity and Access Management (IAM) were born as a result of security models evolving to manage users' digital identities safely across many platforms.

1.2.9 ZERO TRUST (2010S–PRESENT)

The ZT paradigm has become popular in the modern period in response to the changing threat landscape. The idea of implicit trust in networks has traditionally been contested by ZT. It places a focus on constantly verifying all users and devices, whether they are inside or outside the network boundary. This concept emphasizes stringent access controls and has emerged as a crucial tactic in modern cybersecurity.

These phases in the development of security models are a reflection of both the evolving nature of technology and the constant attempt to modify security controls to effectively address new threats and vulnerabilities.

1.3 WHY TRADITIONAL SECURITY MODELS ARE NO LONGER EFFECTIVE

1.3.1 EVOLVING THREAT LANDSCAPE

Once effective, traditional security strategies have had difficulty keeping up with the constantly changing threat landscape. These models were initially developed to defend against relatively simple attacks that mainly targeted network perimeters [6]. The sophistication of modern cyber-attacks has increased substantially, though, and now includes a wide variety of attack vectors, including social engineering, sophisticated malware, ransomware, and persistent, highly targeted threats. These sophisticated threats frequently get beyond the traditional security procedures that were once thought to be adequate.

1.3.2 OUTDATED PERIMETER-CENTRIC APPROACH

Traditional security approaches depended heavily on the idea of a clearly defined network perimeter, presuming that dangers were most likely to exist outside of this barrier. However, the prevalence of mobile devices, cloud computing, and remote work has rendered this perimeter-centric strategy obsolete. Since attacks can come from both inside and outside of conventional borders, the network perimeter is no longer a reliable line of defense.

1.3.3 INSIDER THREATS

The foundation of traditional security models was the belief that people in organizations, particularly employees, were trustworthy. However, the rise of insider threats, which may be purposeful or unintentional, calls into question this trust-based premise. These dangers take advantage of their legitimate access to systems and data, providing a serious risk that conventional approaches could find difficult to handle successfully.

1.3.4 LACK OF ACCESS CONTROL GRANULARITY

The granularity of access control in many conventional security methods is lacking. They frequently use simple password-based authentication and broad permissions, which might result in situations where people have greater access rights than are essential for their roles. Due to the absence of fine-grained control, there may be privileged users with excessive access and potential security flaws.

1.3.5 SLOWER INCIDENT RESPONSE

Incident detection and response times may be slower with traditional security models. Rapid detection and prompt action are essential in today's continuously changing threat environment to reduce the impact of security incidents. Response times that are too slow can result in data breaches, system compromises, and serious harm.

1.3.6 OVEREMPHASIS ON COMPLIANCE

Some conventional security techniques give compliance requirements more weight than proactive threat detection and avoidance. Although compliance is crucial, it isn't always compatible with the changing threat environment. Organizations that rely entirely on compliance requirements may be exposed to new dangers that are not covered by regulatory frameworks.

1.3.7 USER-CENTRIC SECURITY CHALLENGES

Identity and access control are now the most important security concerns as users access resources from various devices and places. Traditional methods might not be able to adequately address these user-centric issues, which could result in identification and access control problems [7]. Organizations are increasingly using proactive and adaptive security strategies, such as the ZT model, in response to these problems. These methods prioritize constant verification and stringent access controls and operate under the presumption that dangers may exist both inside and outside the network. They more closely match the volatility and complexity of the current cybersecurity landscape.

1.4 CORE PRINCIPLES OF ZERO TRUST

1.4.1 VERIFY IDENTITY

The identity verification principle lies at the core of ZT. In a ZT architecture (ZTA) [8], strict authentication is required for all users and devices trying to access resources. Trust is never presumed. To verify that people and devices are who they say they are, powerful authentication techniques like MFA must be used.

1.4.2 LEAST PRIVILEGE ACCESS

The least privileged access is emphasized by ZT. In other words, people, and devices are given the minimal level of access necessary to carry out their particular activities. Restrictions on unnecessary access privileges lower the potential attack surface and limit exposure to dangers.

1.4.3 MICRO-SEGMENTATION

ZT takes network segmentation to a very specific level. Access to resources is carefully managed, and they are kept apart from one another. The network cannot migrate laterally as a result of this strategy, which makes it much harder for attackers to move freely once they have gained access to one region.

1.4.4 CONTINUOUS MONITORING

ZT is fundamentally based on continuous monitoring. Real-time anomaly and suspicious activity detection enable prompt action and proactive threat mitigation. This ongoing watchfulness aids in keeping the organization's security posture.

1.4.5 STRICT ACCESS CONTROLS

A ZT paradigm strongly enforces access controls [9]. These restrictions are dependent on a number of variables, such as user identity, device health, location, and contextual data. Access decisions can be accurate and flexible because access policies are dynamic.

1.4.6 DATA ENCRYPTION

ZT places a premium on data security [10]. Data is encrypted to prevent unauthorized access to it, whether it is in transit or at rest. Sensitive information is kept secure and secret by using robust encryption technologies.

1.4.7 CONTINUOUS AUTHENTICATION

ZT encourages ongoing authentication during a user's session rather than just the initial login. By ensuring that the user's identity is continuously validated after the initial login, security is improved.

1.4.8 *ADAPTIVE SECURITY*

ZT is very responsive and flexible. It can adapt to changes in both the organization's infrastructure and the threat landscape. Policies and controls can be dynamically changed as needed to maintain effectiveness.

1.4.9 *ASSET INVENTORY*

In a ZT environment, it's critical to keep an up-to-date inventory of all assets, including hardware and software. By ensuring that assets can be properly monitored and secured, this inventory helps to minimize vulnerabilities.

These guiding principles can help organizations create a strong ZT security framework that improves their cybersecurity posture [11] and offers thorough defense against a variety of internal and external attacks (Figure 1.1).

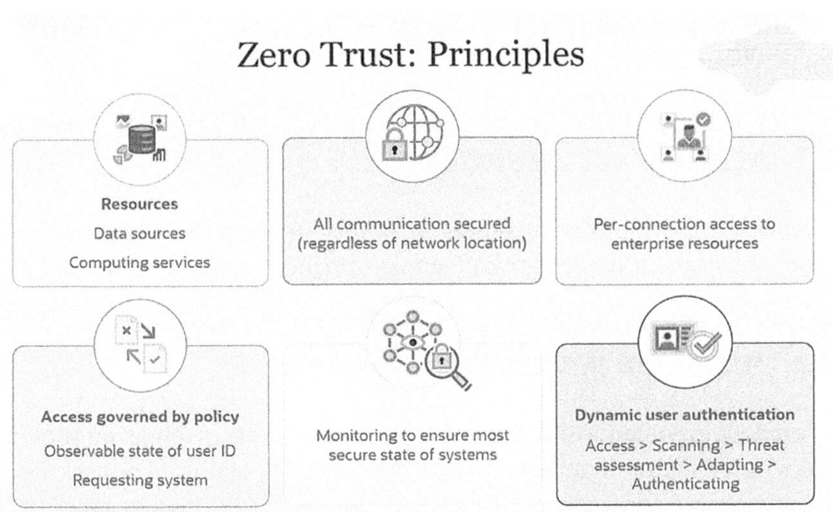

FIGURE 1.1 Key principles of zero trust.

1.5 LEAST PRIVILEGE ACCESS: LIMITING ACCESS TO THE MINIMUM NECESSARY

A security principle known as "least privilege access" states that people, systems, or processes should only be given access to the minimal range of resources, functionality, or data needed to carry out their particular responsibilities or activities. A solid cybersecurity plan must start with this idea. Organizations can decrease the potential attack surface, lower the risk of unauthorized access, and improve overall security by limiting access to the bare minimum required.

The idea of least privilege access is simple: people or organizations should only be granted access that is directly related to their tasks and responsibilities. Beyond what is necessary for them to carry out their job functions properly, no further permissions or privileges should be allowed. This strategy aids in reducing the dangers connected to excessively permissive access, which can result in security breaches, data leaks, and unauthorized system manipulation.

Planning carefully and implementing thorough access control procedures are necessary when implementing least privilege access. Organizations must specify and record each user's roles, duties, and the particular resources or data to which they need access. Then, these access limitations are put into effect using access control lists (ACLs), RBAC, and other techniques.

Businesses can get a number of security advantages [12] by following the least privilege access principle:

1.5.1 REDUCED ATTACK SURFACE

By reducing the number of potential points of entry for attackers, access restrictions make it harder for bad actors to penetrate the system.

1.5.2 MITIGATED INSIDER THREATS

The organization can lower the risk of insider threats, such as unintentional employee abuse or inadvertent data exposure, by limiting access to only what is required for a specific function.

1.5.3 ENHANCED COMPLIANCE

Least privilege access is a requirement for many legislative regimes as part of data protection and privacy safeguards. Following this idea makes it easier to comply with these rules.

1.5.4 *MINIMIZED DAMAGE POTENTIAL*

Attackers with fewer privileges will find it more difficult to steal important information or cause considerable harm in the event of a security breach or unauthorized access. Overall, least privilege access is essential to a comprehensive security plan [13] since it promotes the idea of "only what's necessary, nothing more." It guarantees that people and machines can work efficiently while maintaining a robust defense against any security risks.

1.6 MICRO-SEGMENTATION

A network is divided into smaller, isolated segments or zones, each with its own particular security policies and controls, as part of the network security technique known as micro-segmentation. Micro-segmentation offers a higher level of granularity compared to typical network segmentation, which frequently divides networks into broad divisions [14]. It establishes constrained, secure areas that can be tailored to fit particular devices, workloads, or applications.

Micro-segmentation's main goal is to improve network security by preventing attackers from moving laterally through the network. Following are some crucial features and advantages of micro-segmentation:

1.6.1 *GRANULAR SECURITY CONTROLS*

Organizations can apply more specialized security measures inside each segment via micro-segmentation. This implies that security guidelines, access controls, and firewall configurations can all be properly adapted to the needs of the workloads or applications hosted in that segment.

1.6.2 *REDUCED ATTACK SURFACE*

Micro-segmentation greatly decreases the attack surface within the network by isolating portions. Even if an attacker manages to access one segment, they are separated from the rest of the network, making it challenging for them to travel laterally and increase their level of access.

1.6.3 *ENHANCED DATA PROTECTION*

With rigorous access controls and encryption, sensitive data can be stored in separate, extremely safe segments. By doing this, sensitive data is protected from unauthorized access or data breaches.

1.6.4 *ZERO TRUST ALIGNMENT*

The ZT security concept, which believes that dangers might exist both within and outside the network, is in line with micro-segmentation. Regardless of the source or destination of any communication, it imposes strict access controls and verification.

1.6.5 *APPLICATION-CENTRIC SECURITY*

Applications have the ability to design and implement security policies. As a result, security controls are closely related to particular programs or services, guaranteeing that each application is secured in accordance with its particular needs.

1.6.6 *SIMPLIFIED INCIDENT RESPONSE*

Micro-segmentation can contain the threat within the afflicted segment in the event of a security incident or breach, preventing it from spreading to other areas of the network. The containment of the issue streamlines incident response procedures and reduces possible harm.

Organizations frequently rely on technologies like software-defined networking (SDN), next-generation firewalls, intrusion detection and prevention systems (IDPS), and network virtualization to perform micro-segmentation efficiently [15]. Through the use of these technologies, isolated segments may be created, access controls can be enforced, and network traffic inside and between segments [16] can be monitored.

In summary, by dividing up a network's architecture into smaller, well-defined security zones and lowering its overall attack surface, micro-segmentation is a strong method for boosting network security. It is especially useful in contemporary, complicated network contexts where traditional perimeter defenses might not be able to offer full protection.

1.7 ASSUME BREACH

In cybersecurity and risk management, the idea of "Assume Breach" refers to operating under the presumption that a security breach or compromise is always possible, even when stringent preventive measures are in place. This proactive strategy emphasizes effective planning, detection, and response to security issues while acknowledging the dynamic and persistent nature of cyber threats [17]. The "Assume Breach" mindset's main features and advantages are as follows:

1.7.1 CONTINUOUS VIGILANCE

Businesses that adopt the "Assume Breach" mentality are constantly on notice. They are aware that attackers are tenacious and that fresh dangers can appear at any time. To stay ahead of potential breaches, this approach supports continual surveillance and threat intelligence collection.

1.7.2 PREPAREDNESS

A key component of this approach is being ready for a breach. The development and testing of incident response plans, regular security drills, and team training for handling security issues are all practices carried out by organizations. In the event of a breach, being prepared lessens the effect and recovery time.

1.7.3 EARLY IDENTIFICATION

The "Assume Breach" method places a significant value on early identification of security events rather than just relying on prevention. To quickly spot odd or suspicious activity, this calls for the adoption of powerful IDSs, security information and event management (SIEM) tools, and other monitoring tools.

1.7.4 ADAPTIVE SECURITY

"Assume Breach" security policies are compatible with this attitude. They recognize that sophisticated attacks can get past conventional security measures or that attackers may already be inside the network. Security measures are therefore made to respond quickly to changing threats and adapt as necessary.

1.7.5 DATA SECURITY

"Assume Breach" emphasizes the importance of safeguarding sensitive data. To protect vital information from unauthorized access or exfiltration, organizations deploy strong data encryption, access controls, and data loss prevention (DLP) techniques.

1.7.6 POST-INCIDENT LEARNING

After a breach or security incident, organizations carry out extensive post-incident assessments to comprehend the underlying causes, spot vulnerabilities, and enhance security precautions. This process of learning makes future occurrences of this kind less likely.

1.7.7 RISK MITIGATION

Organizations can actively manage and reduce risks by adopting the "Assume Breach" approach. Based on risk evaluations, threat intelligence, and the potential effects of security incidents, they prioritize security spending.

In conclusion, the "Assume Breach" concept signifies a change in cybersecurity from a solely preventative strategy to a proactive, adaptable, and ready stance. It acknowledges that no security system is impenetrable and that to combat the constant and developing nature of cyber threats, effective cybersecurity necessitates a combination of prevention, detection, response, and continual improvement.

1.8 ZERO TRUST ARCHITECTURE

A modern and effective cybersecurity paradigm called ZTA [18] was created to address the changing landscape of online threats and vulnerabilities. At its core, ZTA promotes a "never trust, always verify" mentality, challenging the conventional assumption of trust within business networks. Whether people, devices, or systems are located inside or outside the organization's network perimeter, trust is never assumed in ZTA. Instead, in order to access network resources, each entity must go through strict and ongoing authentication, authorization, and verification procedures (Figure 1.2).

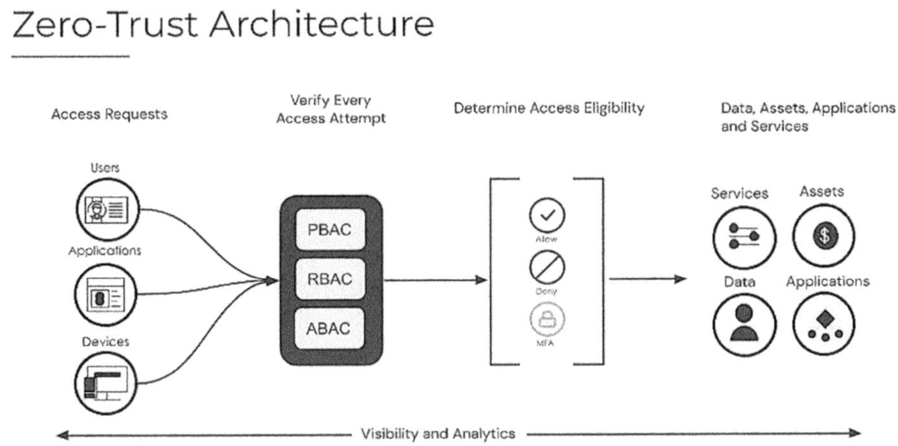

FIGURE 1.2 Architecture of zero trust "never trust, always verify."

Source: Skyflow. https://www.skyflow.com/post/what-is-zero-trust

The idea of least privilege access, where people and devices are given the minimum level of access required for their particular roles and responsibilities, is a basic ZTA philosophy. By drastically reducing the potential attack surface, this strategy greatly lowers the danger of unauthorized access or privilege escalation. ZTA also places a strong emphasis on micro-segmentation, which includes creating separated network portions with strict access constraints. By restricting lateral movement within the network, this isolation makes it extremely difficult for attackers to pass through various parts.

The foundation of ZTA is continuous monitoring, which enables the real-time detection of anomalies and suspicious activity. Adaptive controls that enable dynamic access policy modifications support this vigilance. ZTA places a high priority on data security, encrypting data both in transit and at rest to protect private information from unauthorized access. An essential element is MFA, which raises security by requesting various forms of identification before granting access.

While integrating threat intelligence feeds keeps security measures in line with new threats, centralized policy administration enables the uniform application of access controls throughout the organization. By creating a baseline of typical behavior through user and device profiling, it is possible to quickly identify variations that could indicate security vulnerabilities. ZTA relies heavily on automation to provide quick responses to security events.

One major strength of ZTA is its capacity to adjust to a shifting security environment. It enables dynamic adjustments to policies and procedures to effectively address new risks and problems. ZTA is an all-encompassing cybersecurity strategy that is focused on the future and promotes continuous verification, stringent access controls, and a steadfast dedication to increasing security in the face of ongoing and developing cyber threats.

1.8.1 OVERVIEW OF THE ZERO TRUST ARCHITECTURE

The ZTA redefines conventional ideas of trust within organizational networks as an innovative approach to cybersecurity. ZTA's fundamental operating tenet is "never trust, always verify." Because of this fundamental change, trust is never taken for granted, whether users or devices are connected to the corporate network or not. Instead, to verify their legitimacy and access privileges, every entity requesting access to network resources must go through ongoing authentication and authorization procedures.

Granular identity verification, allowing least privilege access, and micro-segmenting the network to establish isolated zones with stringent access rules are important components of ZTA. ZTA's core functionality is the continuous monitoring of user and device behavior, which enables the real-time detection of anomalies and suspicious activity. Sensitive information is safeguarded by strict enforcement of access limits and data encryption procedures. Continuous authentication ensures that user identities are validated throughout sessions, while MFA adds an additional layer of security.

ZTA's dynamic and adaptive security solution includes centralized policy management, integration of threat intelligence feeds, user, and device profiling, automation for quick incident response, and more. The adaptability of this architecture enables organizations to successfully address changing threats and difficulties, making it a strong and cutting-edge paradigm for boosting cybersecurity in the contemporary digital environment.

1.8.2 COMPONENTS AND BUILDING BLOCKS OF A ZERO TRUST MODEL

A secure and flexible cybersecurity framework is built using a number of fundamental elements and building blocks that make up a zero-trust paradigm. The "never trust, always verify" tenet is upheld by these elements in concert. These are the main components:

1.8.2.1 IDENTITY AND ACCESS MANAGEMENT (IAM)

The foundation of ZT is IAM. It entails meticulously confirming the legitimacy of individuals, devices, and systems making resource access requests. To improve identity security, single sign-on (SSO) and MFA are frequently used.

1.8.2.2 LEAST PRIVILEGE ACCESS

The minimum level of access required for each user's or device's particular duties and responsibilities is guaranteed by this principle. To lessen the potential attack surface, access rights are restricted.

1.8.2.3 MICRO-SEGMENTATION

To generate distinct segments or zones, network segmentation is applied at a finer scale. Strict access controls on each segment prevent possible attackers from moving laterally within the network.

1.8.2.4 USER AND DEVICE PROFILING

By establishing a baseline of typical behavior, user, and device profiling enables the quick identification of variations that can signify security issues.

1.8.2.5 AUTOMATION

Automation is a crucial component that makes it possible to respond quickly to security events. When suspicious activity is noticed, automated reactions may include isolating compromised devices or revoking access privileges.

1.8.2.6 CONTROLS FOR NETWORK SEGMENTATION

SDN and next-generation firewalls (NGFW) are used to manage network segmentation, enabling flexible and granular control over network traffic.

1.8.2.7 CLOUD SECURITY

It's critical to apply ZT concepts to cloud systems. Identity federation systems and cloud access security brokers (CASBs) [19] both contribute to the security of cloud access. The foundation of a ZT model is made up of these elements and building blocks as a whole, allowing organizations to develop a safe, adaptable, and resilient cybersecurity posture that is suitable for the current threat scenario.

1.8.3 HOW ZERO TRUST (ZT) DIFFERS FROM TRADITIONAL NETWORK SECURITY ARCHITECTURE

In several significant aspects, ZT marks a radical break from conventional network security architecture [20]. A comparison of ZT and conventional security concepts is shown below:

1.8.3.1 ASSUMPTION OF TRUST

1. **Traditional Security:** When it comes to traditional security models, it's frequently assumed that once users or devices are connected to a corporate network, one may implicitly trust them. On the basis of network location, trust is often granted, and access controls are laxer inside the perimeter [21].
2. **Zero Trust:** ZT questions this premise and operates under the motto "never trust, always verify." Regardless of whether a user is inside or outside the network, trust is never assumed. Every access request must undergo ongoing verification and authentication.

1.8.3.2 NETWORK PERIMETER

1. **Traditional Security:** The network perimeter is frequently used in traditional models as the main security boundary. At this boundary, security mechanisms like firewalls and intrusion detection systems are focused.
2. **Zero Trust:** The value of the network perimeter is diminished by ZT. Instead, it focuses on using rigorous access controls, encryption, and ongoing monitoring to secure particular assets, regardless of where they are.

1.8.3.3 SEGMENTATION

1. **Traditional Security:** Traditional models' network segmentation frequently divides networks into broad divisions at a coarser level. Different internal groups may have different levels of trust.

2. **Zero Trust:** Micro-segmentation, in which the network is split into tiny, isolated portions with strong access rules, is encouraged by ZT. Each segment might be as specialized as particular devices, workloads, or applications.

1.8.3.4 DATA PROTECTION

1. **Traditional Security:** Traditional models might not fully emphasize data protection. Encryption and other data security measures could be deployed selectively.
2. **Zero Trust:** Data security is a priority for ZT. To prevent unauthorized access or data breaches, data is automatically encrypted both in transit and at rest.

1.8.3.5 CONTINUOUS MONITORING

1. **Traditional Security:** Traditional approaches may not have real-time monitoring capabilities and frequently rely on annual security evaluations.

2. **Zero Trust:** One of the core principles of ZT is the constant observation of user and device behavior. Real-time anomalies and suspicious activity detection allow for quick action.

1.8.3.6 SECURITY POSTURE

1. **Traditional Security:** Traditional methods presuppose a constant security posture, which might not be able to keep up with changing threats.

2. **Zero Trust:** By adopting an adaptable security posture, ZT enables rules and controls to be changed on the fly in response to changing threat intelligence and environmental circumstances.

1.9 CONCLUSION

In conclusion, ZT radically questions the conventional security paradigm by emphasizing continuous verification, least privilege access, micro-segmentation, and data protection in all areas in medicine [22], all the while reducing dependency on the network perimeter. The dynamic and persistent nature of contemporary cyber threats is well-suited for this strategy.

KEYWORDS

- **artificial intelligence**
- **discretionary access control**
- **least privilege access**
- **micro-segmentation**
- **software-defined networking**
- **threat detection**
- **user authentication**
- **zero trust learning**

REFERENCES

1. Rose, O., Borchert, S., Mitchell, S., & Connelly, S. (2020). *Zero Trust Architecture*. NIST Special Publication 800–207. National Institute of Standards and Technology, Gaithersburg, MD, USA.
2. American Council for Technology and Industry Advisory Council (ACT-IAC). (2019). *Zero Trust Cybersecurity Current Trends.*
3. Anderson, B., & McGrew, D. (2017). Machine learning for encrypted malware traffic classification: Accounting for noisy labels and non-stationarity. In *Proceedings of the 23rd ACM SIGKDD International Conference on Knowledge Discovery and Data Mining* (pp. 1723–1732). ACM. https://doi.org/10.1145/3097983.3098163.
4. National Institute of Standards and Technology (NIST). (2004). *Standards for Security Categorization of Federal Information and Information Systems* (FIPS Publication 199). U.S. Department of Commerce, Washington, DC.
5. Gilman, E., & Barth, D. (2017). *Zero Trust Networks: Building Secure Systems in Untrusted Networks* (1st ed.). O'Reilly Media, Inc.
6. Cohen, R., Barabash, K., Rochwerger, B., Schour, L., Crisan, D., Birke, R., Minkenberg, C., Gusat, M., Recio, R., & Jain, V. (2013). An intent-based approach for network

virtualization. In *2013 IFIP/IEEE International Symposium on Integrated Network Management (IM 2013)*. https://doi.org/10.1109/INM.2013.6572981.

7. Office of Management and Budget (OMB). (2018). *Strengthening the Cybersecurity of Federal Agencies by Enhancing the High-Value Asset Program* (OMB Memorandum M-19-03). The White House, Washington, DC.
8. Ferraiolo, D. F., Gavrila, S., & Jansen, W. (2015). *Policy Machine: Features, Architecture, and Specification* (NIST Interagency or Internal Report (IR) 7987, Rev. 1). National Institute of Standards and Technology, Gaithersburg, MD.
9. Brooks, S. W., Garcia, M. E., Lefkovitz, N. B., Lightman, S., & Nadeau, E. M. (2017). *An Introduction to Privacy Engineering and Risk Management in Federal Systems* (NIST Interagency or Internal Report (IR) 8062). National Institute of Standards and Technology, Gaithersburg, MD.
10. Joint Task Force. (2018). *Risk Management Framework for Information Systems and Organizations: A System Life Cycle Approach for Security and Privacy* (NIST Special Publication (SP) 800-37, Rev. 2). National Institute of Standards and Technology, Gaithersburg, MD.
11. Doherty, N. F., Ashurst, C., & Peppard, J. (2012). Factors affecting the successful realization of benefits from systems development projects: Findings from three case studies. *Journal of Information Technology, 27*(1), 1–16. https://doi.org/10.1057/jit.2011.29.
12. The Zero Trust eXtended (ZTX) Ecosystem: Extending zero trust security across your digital business. (n.d.).
13. Taffel, S. (2019). Trusted machines? Machine learning, more-than-human speed and dead labor in platform capitalism. *AoIR Selected Papers of Internet Research, 10.*
14. Albarakati, A., Moussa, B., Debbabi, M., Youssef, A., Agba, B. L., & Kassouf, M. (2018). OpenStack-based evaluation framework for smart grid cybersecurity. In *2018 IEEE International Conference on Communications, Control, and Computing Technologies for Smart Grids (SmartGridComm)*. https://doi.org/10.1109/SmartGridComm.2018.8587434.
15. Bevish Jinila, Y., Prayla Shyry, S., & Christy, A. (2022). A multi-component-based zero trust model to mitigate the threats in the internet of medical things. In *Data Engineering for Smart Systems: Proceedings of SSIC 2021*. Manipal University Jaipur, India.
16. Bien, J., Gaynanova, I., Lederer, J., & Müller, C. L. (2019). Prediction error bounds for linear regression with the TREX. *Test, 28*, 451–474. https://doi.org/10.1007/s11749-018-0584-4.
17. Gao, P., Yan, L., Chen, Z., Wei, X., Guo, L., & Shi, R. (2021). Research on zero-trust-based network security protection for power internet of things. In *2021 IEEE 4th International Conference on Automation, Electronics and Electrical Engineering (AUTEEE), Shenyang, China*. https://doi.org/10.1109/AUTEEE.2021.9497436.
18. Hatakeyama, K., Kotani, D., & Okabe, Y. (2021). Zero trust federation: Sharing context under user control towards zero trust in identity federation. In *2021 IEEE International Conference on Pervasive Computing and Communications Workshops and other Affiliated Events (PerCom Workshops)*, Kassel, Germany. https://doi.org/10.1109/PerComWorkshops51409.2021.9431133.
19. Rodigari, S., O'Shea, D., McCarthy, P., McCarry, M., & McSweeney, S. (2021). Performance analysis of zero-trust multi-cloud. In *2021 IEEE 14th International*

Conference on Cloud Computing (CLOUD), Chicago, IL, USA. https://doi.org/10.1109/CLOUD53861.2021.00043.

20. Samaniego, M., & Deters, R. (2018). Zero-trust hierarchical management in IoT. In *2018 IEEE International Congress on Internet of Things (ICIOT)*, San Francisco, CA, USA. https://doi.org/10.1109/ICIOT.2018.00019.
21. Shan, L., Iqbal, M., & Saxena, N. (2022). Future industry internet of things with zero-trust security. *Information Systems Frontiers*, 1–14. https://doi.org/10.1007/s10796-021-10199-5.
22. Sultana, M., Hossain, A., Laila, F., Abu Taher, K., & Nazrul Islam, M. (2020). Towards developing a secure medical image sharing system based on zero trust principles and blockchain technology. *BMC Medical Informatics and Decision Making, 20*(1), 1–10. https://doi.org/10.1186/s12911-020-01275-y.

CHAPTER 2

Importance of Zero Trust and Transformation of Modern Network Security

S. MURALIMOHAN,[1,*] U. SIVAJI,[2,*] R. THARUN,[3,*] B. RUPA DEVI,[4,*] and T. RAVI KUMAR[5,*]

[1]*Department of ECE, Mother Theresa Institute of Engineering and Technology, Palamaner, Andhra Pradesh, India*

[2]*Institute of Aeronautical Engineering, Department of IT, Dundigal, Hyderabad, Telangana, India*

[3]*Department of CE, Mother Theresa Institute of Engineering and Technology, Palamaner, Andhra Pradesh, India*

[4]*Department of CSE, Annamacharya Institute of Technology and Sciences, Tirupati, Andhra Pradesh, India*

[5]*Department of CSE, Aditya Institute of Technology and Management, Tekkali, Andhra Pradesh, India*

*Corresponding author

ABSTRACT

The imperative for modern network defense systems has become irrefragable in the ever-evolving digital landscape. The implacable wave of cyber hazards, marked by their increasing complexity and frequency, has exposed the shortcomings of conventional security models reliant on perimeter-based defenses and implicit trust. This chapter investigates the compelling rationale behind the need for contemporary network defense mechanisms and explores the vital role played by Zero Trust (ZT) in shaping our modern

Zero-Trust Learning: Applications in Modern Network Security. Chetna Kaushal, Amandeep Kaur, Mohit Angurala, Aryan Chaudhary (Eds.)

cybersecurity framework. ZT, a revolutionary cybersecurity paradigm, challenges established norms by fundamentally reconsidering the concept of trust within network environments. It accomplishes this by placing paramount importance on rigorous access controls, continuous surveillance, and dynamic verification, thereby eliminating the assumption of inherent trust for any entity, irrespective of its position within or beyond the network perimeter. By examining the essential limitations of traditional security models and addressing the ever-shifting threat landscape, this chapter elucidates how ZT offers a flexible, adaptive, and comprehensive approach to modern network defense. It emphasizes the critical importance of embracing ZT as a foundation for strengthening our digitally connected world against numerous evolving cyber threats.

2.1 INTRODUCTION

Traditional network security models have yet to be ascertained as sufficient in an age of increasingly sophisticated cyber threats and a rapidly developing digital landscape. In today's fast-paced and ever-changing world, the perimeter-based security strategy that was once considered the gold standard for safeguarding enterprises is no longer sufficient to protect against many risks. With the increasing sophistication of cyber-attacks [1], adopting a multi-layered approach to security is essential beyond traditional perimeter defenses. As information technology and network security evolve rapidly, protecting digital assets from advanced cyber threats has become increasingly challenging. Recently, the approach to security tactics has undergone a significant transformation, leading to the emergence of the innovative concept of "ZT" [2]. This approach is based on the principles of continuous verification, providing minimal access privileges, and implementing strict segmentation. By doing so, ZT offers a modern way of protecting digital infrastructures against prevalent and evolving security threats [3].

As technology advances, it has become increasingly clear that relying on the traditional security model is no longer enough because cyber attackers constantly search for vulnerabilities to exploit, and the consequences can be devastating. Organizations need more than just traditional security measures like firewalls and VPNs; they must adapt and implement additional robust security measures to defend against cyber threats. The expanding Internet of Things (IoT), cloud computing, and mobile workforce [4] have made the traditional security model of trusting internal networks vulnerable. The static perimeters and "castle-and-moat" security strategies have proven ineffective against advanced threats that infiltrate organizations, often going unnoticed [5]. In the current environment, security is of foremost priority, and this is where the concept of Zero Trust comes into play. Unlike the traditional approach of unquestioningly trusting the internal network, the Zero Trust framework considers every user, device, and application as untrusted until they can prove their authenticity and authorization through continuous verification. This paradigm shift challenges conventional wisdom and requires ongoing authentication, authorization, and validation for entities attempting to access critical resources. With the "never trust, always verify" approach, Zero Trust disrupts traditional trust assumptions and ensures maximum security [6]. Zero Trust is a well-known security architecture utilized in various industries and businesses, including government agencies, financial institutions, healthcare providers, and companies of all sizes [7]. Its adaptable and scalable design makes it an attractive option for modern networks that require flexibility, dynamic access controls, and robust security measures. Organizations often face obstacles when transitioning from legacy systems to ZT frameworks, such as resistance to change and cultural barriers. To successfully implement ZT, it is essential to understand these hurdles and develop effective strategies to overcome them [8].

ZT presents a transformative approach that can bolster defenses against various cyber threats, but this isn't just about protection against breaches and attacks. By embracing ZT, organizations can foster agility, adaptability, and resilience in an increasingly interconnected and dynamic digital landscape. The benefits of ZT are clear, and it's something that companies of all sizes and industries must consider as they strive to protect their data, assets, and reputation in today's complex and ever-evolving threat landscape.

2.2 BACKGROUND

Cyber dangers and extremely sophisticated assaults have rapidly increased in recent years, endangering the security of many international organizations. Traditional security paradigms that depend mainly on perimeter-based defenses and an implicit trust model must be reevaluated to secure sensitive information and vital assets appropriately. ZT has significantly increased its popularity in network security due to the expanding threat landscape and the growing demand for more robust and adaptive security solutions. The ZT idea was initially introduced by Forrester Research in 2010 to address the shortcomings of traditional perimeter-centric security solutions. It radically alters how trust is produced and confirmed within a network [9]. It's crucial to understand that no traffic in a zero-trust architecture can be automatically trusted; therefore, relying on location as a security criterion is pointless. This paradigm adopts a comprehensive strategy and mandates security measures for each entrance attempt. This necessitates the implementation of limited authorization regulations and the use of strong access controls. Additionally, it is necessary to view and analyze all network activity properly. These fundamental concepts starkly contrast the usual perimeter-based security architecture and considerably depart from accepted security practices. As a result, the level of security and resilience to evolving cyber threats is substantially higher [10]. In modern cybersecurity, the perimeter-based security approach must be updated, which typically depends on trustworthy organizations within network borders. Its flaws are exposed when confronted with difficulties like fending off insider attacks, battling masked network activity, and juggling the needs of remote work, mobile devices, and the unrelenting persistence of sophisticated threats. Its flaws include limited scalability and slow incident response, frequently resulting in compliance barriers and the possibility of crucial single points of failure. As a result, corporations are actively researching new security paradigms, such as ZT, to strengthen their defenses in a cybersecurity environment that is still rapidly evolving [7].

2.3 CONCEPT OF ZERO TRUST

Comparing ZT to conventional security approaches, it is a noticeably better cybersecurity strategy. Unlike traditional approaches, which frequently functioned on the premise that everyone in the network was trustworthy and heavily depended on perimeter defenses, ZT adheres to a fundamentally different set of principles, which are summed up in the maxim "never trust, always verify." Using this strategy, users' and devices' identities are dynamically verified while continuously monitored and subject to strict access constraints. The ability of ZT to successfully deal with the difficulties posed by advanced persistent threats (APTs), combat insider threats, and adapt effortlessly to the changing needs of a remote and mobile workforce provides a more robust and durable security approach adapted to the intricacies of today's fast-evolving threat landscape by prioritizing minimum permission and real-time visibility. This reduces the reliance on a potentially susceptible network perimeter (Table 2.1).

TABLE 2.1 Best Parameters for Different Models

Traditional Security Model	Zero-Trust Model
Access control based on the IP (Port, Protocol)	Access control focused on data
Encrypted external / Not encrypted Internal	Encrypted for both traffic
Established guidelines and procedures	Specific and flexible Rules (Requires Security Assessment)
Monitoring and visibility (Individual)	transparency and automation coordination of conduct, tools, services, and security

The ZTA is a coherent framework for system design based on the core idea that risks to a computer network, whether from internal or external sources, must always be foreseen. This groundbreaking idea for a network system architecture necessitates almost continual monitoring and verifying of all network elements, including nodes, services, applications, and user groups. The fundamental problem with this worldview is that it automatically distrusts all entities. The ability to access, use, or alter network resources, such as databases, other network nodes, servers inside the network, and network regulations, is only granted to network components after passing a laborious verification procedure and within a continuously shrinking time window.

2.4 ZERO TRUST FUNDAMENTALS

Firewalls and intrusion prevention systems (IPS) are no longer enough to protect against cybercriminals. Zero Trust's "never trust, always verify" model applies granular security rules to eliminate blind trust and make trust binary. This revolutionary architecture requires a fundamental shift in perspective for organizations to reconsider their approach to security. When embarking on the journey to create a network with Zero Trust DNA, there are several following critical considerations [11].

2.4.1 SECURE DATA ACCESS BASED ON USER AND LOCATION

Zero Trust prioritizes user and location over network or device, ensuring secure data access through strict authentication and monitoring to prevent cyberattacks (Figure 2.1).

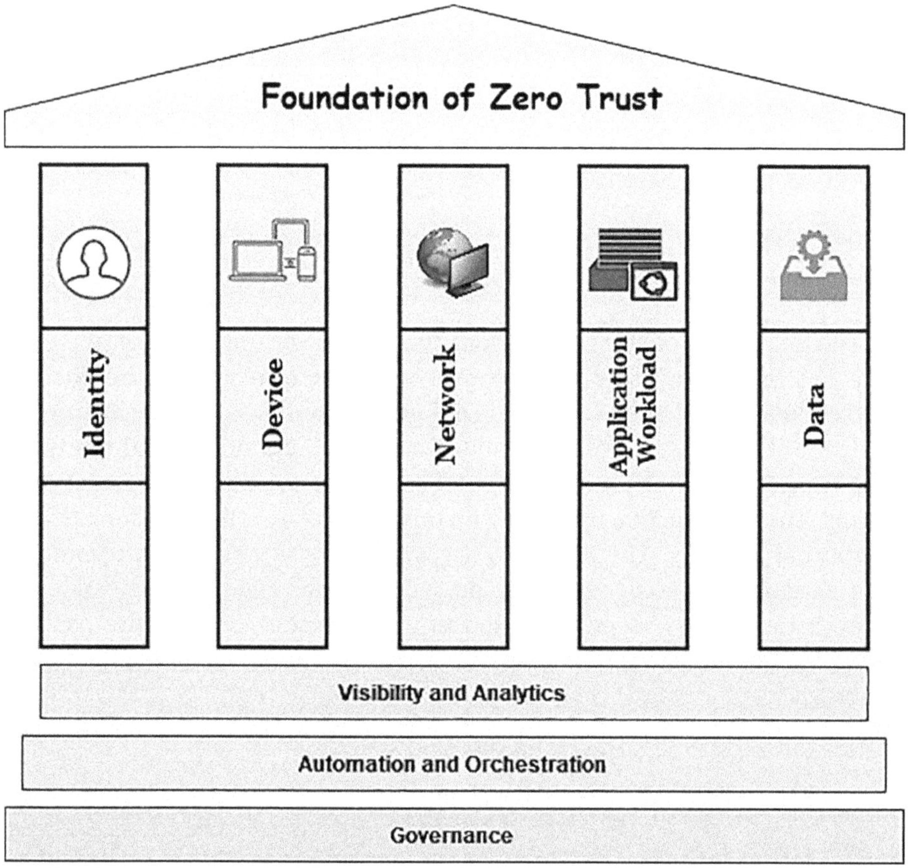

FIGURE 2.1 Major foundation of zero trust.

2.4.2 STRONG EMPHASIS ON ACCESS CONTROL

To ensure security, access control is no longer just a suggestion but a necessary foundation of ZT. The principle of least privilege ensures that only the necessary resources are accessible to users and devices based on their specific tasks.

2.4.3 COMPREHENSIVE TRAFFIC LOGGING AND INSPECTION

Monitoring network traffic is crucial for security. It helps detect potential threats and enables a swift response to anomalies, ensuring a safe environment for users and devices. Figure 2.1 illustrates the major foundation of ZT. In today's world of digital interconnectedness, it's essential to be vigilant about security. ZT is a new way of thinking about trust, emphasizing a continuous and dynamic approach to security that considers each situation's context. By adopting this approach, we can better protect ourselves and our digital assets from potential threats.

2.4.4 APPLICATIONS OF ZERO TRUST

Zero Trust's foundational principle of "never trust, always verify" has rapidly gained importance in cybersecurity due to its adaptability and effectiveness in addressing modern threats and evolving network landscapes. Its applications are various and extend to diverse cybersecurity and network management aspects. Figure 2.2 illustrates the ZT application, and here are some critical applications of ZT that we describe below.

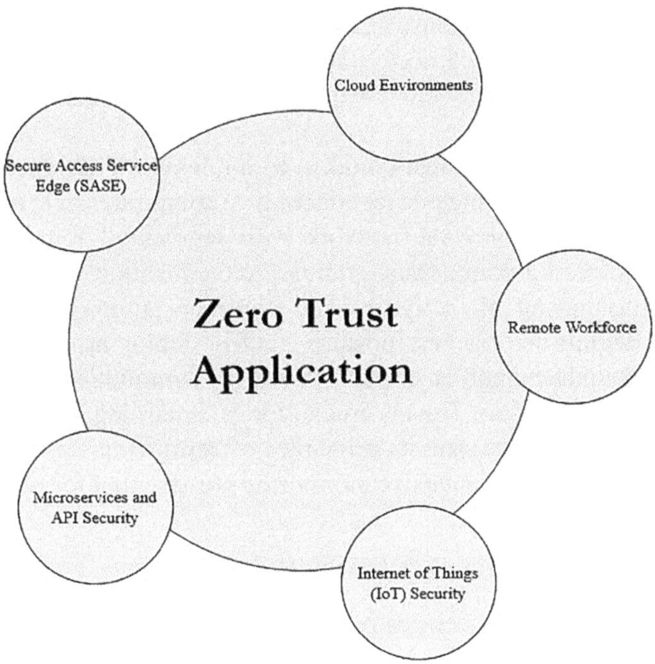

FIGURE 2.2 The application of zero trust.

2.4.4.1 CLOUD ENVIRONMENTS

Zero Trust is an excellent security framework that provides a robust solution for protecting cloud-based assets and data. This approach involves continuous verification and access controls, which helps to mitigate risks and provides granular control over cloud resources.

2.4.4.2 REMOTE WORKFORCE

Zero Trust is an excellent option for companies looking to secure remote workers who need to access corporate resources outside the traditional network perimeter. This approach continuously verifies remote devices and connections, significantly reducing the risk of unauthorized access. It's an innovative and effective way to protect your business from threats.

2.4.4.3 INTERNET OF THINGS (IOT) SECURITY

Securing the growing network of interconnected devices in the IoT ecosystem is crucial, and Zero Trust is instrumental in achieving this. Zero Trust effectively thwarts unauthorized access and mitigates the risks associated with IoT-related cyber threats by enforcing strict access controls and continuous device verification.

Several essential steps must be taken to implement a ZT security model for IoT devices. These include conducting a comprehensive inventory of all IoT devices, designing the network with segmented zones and micro-perimeters, enforcing continuous, granular authentication and authorization protocols, monitoring all activities and communications of IoT devices, adopting a default-deny access posture, swiftly deploying security patches and firmware updates, and ensuring all network communication among IoT devices is encrypted. Zero Trust is crucial for safeguarding IoT devices in the expanding IoT landscape, and its principles of continuous verification, strict access control, and comprehensive monitoring are essential for achieving this.

2.4.4.4 MICROSERVICES AND API SECURITY

Modern application architectures rely heavily on microservices and APIs to ensure that the communication between microservices is secure and that only authorized entities access the APIs. Zero Trust principles can be applied. This approach provides an additional layer of security to safeguard against potential threats and vulnerabilities. By implementing ZT, businesses can rest assured that their data and systems are protected and operating in a safe and secure environment.

2.4.4.5 SECURE ACCESS SERVICE EDGE (SASE)

Zero Trust principles are the way to go when securing access to applications and data in a distributed, cloud-centric network environment. Organizations can significantly enhance their network security posture and adapt to dynamic threat landscapes by effectively implementing Zero Trust, and this aligns closely with the SASE framework, which integrates network and security functions into a unified cloud-based service. Organizations can protect their critical assets from advanced cyber threats with a resilient security framework that fosters trust and confidence in their digital operations. Zero Trust is a powerful tool for safeguarding data and resources in today's ever-changing digital landscape.

2.5 BENEFITS OF ZERO TRUST

Zero Trust is a cybersecurity paradigm that offers many advantages spanning various facets of network security and operational resilience. Foremost among its benefits is the significant reduction in the attack surface it provides. Through the strict enforcement of access controls and the segmentation of networks into discrete zones, Zero Trust effectively limits the lateral movement of threats and confines potential breaches. This pivotal measure bolsters organizations' security posture, ensuring a robust defense against malicious actors. Another crucial advantage of Zero Trust lies in its ability to enhance adaptability in the face of constantly evolving cyber threats. Continuously verifying entities such as users, devices, or applications enables organizations to dynamically respond to potential risks in real-time. Unlike the traditional perimeter-based model, Zero Trust equips organizations with the means to thwart conventional and advanced attacks, resulting in heightened threat detection and response.

2.6 ZERO TRUST PROPOSES A RANGE OF BENEFITS BEYOND ENHANCED SECURITY

2.6.1 IMPROVED SECURITY

Zero Trust provides only authorized users access resources from verified devices, bolstering security.

2.6.2 *GREATER CONTROL*

It deploys granular controls and automated responses based on real-time behavior analysis, enhancing control.

2.6.3 *INCREASED VISIBILITY*

Continuous monitoring delivers better insight into network traffic, aiding anomaly detection and malware mitigation.

2.6.4 *IMPROVED COMPLIANCE*

Zero Trust aligns with compliance standards, making it more comfortable for organizations to meet regulatory requirements.

2.6.5 *INCREASED EFFICIENCY*

Automation decreases manual security tasks, enhancing efficiency and timeliness in maintaining a secure environment.

2.6.6 *MORE SECURE FUTURE*

Investing in ZT protects against breaches and ensures a company's long-term viability by freeing up resources for strategic tasks.

Furthermore, the adoption of ZT aligns seamlessly with contemporary technology paradigms, such as cloud computing and IoT ecosystems. Its adaptable nature facilitates the secure integration of these transformative technologies, fostering an environment conducive to innovation without compromising security. Zero Trust fortifies an organization's defense against cyber threats and empowers it to confidently embrace the latest technological advancements, making it a crucial strategy for modern cybersecurity.

2.7 IMPLEMENTING ZERO TRUST

Several network sensors must thoroughly deploy a zero-trust network to improve situational awareness inside and outside the network. It's crucial that this system functions without impairing performance and concentrates on preventing illegal events. When implementing this approach to a network, extra metrics may be necessary. These may include east-west security measures that partition networks and classify users and nodes according to established access criteria. Furthermore, endpoint detection and response (EDR) tools and the sandboxing of specific network elements, such as virtual machine monitors, can be required. The Zero Trust Architecture (ZTA) core still needs a trust computation system. There are several ways to calculate trust. Still, the algorithm eventually decides whether to authorize or reject a specific access request, especially when using network coffers. This algorithm may be allowed as a function that considers numerous material features. These characteristics could include, among others, the type of access request, the requesting entity's past behavior, past resource operation patterns, any forfeitures incurred in the history, current Identity and Access Management (IAM) programs, the trust position connected to the entity's group class, and the troubled terrain at the time.

2.7.1 MODERN THREATS

The kind and intensity of recently discovered attacks or network abnormalities, as well as the time frames in which they occur, are just some of the many aspects that make up the current trouble terrain. A thorough trouble matrix for the network is constructed using this collection of trouble data as a base. In this case, the trust algorithm is flexible enough to draw widely from the set of risk traits as necessary, integrating particular attributes relevant to the current circumstance.

2.7.2 REQUEST FOR ACCESS

The current access request initiated by the entity to gain entry to a secured element necessitates the addition of supplementary metadata. This new information encompasses crucial details such as the timestamp of the request, the number of previous requests made for the same element, the level of authorization required for access, and the current specifics of the service or operation used to submit the request, as well as logical network identifiers like MAC addresses and IP addresses. These combined attributes contribute to a comprehensive profile of the access request, enhancing the trust computation process.

2.7.3 RESOURCE USE PATTERNS

Encompasses the historical utilization of resources, including database requests, network protocols employed to access managed zones, files, and directories accessed on web servers, and the typical applications the user uses. These elements serve as valuable attributes for the trust calculation process. They impart a historical dimension to trust assessment, offering insights into how specific network resources have been employed in connection with the tasks assigned to a particular entity (Figure 2.3).

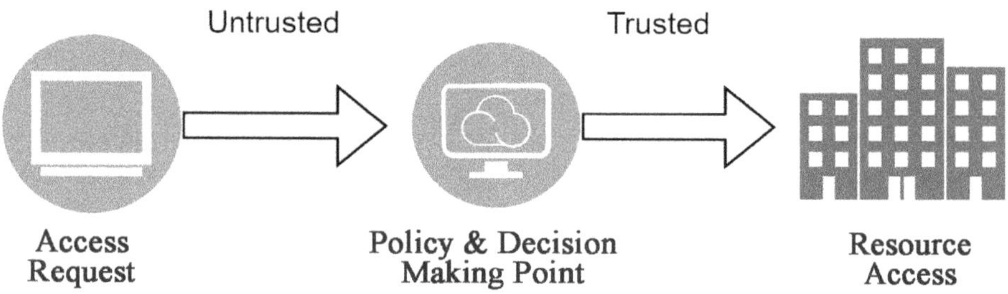

FIGURE 2.3 Graphical representation of zero trust architecture.

2.7.4 RESOURCE CONDITIONS AND STATUS

The security position of specific resources may be adaptively changed within a dynamic network terrain as necessary. This flexibility enables the network to more precisely check access requests aimed at those particular resources. For this case, the authority in charge of the ZTA may specify that the entity requesting access must abide by particular security conditions if the network has lately suffered illegal access or observed aberrant activity. This could include using the most recent firmware and software versions, strictly adhering to secure protocols, and taking responsibility for their conduct once they've gained access. Similar measures are implemented to strengthen security in response to increased threats.

The algorithm can assign fresh characteristics about the network's condition, with the flexibility to attribute varying weights to each trait.

2.8 ZERO TRUST SECURITY

In an era where the threat landscape has evolved to encompass not just external adversaries but also insidious attackers lurking within an organization's digital corridors, the traditional approach to network security has become obsolete. Designing complex systems can be a daunting task, and that's why there are different models like the V-model, waterfall, and spiral model that provide a structured approach to the process. The goal is to ensure that all aspects of the system are comprehended, and potential outcomes for various inputs are considered. That's why subject matter experts are often brought in to handle specific components of the system design, while systems engineers integrate and coordinate the efforts of these experts. Systems engineers also conduct analyses that span different domains, including safety, security, and failure analysis. They employ methods like FMECA, PRA, hazard analysis, and reliability analysis to identify potential issues that could adversely affect the system. Despite all these rigorous methodologies, complex systems can still experience "black swan events" that were either unpredicted or disregarded during the system design phase. Ongoing research aims to address these challenges and enhance the safety and reliability of complex systems.

Zero Trust security is founded on the idea that potential threats can come from inside and outside the network, and this means that trust is no longer automatically granted to users and machines (Figure 2.4).

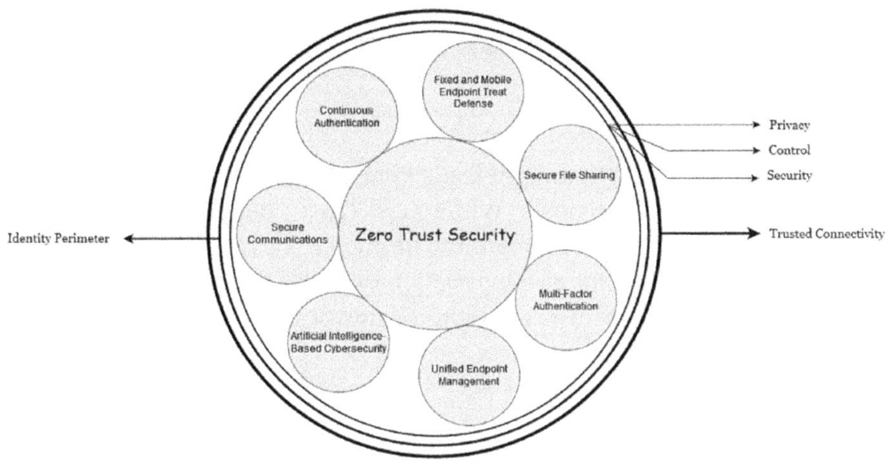

FIGURE 2.4 Graphical representation of zero trust security.

Instead, it revolves around the principle of least privilege, which only grants access to users for what they need, similar to an army general's need-to-know approach. It utilizes micro-segmentation to divide security perimeters into separate zones to maintain strict access control. Overall, the emphasis is on proactive measures, validation, and verification rather than simply granting trust by default. Zero Trust Security is becoming increasingly crucial in modern network security due to several key factors.

Cyberattacks are growing increasingly sophisticated, rendering traditional security models obsolete. Zero Trust's proactive and continuous approach is essential for safeguarding organizations against these evolving threats. Still, Zero Trust Security accommodates this shift by securing access regardless of the user's location or device. This security framework prioritizes data protection at all stages and enforces the principle of least privilege access, decreasing the risk of data breaches and limiting potential damage. Additionally, continuous monitoring, user-centric authentication, and compliance with regulatory requirements are critical aspects of Zero Trust Security. Overall, it enhances business continuity by minimizing the impact of security incidents and offers flexibility and scalability to adapt to an organization's evolving needs and infrastructure. Zero Trust Security's adaptability, user-centric focus, and real-time threat detection make it an indispensable framework for modern network security. Enter Zero Trust Security, a revolutionary paradigm that stands on a singular premise: "Trust No One, Verify Everything."

2.9 CHALLENGES AND FUTURE TRENDS

2.9.1 CHALLENGES

Adopting the Zero Trust paradigm leads to new difficulties that need careful analysis and strategic outcomes. While Zero Trust's benefits are undeniably alluring, successfully overcoming these obstacles is essential for integrating and realizing its full potential. This section explores the many challenges of enacting Zero Trust and offers insightful viewpoints on overcoming these challenges.

2.9.1.1 INTEGRATION AND TECHNICAL DIFFICULTY

Integrating ZT into an organization's framework can be challenging and resource-intensive. Intricate strategic planning and a phased implementation approach may need to be improved by the complexities presented by legacy systems, a variety of remote technologies, and various platforms. The key to unlocking the full benefits that ZT offers is the seamless integration of these dynamic elements, ensuring that they work together harmoniously.

2.9.1.2 SECURITY MEASURES

Balancing the rigorous security measures of ZT with an optimal user experience presents a vital challenge. The continual verification and access controls essential to this approach can potentially lead to user frustration and operational inefficiencies. Addressing this challenge entails developing intuitive interfaces, simplifying authentication procedures, and fine-tuning workflows to guarantee that the bolstered security measures don't inhibit user productivity.

2.9.1.3 SCALABILITY OF ORGANIZATION

The implementation strategies for ZT must be agile enough to accommodate the growth and scalability of the organization. Maintaining the effectiveness of the ZTA throughout the organization's evolution requires meticulous architectural planning and adherence to evolving requirements. In summary, the challenges that come with ZT implementation emphasize the vital need for a comprehensive and strategic approach. By addressing technical, specialized, usability, fiscal, and regulatory considerations, organizations can navigate these challenges and harness the transformative potential of ZT. Successfully overcoming these obstacles empowers organizations to strengthen their security posture and embrace a future where ZT is a flexible and adaptable bulwark against modern cyber threats.

2.10 ENVISIONING THE FUTURE OF ZERO TRUST

The future direction of ZT promises to be both exciting and essential for organizations across various industries. Integrating ZT principles with cutting-edge technologies will be a fundamental trajectory moving forward. The synergy between ZT and emerging technologies such as quantum computing, blockchain, and 5G networks will become increasingly important as the digital landscape evolves. These advancements will enable even more robust and adaptive security postures, allowing organizations to protect their networks and data in a rapidly changing threat environment. Moreover, ZT will likely extend its influence beyond traditional network security and delve deeper into application and device security. With the proliferation of IoT devices and the growing reliance on cloud-native applications, a holistic approach to security is essential. Zero Trust will evolve to encompass these facets, ensuring that every component of an organization's digital ecosystem is subject to the same rigorous security scrutiny. In addition, the role of automation and machine learning in ZT will continue to expand. AI-driven decision-making, coupled with the ability of ZT to analyze context and intent, will further enhance the adaptive nature of cybersecurity. This synergy will result in a security fabric that proactively responds to threats, predicts, and prevents them, reduces reliance on manual intervention, and minimizes the attack surface. In conclusion, the future of ZT security is poised to be a dynamic and transformative journey. Organizations that commit to its principles and adapt to its evolving landscape will be better equipped to navigate the ever-changing cyber terrain with resilience, agility, and unwavering trustworthiness, ensuring the protection of their critical assets and data in an increasingly digital and interconnected world.

2.11 CONCLUSION

The importance of ZT in modern network security cannot be overstated. The ZT paradigm offers a transformative approach to modern cybersecurity, challenging traditional models and providing a forward-looking solution. As our digital landscape continues to develop, so do the threats that seek to compromise our systems and data. Traditional perimeter-based security models can no longer protect against these advanced and persistent threats. A ZT model can minimize the attack surface, reduce the risk of insider threats, and better protect sensitive data and assets.

Furthermore, ZT aligns with the realities of today's interconnected and dynamic IT environments. It acknowledges that threats can come from external and internal sources, requiring constant vigilance and adaptive security measures to counteract them effectively.

Zero Trust is not merely a buzzword; it's a fundamental shift in how we think about and implement network security. Embracing the principles of ZT is a crucial step toward building a robust, resilient, and future-proof security posture that can adapt to the ever-changing threat landscape. It's an investment in any modern organization's long-term success and security.

KEYWORDS

- **access control**
- **advanced persistent threats**
- **authentication**
- **cybersecurity**
- **internet of things**
- **network security**
- **zero trust architecture**
- **zero trust security**

REFERENCES

1. Guembe, B., Azeta, A., Misra, S., Osamor, V. C., Fernandez-Sanz, L., & Pospelova, V. (2022). The emerging threat of AI-driven cyber attacks: A review. *Applied Artificial Intelligence, 36*(1), 2037254. https://doi.org/10.1080/08839514.2022.2037254.
2. Stafford, V. A. (2020). Zero trust architecture. *NIST Special Publication,* 800–207. National Institute of Standards and Technology.
3. Cao, Y., Pokhrel, S. R., Zhu, Y., Doss, R. R. M., & Li, G. (2022). Automation and orchestration of zero trust architecture: Potential solutions and challenges. *Machine Intelligence Research, 21*(10).
4. Ahmad, T., & Zhang, D. (2021). Using the internet of things in smart energy systems and networks. *Sustainable Cities and Society, 68,* 102783. https://doi.org/10.1016/j.scs.2021.102783.
5. Egerton, H., Hammoudeh, M., Unal, D., & Adebisi, B. (2021). Applying zero trust security principles to defense mechanisms against data exfiltration attacks. In *Security and Privacy in the Internet of Things: Architectures, Techniques, and Applications* (pp. 57–89). Springer.
6. Alevizos, L., Ta, V. T., & Eiza, M. H. (2022). Augmenting zero trust architecture to endpoints using blockchain: A state-of-the-art review. *Security and Privacy, 5*(1), e191. https://doi.org/10.1002/spy2.191.
7. Ghasemshirazi, S., Shirvani, G., & Alipour, M. A. (2023a). *Zero Trust: Applications, Challenges, And opportunities.* arXiv preprint arXiv:2309.03582.
8. Astillo, P. V., Choudhary, G., Duguma, D. G., Kim, J., & You, I. (2021). TrMAps: Trust management in specification-based misbehavior detection system for IMD-enabled artificial pancreas system. *IEEE Journal of Biomedical and Health Informatics, 25*(10), 3763–3775. https://doi.org/10.1109/JBHI.2021.3071959.
9. Teerakanok, S., Uehara, T., & Inomata, A. (2021). Migrating to zero trust architecture: Reviews and challenges. *Security and Communication Networks, 2021,* 6692882. https://doi.org/10.1155/2021/6692882.

10. He, Y., Huang, D., Chen, L., Ni, Y., & Ma, X. (2022). A survey on zero trust architecture: Challenges and future trends. *Wireless Communications and Mobile Computing, 2022*, 5263467. https://doi.org/10.1155/2022/5263467.
11. Kindervag, J., et al. (2010). *Build Security into Your Network's DNA: The Zero Trust Network Architecture.* Forrester Research Inc. 27.

CHAPTER 3

Advantages of Zero-Trust Security

NEHA SHARMA,[1] SHUBHANGI MISHRA,[1] MOHIT ANGURALA,[2,*] and YASHIKA SHARMA[1]

[1]*Apex Institute of Technology (CSE), Chandigarh University, Mohali, Punjab, India*

[2]*Department of Computer Science, Guru Nanak Dev University College, Pathankot (Constituent College of Guru Nanak Dev University, Amritsar), Punjab, India*

*Corresponding author

ABSTRACT

In today's interconnected digital environment, traditional security models based on network perimeters face significant limitations against evolving cyber threats. This chapter introduces the Zero-Trust Security framework, which challenges the conventional notion of inherent trust within networks. Adopting a "never trust, always verify" philosophy, Zero Trust continuously validates users, devices, and applications, ensuring secure access only after rigorous checks. The discussion emphasizes core aspects of Zero Trust, including identity-focused verification, micro-segmentation to prevent lateral attacks, minimal privilege allocation, ongoing activity monitoring, and the integration of multi-factor authentication. These components work cohesively to mitigate risks and establish robust defenses. The framework is particularly effective in addressing insider threats, protecting sensitive data, and ensuring compliance with regulations such as GDPR and HIPAA.

3.1 INTRODUCTION

The idea of security has changed significantly in the quickly changing digital environment of today. The increasingly sophisticated attacks that

Zero-Trust Learning: Applications in Modern Network Security. Chetna Kaushal, Amandeep Kaur, Mohit Angurala, Aryan Chaudhary (Eds.)

corporations and organizations must contend with have shown that traditional perimeter-based security strategies fall short of securing critical data and systems. The concept of Zero Trust security becomes revolutionary in this situation.

The way we approach cybersecurity has changed, and Zero Trust security is one example of this. Zero Trust operates under the presumption that no entity, whether inside or outside the network, can be trusted by default, unlike traditional security models that depend on the idea of trust within the network perimeter. Until otherwise demonstrated, every person, device, and application is considered potentially unreliable [1].

3.1.1 BRIEF RECAP OF WHAT ZERO TRUST SECURITY IS?

Zero Trust security is a methodology and concept for cybersecurity that operates under the guiding principle of "never trust, always verify." Whether a user, device, or program is inside the network perimeter or not, in a zero-trust paradigm, trust is never inferred. Instead, before having access to resources, each entity is continuously verified [1].

Zero Trust security's essential components included here.

3.1.1.1 IDENTITY-CENTRIC SECURITY

Zero Trust places a strong emphasis on confirming the identities of people and devices through authentication, authorization, and ongoing monitoring.

3.1.1.2 MICRO-SEGMENTATION

In order to prevent lateral movement by possible attackers and to ensure that even if one segment of a network is compromised, access to the rest of the network is not immediately granted, networks are segmented into smaller sections.

3.1.1.3 LEAST PRIVILEGE ACCESS

In order to minimize the attack surface, users, and systems are only given the access necessary to carry out their particular tasks.

3.1.1.4 CONTINUOUS MONITORING

In order to identify anomalies or threats in real-time, it is crucial to continuously monitor network traffic and user activity.

3.1.1.5 MULTI-FACTOR AUTHENTICATION (MFA)

Frequently, MFA is used to increase security by requiring several forms of identification before providing access.

Overall, Zero Trust security aims to reduce the risk of unauthorized access and data breaches by constantly vetting the legitimacy of entities trying to access resources. It is designed to improve an organization's cybersecurity posture by assuming that threats exist both inside and outside the network [2].

Figure 3.1 shows that zero-trust security model.

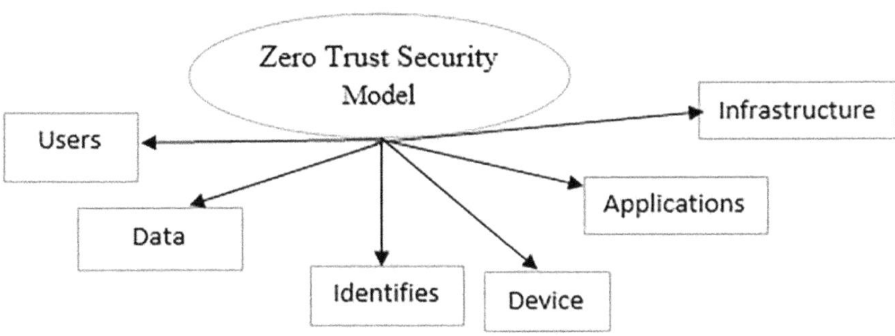

FIGURE 3.1 Zero trust security model.

3.2 THE IMPORTANCE OF UNDERSTANDING THE ADVANTAGES OF ZERO TRUST

In the current cybersecurity environment, understanding the benefits of Zero Trust security is listed here.

3.2.1 ENHANCED SECURITY POSTURE

A proactive strategy called "Zero Trust" lowers the attack surface by avoiding the assumption of trust. Organizations can improve their security posture by strengthening their understanding of its benefits, which makes it much more difficult for cyber threats to access networks and compromise data.

3.2.2 PROTECTION AGAINST INSIDER THREATS

Insider threats, which are frequently a major security risk, can be reduced by businesses by adopting a Zero Trust mentality. Constantly monitoring both internal and external activity enables early detection of suspect conduct and rapid remediation.

3.3.3 ADAPTATION TO MODERN WORK ENVIRONMENTS

Traditional perimeter-based security solutions are being rendered obsolete by the growth of remote work and cloud computing. Organizations can adapt to these contemporary work contexts, where personnel and resources are dispersed across numerous locations, by understanding zero trust [3].

3.3.4 DATA PRIVACY AND COMPLIANCE

For many firms, adherence to data protection laws like GDPR and HIPAA is essential. Controlling access and keeping track of data flows helps Zero Trust maintain data privacy and security requirements, lowering the likelihood of regulatory penalties and legal implications.

3.3.5 REDUCED CYBERSECURITY RISKS

Reducing cybersecurity risks, such as data breaches and cyberattacks, is one of Zero Trust's benefits. Understanding these advantages can help firms save a lot of money while safeguarding their reputations.

3.3.6 FLEXIBILITY AND SCALABILITY

Zero Trust is not a universally applicable solution. Understanding its benefits helps organizations, whether they are small businesses or large enterprises, to customize their security measures to their own demands.

3.3.7 RESILIENCE TO EVOLVING THREATS

The cybersecurity environment is ever-changing, with new threats appearing frequently. With its constant monitoring and adaptation, Zero Trust supports enterprises in remaining resilient in the face of changing cyber threats.

3.3.8 STRATEGIC ADVANTAGE

Businesses that adopt Zero Trust security can establish themselves as industry leaders in cybersecurity, thereby obtaining a competitive advantage and gaining the confidence of clients and partners.

3.3.9 ENHANCED DATA PROTECTION

Data is kept secure even in the case of network intrusions thanks to Zero Trust security. Particularly when working with sensitive consumer information and legal compliance standards like GDPR or HIPAA, this is essential.

3.3.10 REDUCED ATTACK SURFACE

Micro-segmentation is frequently used in Zero Trust implementation to lower the attack surface within the network. As a result, if one section of the network is hacked, access to other portions of the network is not immediately granted.

3.3.11 BUSINESS CONTINUITY

Organizations may guarantee business continuity and avoid expensive disruptions brought on by data breaches by reducing the likelihood of successful cyberattacks.

3.3.12 COST SAVINGS

Zero Trust may need to be implemented initially, but over time it can save money by preventing expensive security breaches and the cleanup and legal costs that come with them.

3.3.13 *COMPETITIVE ADVANTAGE*

Zero Trust security can give organizations a competitive edge if they adopt it. Customers and partners frequently prefer doing business with organizations that put data security and privacy first.

In conclusion, it is critical to comprehend the benefits of Zero Trust security because it offers a strategic framework for safeguarding against a variety of cyber threats, responding to shifting work settings, and guaranteeing regulatory compliance. It is a proactive strategy that can greatly improve an organization's cybersecurity posture in a world that is becoming more connected and digital [4].

3.4 ENHANCED SECURITY

In order to provide a better level of protection against various threats and dangers, enhanced security refers to a strategy or collection of actions that go above and beyond the fundamental or industry-standard security standards. Standard practices for enhanced security implementation are included here.

3.4.1 *INFORMATION SECURITY*

Information security in the modern era is essential. Advanced encryption methods, MFA, intrusion detection and prevention systems (IDPS), and regular security audits and assessments are some examples of enhanced security measures.

3.4.2 *PHYSICAL SECURITY*

Protecting physical assets, including buildings, data centers, and crucial infrastructure, requires the employment of enhanced physical security measures. This can involve security guards, alarm systems, surveillance cameras, and biometric access control.

3.4.3 *NETWORK SECURITY*

Data must be protected when traveling through networks, which requires network security. Firewalls, intrusion detection systems (IDS), intrusion prevention systems (IPS), and cutting-edge threat intelligence are frequently used to enhance network security.

3.4.4 CYBERSECURITY

Threat hunting, threat intelligence sharing, and the use of cutting-edge technologies and procedures to detect and address cyber threats are some examples of better security practices in the field of cybersecurity.

3.4.5 DATA SECURITY

Protecting sensitive data from unwanted access or disclosure is the main goal of improved data security. Data encryption, data masking, data loss prevention (DLP) technologies, and strong access controls may all be used in this.

3.4.6 APPLICATION SECURITY

For software and web applications to be protected from flaws and assaults, enhanced application security is crucial. Penetration testing, secure coding techniques, and frequent security updates are all part of it [5].

3.5 EXPLANATION OF HOW ZERO TRUST ELIMINATES THE ASSUMPTION OF TRUST

A cybersecurity methodology and idea called "Zero Trust" does away with the conventional premise of trust in the security architecture of a network or organization. In the past, network security was predicated on a perimeter-based concept, in which most devices and people inside the network perimeter were trusted. Once a user or device had network access, they were frequently given unrestricted access to systems and resources based on this initial trust. The insider threat, advanced persistent threat (APT), and sophisticated external attackers are only a few examples of the modern cybersecurity risks that have shown that this strategy is insufficient to protect against them.

Zero Trust does away with the notion of trust in the following ways.

3.5.1 AUTHENTICATION AND VERIFICATION

- The foundation of Zero Trust is the maxim "never trust, always verify." It necessitates constant authentication and verification of all users and devices, independent of their location or manner of network access, as opposed to assuming trust based on network segment or location.
- Multiple factors are used to authenticate users and devices, including things users know (passwords), have (smart cards, tokens), and are (biometrics). Only parties with authorization can access the system thanks to the MFA [6].

3.5.2 MINIMUM PRIVILEGE ACCESS

- Zero Trust imposes access restrictions on resources based on the least privilege principle. This means that users and devices are only given access that is absolutely necessary for them to carry out their tasks.
- According to shifting circumstances, user roles, and device positions, access is dynamically changed. Users are only given access to the resources they need to do their jobs, which lessens the chance of lateral attacker movement.

3.5.3 CONTINUOUS MONITORING

- The behavior of users and devices is continuously monitored by Zero Trust. In order to do this, data on user behavior, device health, network activity, and application usage must be gathered and analyzed.
- Alerts are sent out in reaction to anomalies and suspicious conduct, and automated responses can be started to neutralize any threats. Real-time detection and response to security incidents are made possible through continuous monitoring [7].

3.5.4 MICRO-SEGMENTATION

- Segmenting networks is a crucial component of Zero Trust. Firewalls, network access controls, and other security measures are used to segment networks into smaller, isolated pieces known as micro-segments.

- In order to prevent attackers from simply moving to other portions of the network even if they obtain access to one segment, traffic between segments is tightly monitored, and restrictions are enforced at the application level.

3.5.5 DATA ENCRYPTION

- Zero Trust stresses the use of encryption to safeguard data while it is in use and at rest. To avoid data theft and eavesdropping, all communication between systems and devices should be encrypted.
- Encrypting data on endpoints, as it travels over networks, as well as while it is stored in databases or cloud storage, is part of this [8].

3.6 WHY ARE TRADITIONAL SECURITY MODELS VULNERABLE TO INSIDER THREATS AND LATERAL MOVEMENT?

Traditional security models are susceptible to lateral movement and insider threats for a variety of reasons are listed here.

3.6.1 TRUST-BASED MODEL

- Traditional security models frequently adopt a trust-based strategy, whereby once a user or device has access to a network or specific system, they are given a high degree of trust and unrestricted access to resources. Users and equipment inside the network perimeter are also covered by this trust.
- Insiders, especially workers and contractors, frequently take advantage of this trust because they already have legal access to specific resources, making it simpler for them to engage in malevolent acts without immediately raising suspicion.

3.6.2 INSUFFICIENT ACCESS CONTROLS

- Strong access restrictions may be missing from traditional security methods. Users frequently receive a lot of extra permissions to make their jobs easier, which might be abused by insiders.

- Insiders with valid access might misuse their control over private information and systems, resulting in data breaches, data theft, and other harmful behavior [9].

3.6.3 *LIMITED MONITORING*

- Traditional security methods often place a greater emphasis on perimeter defense and external threats than they do on internal network activities, which results in inadequate monitoring.
- This scant surveillance presents an opportunity for insiders to launch assaults undetected. For instance, they can gradually siphon off data over a long period of time, making it challenging to pinpoint their activity.

3.6.4 *WEAK AUTHENTICATION METHODOLOGIES*

- Many conventional security methods rely heavily on usernames and passwords for authentication, which can be exploited by a number of techniques, such as social engineering, phishing, or credential theft.
- Once an insider has access to another user's login information, they can assume that user's identity and move covertly throughout the network.

3.6.5 *SEGMENTATION ISSUES*

- Traditional networks frequently lack effective segmentation, which means that once an insider has access to one region of the network, they might have fairly simple access to other parts of the network.
- Because there is no segmentation, insiders can roam more easily between systems, potentially accessing confidential information or causing network disturbances.

3.6.6 *INSUFFICIENT INSIDER THREAT DETECTION*

- Dedicated tools for identifying insider threats or atypical user behavior may not exist in traditional security frameworks.

- Insiders can avoid discovery until major damage has already been done if they gradually ramp up their operations or conceal their actions inside regular network traffic [10].

3.7 HOW DOES ZERO TRUST HELP IN REDUCING THE ATTACK SURFACE?

The network and systems of an organization play a significant role in lowering their attack surface thanks to Zero Trust. The total number of potential access points and vulnerabilities that can be exploited by attackers is referred to as the attack surface. Organizations can reduce their attack surface in several ways by applying Zero Trust concepts and strategies.

3.7.1 LEAST PRIVILEGE AND ACCESS CONTROL

Zero Trust upholds the least privilege concept, which states that users and devices are only given the minimal amount of access necessary to complete their jobs. As a result, reducing unnecessary access to important resources decreases the possible attack surface. It will be more difficult for attackers to move laterally and access crucial systems or data, even if they manage to obtain a user's login credentials.

3.7.2 MICRO-SEGMENTATION

Zero Trust divides the network into smaller, isolated portions through the use of network segmentation. Typically, these parts are created using the least privilege principle, and communication between them is strictly regulated. This reduces the attack surface and makes lateral movement more challenging by restricting the channels that attackers can utilize to travel around the network.

3.7.3 CONTINUOUS MONITORING

Zero Trust emphasizes continuous monitoring of user and device behavior. This means that any unusual or suspicious activity can be detected and responded to in real-time. By closely monitoring network traffic and user behavior, organizations can identify potential threats early and reduce the window of opportunity for attackers to exploit vulnerabilities [11].

3.7.4 STRONG AUTHENTICATION AND MULTI-FACTOR AUTHENTICATION (MFA)

Zero Trust requires strong authentication practices, including the use of MFA. This ensures that only authorized users gain access to systems and data, reducing the risk of credential theft and unauthorized access.

3.7.5 APPLICATION LAYER SECURITY

Zero Trust frequently applies security controls at the application layer. Access controls and application-level firewalls are included in this. Organizations can reduce the attack surface associated with each application and stop attackers from taking advantage of application vulnerabilities by securing individual applications.

3.7.6 ENDPOINT SECURITY

Zero Trust imposes security policies on endpoints, such as laptops, smart-phones, and IoT devices, to further extend protection. Ensuring that endpoints are routinely patched, updated, and in compliance with security standards, lowers the possible attack surface.

3.7.7 DATA ENCRYPTION

Zero Trust emphasizes data encryption, both in transit and at rest. Encrypting data reduces the risk of data interception and tampering, further narrowing the attack surface.

3.7.8 REMOTE ACCESS CONTROLS

In a zero-trust model, remote access to the network and systems is carefully controlled and monitored. This reduces the potential exposure to threats coming from external networks or remote devices.

3.7.9 DYNAMIC ACCESS POLICIES

Access policies in a zero-trust environment are dynamic and can change in response to the situation. For instance, based on the user's role, device posture, location, and other contextual criteria, access may be denied or permitted. Tailoring access in real-time to the user's current demands and level of trustworthiness lessens the attack surface [12].

3.8 IMPROVED DATA PROTECTION

Modern cybersecurity methods must include improved data protection. It entails putting policies and procedures in place to protect delicate and priceless data from illegal access, theft, loss, or corruption. In a world that is becoming more digital and linked, improved data protection is crucial to preserving the privacy, integrity, and confidentiality of data. The following are some crucial components and tactics for enhancing data protection.

3.8.1 DATA ENCRYPTION

Data encryption is a fundamental data security strategy that encrypts data both in transit and at rest. Without the necessary decryption keys, encryption ensures that even if unauthorized people or organizations access data, they cannot read or understand it.

3.8.2 ACCESS CONTROL

To limit data access to approved individuals and devices, implement strong access controls and authorization methods. In order to ensure that users and systems only have access to the data necessary for their unique responsibilities or duties, the principle of least privilege should be observed.

3.8.3 AUTHENTICATION AND IDENTITY MANAGEMENT

Implement robust authentication techniques, such as MFA, to confirm the identities of users and devices accessing data. Identity and Access Management (IAM) systems that are effective aid in ensuring that only authorized users have access to data.

3.8.4 DATA CLASSIFICATION

Arrange data according to its significance and sensitivity. Due to the ability to prioritize protection efforts, businesses can apply stricter security controls to extremely sensitive data.

3.8.5 DATA LOSS PREVENTION (DLP)

Use DLP systems to track and stop the unlawful transfer of sensitive data or its leakage. Data transmission violations can be found and stopped using DLP tools.

3.9 DATA MASKING AND REDACTION

Use data masking or redaction techniques to hide or remove sensitive information while maintaining the usefulness of the data when sharing it with others or using it for testing.

3.9.1 REGULAR BACKUPS

Implement a reliable backup method on a regular basis to assure data availability in the event of data loss, corruption, or cyberattacks. Regular backups should be made and safely stored.

3.9.2 SECURE DEVELOPMENT TECHNIQUES

Adhere to secure coding techniques while creating applications and systems that handle sensitive data to avoid security holes that could be exploited by hackers.

3.9.3 SECURITY PATCH MANAGEMENT

Maintaining the most recent security patches on all software and systems will protect them from known vulnerabilities that might be used to access data without authorization.

3.9.4 SECURITY AWARENESS TRAINING

Inform users and workers about security best practices, such as the value of secure passwords, how to spot phishing scams, and how to play a part in data protection [13].

3.9.5 INCIDENT RESPONSE PLAN

Create and keep up-to-date an incident response strategy that specifies how to respond quickly to security issues or data breaches. Damage and data exposure can be reduced with a well-planned response.

3.9.6 COMPLIANCE WITH DATA PRIVACY RULES

To preserve individuals' privacy rights and prevent legal repercussions, ensure compliance with data privacy rules, such as the General Data Protection Regulation (GDPR) or the California Consumer Privacy Act (CCPA).

3.9.7 ENCRYPTION OF DATA ON THE CLOUD

Implement encryption and access controls to secure data stored in cloud environments if you're using cloud services. Numerous cloud service providers offer products and services that enhance data protection.

3.9.8 DATA AUDITING AND MONITORING

Implement auditing and monitoring systems to track who accesses data and when, as well as to spot unusual or suspicious activity that could be a sign of a security incident [14].

3.10 HOW ZERO TRUST HELPS IN SECURING SENSITIVE DATA?

Secure sensitive data using the cybersecurity framework and method known as "Zero Trust." It is based on the fundamental tenet of "never trust, always verify," which states that no user or device can be trusted without first being authenticated and authorized, even if they are already on the network. Zero Trust helps protect sensitive data in the following ways.

3.10.1 GRANULAR ACCESS CONTROL

Zero Trust implements granular access controls to ensure that individuals and devices only have access to the specific information and resources they need to complete their activities. The risk of sensitive information being disclosed is reduced thanks to the least privilege concept.

3.10.2 CONTINUOUS AUTHENTICATION

Throughout their sessions, users, and devices are continuously authenticated and re-authenticated. As a result, there is a lower chance of unwanted access caused by stolen credentials or session hijacking.

3.10.3 STRONG AUTHENTICATION

MFA, which adds an additional layer of protection beyond passwords, is one method of strong authentication that Zero Trust emphasizes. As a result, there is a far lower chance of sensitive data being accessed by unauthorized parties.

3.10.4 MICRO-SEGMENTATION

A crucial component of Zero Trust is network segmentation. Smaller, isolated network parts decrease the attack surface for potential attackers. Sensitive data can be contained in segments with strict access controls.

3.10.5 DATA ENCRYPTION

Zero Trust supports the use of data encryption for sensitive information, both in transit and at rest. Data that has been encrypted is protected such that, even if it is intercepted, it cannot be decoded without the right decryption keys [15].

3.10.6 USER AND ENTITY BEHAVIOR ANALYTICS (UEBA)

Zero Trust frequently uses UEBA technologies to track and examine user and device behavior. Real-time detection and response are provided for anomalous or suspicious behaviors that can point to unauthorized access to sensitive data.

3.10.7 DYNAMIC ACCESS POLICIES

In a Zero Trust architecture (ZTA), access policies are flexible and context-sensitive. Decisions about access are made in light of factors like location, device health, user roles, and more. The risk of unwanted data access is decreased due to this dynamic approach, which ensures that access rights adjust to changing conditions.

3.10.8 DATA LOSS PREVENTION (DLP)

To track and stop the improper transmission or leaking of sensitive data, Zero Trust can use DLP solutions. To safeguard sensitive data assets, DLP policies can be applied at multiple locations throughout the network.

3.10.9 ENDPOINT SECURITY

Zero Trust imposes security policies on endpoints, such as laptops and cellphones, to further extend protection. As a result, sensitive data is safeguarded on the computers and other devices where it is accessed and used [16].

3.11 DATA CLASSIFICATION AND ENCRYPTION IN A ZERO TRUST MODEL

To safeguard sensitive information and reduce the risk of data breaches, data classification and encryption are crucial elements of a zero-trust architecture. The following describes how data encryption and classification work within a zero-trust architecture.

3.11.1 CLASSIFICATION OF DATA

Data categorization entails grouping data according to its significance, value, and sensitivity to the organization. Security precautions and access controls can be prioritized with the aid of this classification in a model of Zero Trust.

3.11.1.1 IDENTIFICATION OF SENSITIVE DATA

The first step is to identify sensitive data. This can include any information that is essential to the organization, such as personally identifiable information (PII), financial information, intellectual property, and others.

3.11.1.2 DATA LABELING

After sensitive information has been located, it is usually marked with the appropriate classification level. Examples of classifications for data are "Confidential," "Restricted," and "Public."

3.11.1.3 POLICY-BASED ACCESS CONTROL

Data access is controlled by policies that are in line with the classification level of the data. Zero Trust upholds these rules by routinely confirming the legitimacy of users, devices, and their access rights. Users are only given access to data that corresponds to their level of clearance and the needs of their jobs.

3.11.1.4 DYNAMIC ACCESS POLICIES

In a Zero Trust paradigm, access policies are flexible and can change in response to new information. For instance, if an employee changes jobs or locations, their access to sensitive information may be restricted or canceled, ensuring that access is consistent with the principle of least privilege.

3.11.1.5 LOGGING AND AUDITING

To track who accessed the data and when, activities involving sensitive data are meticulously logged and audited. Alerts and actions are triggered by any suspicious or unauthorized access attempts [17].

3.11.2 ENCRYPTION OF DATA

By using data encryption, you can make sure that even if someone gains access to your data without your permission, they won't be able to read or understand it without the right decryption keys. A Zero Trust framework's data protection strategy must include encryption.

3.11.2.1 ENCRYPTION AT REST

Sensitive data should be encrypted before being kept in databases, file servers, or other types of storage. Information is protected from unauthorized access if the physical storage media are compromised by data encryption at rest.

3.11.2.2 ENCRYPTION IN TRANSIT

Data transported across networks should also be encrypted in order to prevent listening in and intercepting. To secure data in transit, VPNs, and secure protocols like HTTPS are frequently employed.

3.11.2.3 ENDPOINT ENCRYPTION

To safeguard data in the event that a device is lost or stolen, storage drives on devices (such as laptops and mobile devices) that access sensitive information should be encrypted.

3.11.2.4 APPLICATION-LEVEL ENCRYPTION

Encryption can be used at the application layer for programs that deal with sensitive data. This makes sure that data is secure, even when it is stored in the application's databases.

3.11.2.5 KEY MANAGEMENT

For encryption, sound key management is crucial. To prevent unwanted access or key loss, encryption keys should be carefully handled and preserved.

3.11.2.6 ACCESS CONTROLS AND ENCRYPTION KEYS

In accordance with the zero-trust philosophy, access to decryption keys should be rigorously regulated. The keys required to decrypt sensitive data should only be accessible to systems or users with the proper authorization [18].

3.12 CASE STUDIES OR EXAMPLES ILLUSTRATING DATA BREACHES PREVENTED BY ZERO TRUST

There are numerous case studies and examples showing how businesses applied the Zero Trust principles and either prevented or significantly reduced the effect of security breaches. Remember that the specific implementation and the changing threat landscape may have an impact on how effective Zero Trust is. Here are a few illustrations of how access to the keys required to decode sensitive information should be managed.

3.12.1 IMPLEMENTATION OF BEYONDCORP BY GOOGLE

- With its BeyondCorp concept, Google is a pioneer in the application of Zero Trust. With Google's strategy, the idea of a trusted network is completely gone.
- In one case, Google used the Zero Trust principles to stop a data leak. An employee unintentionally handed over their login information to an attacker after falling victim to a phishing scam. However, the attacker was unable to access private resources due to Google's Zero Trust design.
- At Google, the concept of Zero Trust meant that even with compromised credentials, access to confidential systems and data was not possible. The attacker's advancement was successfully stopped.

3.12.2 DATA BREACH MITIGATION BY ANTHEM

- One of the biggest health insurance providers in the country, Anthem, experienced a significant data breach in 2015 that exposed 10 million of records.
- Anthem deployed Zero Trust security measures as a result of the hack. This featured rigorous access limits, ongoing monitoring, and MFA for all users.

- Anthem used a Zero Trust strategy to ensure that even if attackers were to obtain initial access, it would be very challenging for them to travel laterally within the network and access private healthcare records.

3.12.3 ZERO TRUST GOVERNMENT AGENCIES

- To improve security, numerous government organizations have adopted the Zero Trust philosophy.
- As an example, the "Zero Trust Reference Architecture" was developed by the US Department of Defense (DoD) to safeguard sensitive military data.
- The DoD hopes to increase security and lower the possibility of data breaches that could jeopardize national security by introducing Zero Trust.

3.12.3 ADOPTION OF ZERO TRUST BY CISCO

- Zero Trust was implemented by the multinational technology company Cisco as part of their cybersecurity plan.
- According to Cisco, the use of Zero Trust principles enabled them to identify and contain a third-party vendor-related data compromise. Due to the establishment of access controls and ongoing monitoring, the incident was contained and its scope was restricted [19].

3.12.4 STRONGER AUTHENTICATION

Modern cybersecurity strategies, intended to improve the security of systems, apps, and sensitive data, must include stronger authentication as a key component. In order to guarantee that only authorized users or systems obtain access, it includes implementing authentication techniques that go beyond conventional username and password combinations. To lessen the dangers connected with hacked or weak credentials, stronger authentication mechanisms are put in place. Stronger authentication has the following essential components.

3.12.4.1 MULTI-FACTOR AUTHENTICATION

One of the most popular and efficient ways to strengthen authentication is through MFA. Users must supply two or more authentication factors, often from the following groups:

- Something you are aware of (like a PIN or password).
- A possession (such as a token, smart card, or mobile device).
- Something you are, like iris scans, facial recognition, or biometric information like fingerprints.

Even if an attacker were to learn a user's password, they would still require the extra factor(s) to get access, making MFA a considerable security improvement.

3.12.4.2 BIOMETRIC AUTHENTICATION

Biometric authentication techniques use a person's distinctive biological traits, like their fingerprints, retinal scans, or facial recognition, to confirm their identification. These techniques offer a high level of security and are challenging to forge.

3.12.4.3 SMART CARDS AND TOKENS

Users carry around real tokens called smart cards. For added protection, they generate one-time passwords or cryptographic keys for authentication.

3.12.4.4 PUSH NOTIFICATIONS

When a person tries to log in using a mobile app, a push notice is sent to their device. Even if credentials are stolen, unlawful access can be stopped if the user approves the login request.

3.12.4.5 TIME-BASED ONE-TIME PASSWORDS (TOTP)

TOTP is a technique that creates one-time passwords with a time constraint. To complete authentication, users must input the most recent TOTP that is presented on their device.

3.12.4.6 PUBLIC KEY INFRASTRUCTURE (PKI)

PKI authenticates users using pairs of public and private keys. It is frequently utilized in VPNs, digital signatures, and encrypted email communication.

3.12.5 BEHAVIORAL AUTHENTICATION

To confirm identity, behavioral biometrics examine user behavior patterns such typing speed, keystroke dynamics, and mouse movements.

3.12.6 RISK-BASED AUTHENTICATION

Using information about the user's location, device, and activity, this approach evaluates the risk involved with a login attempt. Additional authentication processes may be necessary if the risk is judged to be high.

3.12.7 PASSWORDLESS AUTHENTICATION

Passwordless authentication eliminates the necessity for conventional passwords by using techniques such as biometrics, hardware tokens, or mobile app-based authentication to confirm users' identities.

3.12.8 IDENTITY FEDERATION AND SINGLE SIGN-ON (SSO)

Users who use SSO only need to input their credentials once to access several applications. While maintaining strong authentication, identity federation expands this idea to include external systems and services.

3.13 THE ROLE OF MULTI-FACTOR AUTHENTICATION (MFA) IN ZERO TRUST

The implementation of a Zero Trust security paradigm relies heavily on MFA. In a cybersecurity paradigm known as "Zero Trust," no one is trusted, not even users and devices that are inside the network perimeter. Zero Trust relies on MFA because it adheres to the guiding philosophy of "never trust, always verify," how MFA strengthens and supports a zero-trust architecture.

3.13.1 IDENTIFICATION CONFIRMATION

- MFA is used in a zero-trust approach to continuously verify the identities of users and devices. It confirms the identity of the entity attempting to access a system or resource.
- MFA calls for users to submit multiple kinds of identification, usually something they are aware of (like a password or PIN) and something they own (like a mobile device or hardware token). This makes it far more difficult for attackers using stolen credentials or unauthorized users to obtain access.

3.13.2 ADDITIONAL ACCESS CONTROL

- MFA supports access control by introducing an additional layer of protection. An attacker would still require the second factor (e.g., a smartphone with an authentication app) to successfully authenticate even if the user's password was compromised.
- Zero Trust access policies can be set up to demand MFA for particular resources or depending on contextual considerations like the user's location or the importance of the material being accessed. As a result, the attack surface is further reduced and adheres to the principle of least privilege.

3.13.3 PERMANENT AUTHENTICATION

- Continuous monitoring and authentication are mandated by Zero Trust. MFA solutions can assist with this by occasionally asking users for additional verification throughout their sessions or in reaction to changes in the environment.
- MFA can be used to re-verify a user's identity in cases when their behavior or context raises suspicion (e.g., if they unexpectedly log in from a foreign country). This lowers the danger of unwanted access [21].

3.13.4 ADAPTIVE SECURITY

- Contextual data, such as the user's device posture, location, and network circumstances, can be incorporated with MFA solutions. This makes it possible to implement adaptive authentication policies that can change the level of authentication needed on the fly in response to risk considerations.

- For instance, a user connecting from a trusted corporate device inside the workplace might need fewer authentication factors than a person connecting from an unknown device or location.

3.13.5 CREDENTIAL-BASED ATTACK PREVENTION

- Zero Trust recognizes the prevalence of vulnerabilities posed by credential-based assaults, including phishing and credential stuffing. Even if an attacker obtains login credentials, they won't be able to access resources without the additional verification factor, making MFA an effective defense against these attacks.
- By doing this, attacks that only use stolen passwords become less effective.

3.13.6 PRIVACY PROTECTION

MFA improves data security in a Zero Trust environment. It provides an additional layer of protection to safeguard sensitive data by guaranteeing that only authorized and authenticated individuals can access it [22].

3.14 HOW DOES STRONG AUTHENTICATION REDUCE THE RISK OF UNAUTHORIZED ACCESS?

Strong authentication adds levels of security beyond conventional username and password combinations, considerably lowering the risk of unauthorized access. It accomplishes this by requesting various forms of identity proof from users. Strong authentication can reduce the danger of unauthorized access in the following ways.

3.14.1 AVOIDING THE USE OF STOLEN OR COMPROMISED CREDENTIALS

- The compromising of user credentials (such as usernames and passwords) is one of the main dangers associated with unauthorized access. Passwords that are overused or weak can be readily guessed or stolen through phishing, credential stuffing, and other techniques.

- Strong authentication measures, such as MFA, ensure that even if an attacker manages to get their hands on a user's password, they will still be unable to access the system without the extra authentication factors (such as a one-time code from a mobile app).

3.14.2 REDUCED PHISHING ATTACK RISK

Users are frequently tricked by phishing attacks into providing their login information to phony websites or criminal actors. Because attackers cannot access accounts or systems using only stolen credentials, strong authentication techniques like MFA are effective in thwarting phishing efforts.

3.14.3 CREDENTIAL STUFFING PREVENTION

Using stolen credentials to obtain unwanted access to numerous accounts or services where users have repeated passwords is known as a "credential stuffing attack." These attacks are stopped by strong authentication measures, such as MFA, which need additional verification factors.

3.14.4 ENHANCED ACCESS CONTROL

- Access control mechanisms are strengthened with strong authentication. It guarantees that users can only access resources, applications, or systems if they have the required authentication factors.
- Even if a user's password is stolen, illegal access is still prevented since the attacker lacks the extra authentication elements.

3.14.5 PROTECTION FROM BRUTE FORCE ATTACKS

Brute force attacks entail making repeated guesses at passwords. Because each authentication attempt needs a different code, strong authentication techniques, particularly those that use hardware- or time-based one-time codes, are immune to brute force assaults [23].

3.14.6 CONTINUOUS CHECKING

Throughout a user's session, continuous verification is supported by some strong authentication technologies. This lowers the danger of unauthorized access caused by a hijacked session even after the user has logged in for the first time by prompting them to re-authenticate on a regular basis or in reaction to specific occurrences.

3.14.7 AUTHENTICATION IN THE CONTEXT

Strong authentication may take into account the user's location, device, and network in addition to general context. Additional authentication factors might be needed if the authentication context changes (e.g., if the user signs in from a strange place), making it harder for unauthorized people to acquire access.

3.14.8 SECURITY OF SENSITIVE INFORMATION

Sensitive data is further secured by strong authentication. It minimizes the risk of data breaches by ensuring that only authorized users with the appropriate authentication factors can access and interact with sensitive information.

3.14.9 LIMITING ACCESS TO UNAUTHORIZED DEVICES

Strong authentication techniques can be employed in addition to user authentication to confirm the authenticity of devices. This makes it more difficult for unapproved devices to access business networks or resources.

3.14.10 CONDITIONS FOR COMPLIANCE

For access to sensitive data or systems, several regulatory standards and compliance frameworks demand strong authentication. Strong authentication can help firms comply with regulations while lowering the danger of illegal access [19].

3.15 REAL-WORLD EXAMPLES OF AUTHENTICATION-RELATED BREACHES AND HOW ZERO TRUST COULD HAVE PREVENTED THEM. HERE ARE A FEW ILLUSTRATIONS

3.15.1 *EQUIFAX DATA BREACH (2017)*

Attackers used a known vulnerability in an Apache Struts web application during the Equifax breach to obtain unauthorized access to sensitive customer data. Poor authentication procedures made the breach worse because the corporation neglected to repair the flaw and did not use strong authentication to sufficiently safeguard a crucial database. By implementing stringent access restrictions, ongoing monitoring, and adaptive authentication standards, Zero Trust might have avoided this breach. Even if attackers had successfully exploited the vulnerability, accessing sensitive databases would have required a robust authentication process, decreasing the likelihood of unauthorized access.

3.15.2 *TARGET DATA BREACH (2013)*

In the Target data breach, hackers stole login information from a third-party HVAC contractor and used it to access the company's point-of-sale (POS) systems. Customer credit card information was obtained as a result of the stolen credentials granting illegal access to Target's network. By introducing MFA and ongoing authentication for external suppliers accessing the network, Zero Trust principles could have minimized this breach. The POS systems could not have been accessed using just the stolen passwords due to strict authentication requirements and constant verification.

3.15.3 *YAHOO DATA BREACH (2013–2016)*

Numerous data breaches at Yahoo resulted in the exposure of billions of user accounts. These hacks utilized stolen or reused credentials that were obtained in a variety of ways, such as account takeovers.

Trust's implementation of MFA, ongoing surveillance for irregularities, and contextual authentication may have contributed to the prevention of these incidents. For instance, it would have been more difficult for attackers to effectively utilize stolen credentials if login attempts from unknown devices or locations prompted additional authentication checks.

3.15.4 MARRIOTT INTERNATIONAL DATA BREACH (2018)

In the Marriott data breach, hackers used stolen credentials to access the company's Starwood guest reservation database. Millions of visitors' personal information was exposed due to the hack. Strong authentication and ongoing monitoring are two zero-trust concepts that may have lowered the risk of illegal access. Continuous authentication and behavioral analysis may have found anomalous activity even if attackers had obtained valid credentials, setting off alerts and a security response.

3.16 OPM DATA BREACH (2015)

Millions of government workers and contractors had their sensitive personal information compromised by the U.S. Office of Personnel Management (OPM) breach. Through credentials that were stolen, attackers gained entry.

UEBA, stringent access rules, and strong authentication techniques may all be used to implement zero trust in government networks. The illegal access and data exfiltration linked to this attack may have been discovered with continuous monitoring for anomalies [24].

KEYWORDS

- **access control**
- **attacks**
- **biometric authentication**
- **data encryption**
- **decryption keys**
- **multi-factor authentication**
- **remote access**
- **security model**
- **zero-trust architecture**

REFERENCES

1. Anderson, B., & McGrew, D. (2017). Machine learning for encrypted malware traffic classification: Accounting for noisy labels and non-stationarity. In *Proceedings of the 23rd ACM SIGKDD International Conference on Knowledge Discovery and Data Mining* (pp. 1723–1732). https://doi.org/10.1145/3097983.3098163.
2. Samaniego, M., & Deters, R. (2018). Zero-trust hierarchical management in IoT. In *2018 IEEE International Congress on Internet of Things* (pp. 88–95). https://doi.org/10.1109/ICIOT.2018.00019.
3. Tao, Y., Lei, Z., & Ruxiang, P. (2018). Fine-grained big data security method based on zero trust model. In *2018 IEEE 24th International Conference on Parallel and Distributed Systems* (pp. 1040–1045). https://doi.org/10.1109/PADSW.2018.8644614.
4. American Council for Technology and Industry Advisory Council (ACT-IAC). (2019). *Zero Trust Cybersecurity Current Trends*. American Council for Technology and Industry Advisory Council.
5. Zaheer, Z., Chang, H., Mukherjee, S., & van der Merwe, J. (2019). eZTrust: Network-independent zero-trust perimeterization for microservices. In *Proceedings of the 2019 ACM Symposium on SDN Research* (pp. 49–61). https://doi.org/10.1145/3314148.3314349.
6. Dou, Z., Xu, G., Chen, X. B., & Yuan, K. G. (2019). Rational non-hierarchical quantum state sharing protocol. *Computational Materials Continua, 58*(2), 335–347. https://doi.org/10.32604/cmc.2019.06026.
7. Xiaojian, Z., Liandong, C., Jie, F., Xiangqun, W., & Qi, W. (2021). Power IoT security protection architecture based on zero trust framework. In *2021 IEEE 5th International Conference on Cryptography, Security, and Privacy (CSP)* (pp. 166–170). https://doi.org/10.1109/CSP51677.2021.9357607.
8. Yan, X., & Wang, H. (2020). Survey on zero-trust network security. In X. Sun, J. Wang, & E. Bertino (Eds.), *Communications in Computer and Information Science: Artificial Intelligence and Security* (Vol. 1252, pp. 50–60). Springer. https://doi.org/10.1007/978-981-15-8083-3_5.
9. Yao, Q., Wang, Q., Zhang, X., & Fei, J. (2020). Dynamic access control and authorization system based on zero-trust architecture. In *2020 International Conference on Control, Robotics, and Intelligent System* (pp. 123–127). https://doi.org/10.1145/3437802.3437824.
10. Rose, O., Borchert, S., Mitchell, S., & Connelly, S. (2020). *Zero Trust Architecture. NIST Special Publication 800–207.* National Institute of Standards and Technology. https://doi.org/10.6028/NIST.SP.800-207.
11. Office of Management and Budget (OMB). (2018). *Strengthening the Cybersecurity of Federal Agencies by Enhancing the High-Value Asset Program* (OMB Memorandum M-19-03). The White House.
12. Brooks, S. W., Garcia, M. E., Lefkovitz, N. B., Lightman, S., & Nadeau, E. M. (2017). *An Introduction to Privacy Engineering and Risk Management in Federal Systems* (NIST Interagency or Internal Report 8062). National Institute of Standards and Technology. https://doi.org/10.6028/NIST.IR.8062.
13. Joint Task Force. (2018). *Risk Management Framework for Information Systems and Organizations: A System Life Cycle Approach for Security and Privacy* (NIST Special Publication 800-37, Rev. 2). National Institute of Standards and Technology. https://doi.org/10.6028/NIST.SP.800-37r2.

14. Samaniego, M., & Deters, R. (2018). Zero-trust hierarchical management in IoT. In *2018 IEEE International Congress on Internet of Things (ICIOT)*. IEEE. https://doi.org/10.1109/ICIOT.2018.00019.
15. Bien, J., Gaynanova, I., Lederer, J., & Müller, C. L. (2019). Prediction error bounds for linear regression with the TREX. *Test, 28*(2), 451–474. https://doi.org/10.1007/s11749-018-0584-4.
16. Gao, P., Yan, L., Chen, Z., Wei, X., Guo, L., & Shi, R. (2021). Research on zero-trust-based network security protection for power internet of things. In *2021 IEEE 4th International Conference on Automation, Electronics, and Electrical Engineering (AUTEEE)*. IEEE.
17. Hatakeyama, K., Kotani, D., & Okabe, Y. (2021). Zero trust federation: Sharing context under user control towards zero trust in identity federation. In *2021 IEEE International Conference on Pervasive Computing and Communications Workshops and other Affiliated Events (PerCom Workshops)*.
18. Rodigari, S., O'Shea, D., McCarthy, P., McCarry, M., & McSweeney, S. (2021). Performance analysis of zero-trust multi-cloud. In *2021 IEEE 14th International Conference on Cloud Computing (CLOUD)*.
19. Shan, L., Iqbal, M., & Saxena, N. (2022). Future industry internet of things with zero-trust security. *Information Systems Frontiers*, 1–14. https://doi.org/10.1007/s10796-021-10199-5.
20. Bevish Jinila, Y., Prayla Shyry, S., & Christy, A. (2022). A multi-component-based zero trust model to mitigate the threats in the internet of medical things. In *Data Engineering for Smart Systems: Proceedings of SSIC 2021*. Manipal University Jaipur, India.
21. Arai, K., Kapoor, S., & Bhatia, R. (2020). Zero trust in the context of the utility industry. In *Proceedings of the Future Technologies Conference (FTC) 2020* (Vol. 3).
22. Yan, X., & Wang, H. (2020). Survey on zero-trust network security. In *6th International Conference on Artificial Intelligence and Security*.
23. Samaniego, M., & Deters, R. (2018). Zero-trust hierarchical management in IoT. In *2018 IEEE International Congress on Internet of Things (ICIOT)*, San Francisco, CA, USA. https://doi.org/10.1109/ICIOT.2018.00019.
24. Sultana, M., Hossain, A., Laila, F., Abu Taher, K., & Nazrul Islam, M. (2020). Towards developing a secure medical image sharing system based on zero trust principles and blockchain technology. *BMC Medical Informatics and Decision Making, 20*(1), 1–10. https://doi.org/10.1186/s12911-020-01275-y.

CHAPTER 4

A Comprehensive Study on Zero-Trust Frameworks

POONAM DHIMAN[1,*] and AMANDEEP KAUR[2]

[1]*Government PG College, Ambala Cantt, Haryana, India*

[2]*Chitkara University Institute of Engineering and Technology, Chitkara University, Punjab, India*

*Corresponding author

ABSTRACT

The Zero Trust (ZT) security architecture has emerged as a compelling method to address the limitations of traditional network security solutions. This comprehensive survey study offers a detailed elucidation of the fundamental principles of Zero Trust, along with an evaluation of various tactics and potential avenues for its implementation. The study commences by analyzing the significance of authentication and access control in the context of Zero Trust principles. It then proceeds to explore novel authentication and access control solutions in various circumstances. This study delves further into conventional encryption approaches, micro-segmentation, and security automation, highlighting their significance in attaining a secure Zero Trust environment. The concept of Zero Trust Architecture (ZTA) is elucidated in terms of the advantages it offers and the network prerequisites necessary for implementing Zero Trust Network Features. This chapter offers valuable insights into the Zero Trust paradigm, encompassing its various approaches, challenges, and future research goals.

4.1 INTRODUCTION

The infrastructure of a typical organization has experienced a notable increase in complexity. The increasing complexity of modern enterprise networks has

Zero-Trust Learning: Applications in Modern Network Security. Chetna Kaushal, Amandeep Kaur, Mohit Angurala, Aryan Chaudhary (Eds.)

rendered traditional perimeter-based network security solutions inadequate, as the absence of a clearly defined and easily identifiable perimeter poses significant challenges. The effectiveness of perimeter-based network security has been demonstrated to be inadequate, as it fails to prevent unrestricted lateral movement by attackers once they have successfully breached the perimeter. The intricate nature of this endeavor has resulted in the emergence of a novel cybersecurity framework referred to as "zero trust" (ZT). The zero trust security models operate under the assumption that there is a potential attacker within the system, treating an enterprise-owned environment as equally untrustworthy as any non-enterprise-owned environment. In this emerging framework, it is imperative for an organization to adopt a stance of zero implicit trust and consistently assess and appraise the vulnerabilities posed to its resources and operational processes. Subsequently, the organization must implement safeguards to minimize the impact of these risks. In the context of zero trust, the typical approach to safeguarding resources, including data, compute resources, and applications/services, involves restricting access to only those subjects and assets that have been identified as requiring access [1]. ZTA refers to an enterprise cybersecurity framework that is founded upon the concept of zero trust. Its primary objective is to mitigate the risk of data breaches and restrict internal lateral movement within an organization. This scholarly article examines the concept of ZTA, delving into its logical constituents, potential deployment configurations, and associated security risks [2].

4.2 ELEMENTS OF ZERO-TRUST ARCHITECTURE

The zero-trust architecture is constructed based on a foundational set of concepts that facilitate the establishment of a secure and robust environment. These principles guarantee the consistent application and adaptation of security measures to effectively meet the ever-changing threat scenario.

1. Rather than depending on confidence established through network location or prior contacts, it is imperative to authenticate, authorize, and validate each access request prior to providing access. The concept of least privileged access refers to the practice of granting individuals or entities only the minimum level of access rights necessary to perform their authorized tasks or functions [3].
2. The integration and automation of processes have become increasingly prevalent in several industries. Incorporate security

technologies, methods, and solutions to facilitate uninterrupted communication and cooperation.

3. Micro-segmentation involves the partitioning of a network into smaller segments, each of which is assigned its own set of security policies and access controls. This methodology effectively mitigates the risk of unauthorized access to confidential information or valuable assets while also minimizing the possible harm that may arise in the event of a security breach.
4. The concept of context-aware access controls refers to the implementation of access control mechanisms that take into consideration the contextual information around a particular access request. This approach aims to enhance the access decisions that should be formulated by taking into consideration contextual information, including but not limited to the user's role, device posture, location, and risk considerations. This methodology guarantees that access restrictions can adjust to the present circumstances and consider possible risks or weaknesses.
5. Multi-factor authentication (MFA) is a security protocol that necessitates clients to present various kinds of identity as a prerequisite for gaining access to a network. In order to allow access, users are mandated to furnish numerous means of verification to authenticate their identity. It mitigates the potential threat of unauthorized access resulting from compromised credentials by imposing the necessity for an attacker to circumvent multiple stages of verification [4].
6. The area of discussion pertains to the concepts of visibility and analytics. Obtain a thorough understanding of user actions, device condition, and application performance throughout the entirety of the system. Analytics can be employed to discern patterns, recognize possible risks, and provide insights for the purpose of shaping security policy and implementing controls.
7. The practice of continuous monitoring and validation refers to the ongoing process of systematically observing and assessing the performance and accuracy of a system or process. Real-time monitoring is employed to detect potential dangers or anomalies by closely observing user behaviors, device health, and application behavior. It is imperative to consistently assess the efficacy of security controls, policies, and configurations to effectively counter emerging threats [5].

8. To ensure the security of sensitive data in various states, such as at rest, in transit, and during processing, it is imperative to employ robust measures such as encryption, tokenization, and other approaches for data protection. To mitigate the risk of unauthorized access to sensitive data, it is imperative to establish robust access controls and implement comprehensive monitoring mechanisms.
9. User and Entity Behavior Analytics (UEBA) is an academic discipline that centers on the examination and analysis of user and entity behavior within a given system. It involves the use of advanced algorithms and machine learning techniques to detect and identify patterns, anomalies, and potential threats.
10. It is important to note that the ZT at hand holds significant relevance in the academic sphere. The objective is to conduct an analysis of user and entity behavior in order to discover recurring trends and detect any irregularities that could potentially signal threats or malicious activity. UEBA allows organizations to take a proactive approach in addressing possible security incidents and mitigating the potential harm that may result from them [6].

4.3 PRINCIPLES OF ZERO TRUST

Numerous definitions and scholarly discussions pertaining to Zero Trust emphasize the notion of eliminating expansive perimeter defenses, such as enterprise firewalls, as a significant aspect. Nevertheless, the majority of these definitions persist in defining themselves in reference to perimeters to varying extents. For instance, terms like micro-segmentation or micro-perimeters are often used to describe the functional capabilities of a ZTA. The subsequent endeavor aims to establish a comprehensive definition of ZT and ZTA by focusing on the fundamental principles that should be encompassed, rather than emphasizing what is omitted. The principles represent the desired objective, but it is important to recognize that not all principles may be entirely realized in their utmost essence for a particular strategy [7].

1. A network has the potential to consist of various sorts of devices. In addition to its primary components, a network may encompass compact devices that transmit data to loading systems, software-as-a-service platforms, systems responsible for issuing directions to controllers, and various other functionalities. Moreover, an organization might opt to categorize privately owned devices as assets

if they possess the capability to access resources controlled by the organization [8].

2. Irrespective of the network location, all forms of communication are effectively safeguarded. The mere presence of a network site does not necessarily indicate trustworthiness. Admittance appeals originating from resources situated within the enterprise-preserved network substructure, such as those within an inherited network edge, are subject to the same security standards as access requests and interactions originating from another network that is not owned by the enterprise. In essence, it is imperative to refrain from bestowing trust automatically solely on the grounds that a device is connected to the company network infrastructure [9].
3. Exposure to organizational assets is granted on a session basis. The evaluation of the requester's credibility is performed prior to authorizing access. The minimal amount of access rights required to complete the task should be granted. This could potentially indicate a recent occurrence, namely in relation to a transaction, and may not necessarily take place immediately prior to the initiation of a session or the execution of a transaction with a resource. Nevertheless, the act of authenticating and authorizing an individual for one specific resource does not inherently confer access to any other distinct resource [10].
4. Access to resources is decided by a dynamic policy that considers several factors, such as the observable status of client identification, application or service, and the requesting asset. Additionally, other qualities related to behavior and the environment may also be considered [11]. An organization safeguards its resources by establishing a comprehensive framework that encompasses the identification and categorization of its resources, the delineation of its membership, and the determination of the specific resource access requirements for each member.
5. There is no inherent trust associated with any asset. When assessing a resource request, the business conducts an evaluation of the asset's security posture. To effectively deploy a ZTA, it is vital for an organization to build a system for continuous diagnostics and mitigation (CDM) or a comparable mechanism. This might also be applicable to affiliated devices that may be granted access to certain resources while being restricted from accessing others. In addition, the establishment of a comprehensive monitoring and reporting system is

necessary to furnish valuable data pertaining to the present condition of organizational assets [12].

6. The process described involves a perpetual sequence of acquiring authorization, conducting scans and evaluations of potential risks, adjusting strategies accordingly, and consistently reassessing the level of trust in continuous communication.
7. It is imperative for a company to gather data pertaining to the security status of its assets, network traffic, and access requests. This data should then be analyzed and utilized to enhance the formulation and implementation of policies. This data can additionally serve the purpose of offering contextual information for access requests made by individuals.

4.4 ADVANTAGES OF IMPLEMENTING ZERO-TRUST ARCHITECTURE

The implementation of zero-trust architecture inside the Web ecosystem has a multitude of benefits that can assist organizations in constructing safe and resilient decentralized apps, networks, and services. There are several significant advantages associated with the implementation of zero-trust principles include [13]:

1. By embracing the philosophy of "never trust, always verify," diverse applications can actively mitigate threats and reduce the likelihood of unauthorized access, data breaches, and other security problems.
2. The concept of "Reduced Attack Surface" refers to the practice of minimizing the potential vulnerabilities and entry points that might be exploited by malicious actors in a system or network [14].
3. Micro-segmentation and the implementation of least privilege access controls serve to restrict an attacker's capacity to traverse horizontally across the network, so diminishing the total scope of attack and confining the possible harm resulting from a security breach [15].
4. Zero-trust architecture supports the continued relevance and effectiveness of security measures in diverse settings by considering context and risk variables when making access choices.
5. The concept of decentralized identity management refers to the practice of distributing and dispersing the control and ownership

of personal identity information across multiple entities or systems [16].

6. The use of zero-trust principles across many applications facilitates the incorporation of decentralized identification solutions, hence enhancing user privacy and granting individuals greater authority over their personal data. Simultaneously, this approach upholds strong access control measures [17].
7. The topic of concern in this discussion is the preservation of data and the safeguarding of individuals' privacy.
8. The zero-trust architecture places significant emphasis on safeguarding data at all stages of its existence, including when it is at rest, in transit, and being processed, to ensure the protection of sensitive information. This aspect holds significant importance in the realm of the World Wide Web, as data is frequently dispersed among several nodes and platforms [19].
9. The concepts of scalability and flexibility are crucial in various domains and industries. Zero-trust concepts can be implemented across different facets of the Web environment, encompassing user and device authentication, ensuring the security of smart contracts, and establishing access control methods for decentralized applications. The ability to adjust security measures in accordance with the expansion and development of cloud infrastructure empowers organizations with flexibility [20].
10. Interoperability refers to the ability of different systems, devices, or software applications to exchange information. The integration and collaboration of security tools and procedures are essential in the edge computing ecosystem, where many technologies and platforms coexist. Zero-trust architecture plays a significant role in facilitating this integration and collaboration. The ability to interoperate allows organizations to establish comprehensive and uniform security protocols throughout their decentralized infrastructure [21].
11. The concept of automated security refers to the use of technology and algorithms to protect systems, networks, and data against unauthorized access and breaches. The integration of automation into a zero-trust architecture facilitates the optimization of security operations, bolstering the ability to promptly address potential risks and mitigating the likelihood of human mistakes. Efficient and effective security solutions are particularly advantageous within the context of blockchain, given the rapid speed of innovation [22].

12. The topic of interest pertains to the concepts of visibility and analytics. The acquisition of a full understanding of user activity, device health, and application behavior inside the web environment allows organizations to recognize patterns, identify potential risks, and make informed decisions regarding security policies and controls.
13. The implementation of proactive measures to mitigate potential threats is essential. Continuous monitoring, real-time threat detection, and TOTP are essential tools that enable organizations to take a proactive approach in responding to possible security incidents within the web ecosystem. By employing these measures, organizations may effectively minimize the potential harm caused by such incidents and significantly decrease the possibility of successful assaults [23].

4.5 TECHNOLOGICAL PREREQUISITES NECESSARY FOR FACILITATING ZERO TRUST ARCHITECTURE

The network prerequisites are necessary for facilitating ZTA implementation such as:

1. In order to effectively implement ZTA, certain network requirements must be met. These requirements are essential for supporting the principles and objectives of ZTA [24].
2. Enterprise assets possess fundamental network connectivity. The local area network, whether under company control or not, serves as a fundamental framework for routing and architecture. The utilization of all infrastructure services by the remote company asset is not necessarily guaranteed [25].
3. The organization must possess the capability to differentiate between the assets that are owned or controlled by the enterprise and the present security status of the devices. The determination is contingent upon credentials issued by the enterprise, rather than relying on unverifiable information [26].
4. The organization has the capability to monitor and analyze every data transmission over the network. The enterprise captures packets observed on the data plane, even if it lacks the capability to perform application layer analysis on every packet. The organization selectively excludes metadata pertaining to the connection, such as

destination, time, and device identity, to continuously update rules. This information is then communicated to the policy evaluator (PE) during the assessment of access requests.

5. Access to enterprise resources should be restricted unless a proper Policy Enforcement Point (PEP) is utilized. Enterprise resources typically do not permit indiscriminate incoming connections from the internet. Resources will only allow connections that have been custom configured when a client has successfully completed the process of authentication and authorization. The establishment of these communication pathways is facilitated by the PEP. The discovery of resources may be hindered in the absence of access to the PEP [27].
6. The PEPs are required to possess the capability to transmit and receive signals originating from both the data and control planes. The PEP component can be accessed by enterprise assets. To obtain access to resources, enterprise subjects are required to have the capability to access the PEP component.
7. The PEP is the sole element inside the business flow that interacts with the policy administrator. Every PEP that operates within the enterprise network is connected to the policy executive to build communication pathways between trades and assets. All communication related to enterprise business processes is routed through one or more PEPs [28].
8. The ability for remote enterprise assets to directly access enterprise resources, without the requirement of traversing the company network infrastructure, is essential. An instance might be cited where it is deemed unnecessary for a distant entity to establish a connection with the organizational network to avail services offered by a public cloud provider (such as email) that are utilized by the enterprise [29].
9. The scalability of the infrastructure utilized for supporting the ZTA access determination process should be ensured to accommodate potential fluctuations in process demand. The physical entities, physical activities, and physical entity processes employed within a Zero Touch Automation framework assume critical roles as fundamental elements in various organizational workflows [30].
10. Enterprise assets may encounter limitations in accessing certain politically exposed persons because of legislative restrictions or other identifiable circumstances. An illustrative instance is the presence

of a policy that restricts the accessibility of specific resources to mobile assets situated beyond the geographical boundaries of the enterprises. These aspects may encompass geolocation, network location, device type, or other relevant characteristics.

4.6 THE POTENTIAL RISKS AND VULNERABILITIES

Zero Trust is a security structure that aims to enhance network security by assuming that all users, devices, and network components are potentially unreliable. As such, it seeks to mitigate the risks and threats that can arise in a network environment. These threats in Architecture: It is impossible for any organization to completely eradicate the risk associated with cybersecurity. Nevertheless, certain dangers exhibit distinct characteristics during the implementation of a ZTA.

4.6.1 MODIFICATION OF THE ZTA

The policy engine and policy administrator play essential roles within the ZTA framework. This implies that it is necessary to ensure that these components are appropriately configured and regularly maintained. Enterprise operations can be disrupted if an enterprise administrator, who has configuration access to the PE's rules, performs unapproved changes or makes mistakes. Similarly, compromised privileged access could potentially grant unauthorized access to resources that would otherwise be restricted. In order to mitigate the risks that are connected, it is imperative to ensure that the Physical Environment and PA components are appropriately set and monitored. Additionally, any modifications made to the configuration should be reported and subjected to audit procedures [31].

4.6.1.1 UNAUTHORIZED ACQUISITION OF CREDENTIALS

The issue of stolen credentials and insider threats is a significant concern in the realm of cybersecurity. The effective implementation of ZT, along with robust information security and resiliency policies, as well as adherence to best practices, can significantly mitigate the risk of unauthorized access by attackers using stolen credentials or insider attacks. A well-designed and effectively executed ZTA should effectively mitigate the risk of a compromised account or asset gaining unauthorized access to resources beyond its designated scope or established access patterns. Consequently, attackers are inclined to prioritize accounts that possess access policies pertaining to resources of interest.

4.6.1.2 ACCESSIBILITY WITHIN THE NETWORKS

An organization that lacks the capability to do deep packet inspection or analyze encrypted data must rely on alternative methodologies to evaluate potential network intruders. This does not imply that the company lacks the capability to analyze encrypted communication observed on the network. The organization has the capability to gather metadata pertaining to encrypted traffic, which can be utilized to identify the presence of an active attacker or potential malware engaged in communication within the network. Machine learning methodologies can be employed to analyze encrypted and inaccessible traffic for examination purposes. The utilization of this machine learning technique would enable the organization to classify network traffic as either legitimate or potentially hostile, hence necessitating appropriate remedial actions.

4.6.1.3 DENIAL-OF-SERVICE

The occurrence of a Denial-of-Service (DoS) or network disruption event is a significant concern in the field of computer security. Within the context of ZTA, the Privileged Access (PA) assumes a pivotal role in facilitating resource accessibility. The interconnectivity of enterprise resources is contingent upon obtaining authorization from the network administrator, typically the PA, and may require additional configuration measures. In instances of this nature, just a subset of business-related topics is impacted. This capability is also feasible in traditional remote-access virtual private networks and is not exclusive to ZTA [9].

4.6.1.4 POTENTIAL THREAT POSED BY INDIVIDUALS WITH INSIDER ACCESS

Adversaries have the potential to employ various techniques, such as phishing, social engineering, or a combination thereof, in order to illicitly acquire the login credentials associated with high-value accounts. The interpretation of "valuable" can vary depending on the underlying intent of the attacker. For example, enterprise administrator accounts possess significant value.

4.6.1.5 STORAGE AND RETRIEVAL OF NETWORK INFORMATION

One additional concern pertaining to the monitoring and analysis of network traffic within enterprises is the analysis component itself. When monitoring scans, network traffic, and metadata are retained for the purpose of constructing contextual policies, conducting forensics, or facilitating subsequent analysis, this material becomes susceptible to targeting by malicious actors. Similar to network diagrams, configuration files, and various other network architecture documents, it is imperative to ensure the protection of these resources.

4.6.1.6 UNAUTHORIZED ACCESS

If an unauthorized individual manages to obtain access to this data, there exists the potential for them to acquire knowledge regarding the structure of the organization's infrastructure and identify valuable resources for subsequent reconnaissance and exploitation. One additional means through which an attacker can obtain reconnaissance information within a ZT company is through the utilization of the management tool employed for encoding access policies. Similar to a repository of traffic data, this particular element encompasses access regulations pertaining to resources and has the potential to provide an attacker with insights into which accounts hold the highest value for exploitation [10].

4.6.1.7 UTILIZATION OF EXCLUSIVE DATA FORMATS

ZTA utilizes a diversity of data sources that encompass details pertaining to the subject making the request, the asset being utilized, enterprise, and external intelligence, as well as threat assessments. Frequently, the resources employed for the storage and processing of this data lack a universally accepted and publicly accessible framework for facilitating interaction and information sharing. This situation may arise when a business becomes restricted to a limited number of service providers as a result of challenges related to interoperability.

4.6.1.8 SECURITY BREACH

If a single provider encounters a security breach or experiences a disruption, an organization may face significant challenges in transitioning to an alternative provider. This may result in substantial expenses, such as the need to replace several assets or necessitate the implementation of a lengthy transition program. To address potential risks, it is advisable for organizations to conduct a comprehensive evaluation of service providers. These considerations should be taken into account alongside more conventional factors such as performance and stability.

4.6.1.9 ARTIFICIAL INTELLIGENCE AND SOFTWARE-BASED AGENTS

The utilization of artificial intelligence (AI) and software-based agents is being implemented for the purpose of overseeing security concerns within enterprise networks. The interaction between these components and the management components of ZTA is necessary, occasionally replacing the role of a human administrator. The authentication process of these components within a business that implements a ZTA remains an unresolved matter. It is generally presumed that the majority of automated technology systems will employ certain methods of authentication when accessing resource components through an API. The reduction of errors and enhancement of decision-making can be achieved through the implementation of a periodic recalibration study [32].

The potential risk lies in the possibility that an assailant may manipulate or pressure a Non-Privileged Entity into executing a job that exceeds the attacker's authorized privileges. The software agent's threshold for authentication to carry out administrative or security-related tasks may be less stringent in comparison to that of a human user. If an adversary is able to engage with the agent, there exists the potential for the adversary to deceive the agent into granting them increased privileges or to manipulate the agent into executing actions on behalf of the adversary. There exists a potential danger wherein a malicious actor may acquire the authentication details of a software agent, thereby assuming the identity of the said agent throughout the execution of various activities.

4.7 CONCLUSION

The present study provides a thorough analysis of contemporary zero-trust technology implementations, encompassing an extensive examination of their respective advantages and disadvantages. The Zero Trust approach offers a finely detailed and context-specific resolution to network security challenges across many applications. The technique exhibits a high level of agility; yet, additional research and practical application are necessary to draw thorough conclusions regarding its efficacy in real-world implementations. The focus of our study is restricted to research initiatives that are publicly accessible. This chapter provides an in-depth analysis of the characteristics of network design based on the zero-trust model and the frameworks and proofs-of-concept utilized for enhancing security in networks.

KEYWORDS

- **authentication**
- **cloud computing**
- **cybersecurity framework**
- **denial-of-service**
- **network security**
- **zero trust**
- **zero trust architecture**

REFERENCES

1. Stafford, V. A. (2020). *Zero Trust Architecture. NIST Special Publication,* 800, 207.
2. Dimitrakos, T., Dilshener, T., Kravtsov, A., La Marra, A., Martinelli, F., Rizos, A., & Saracino, A. (2020). Trust aware continuous authorization for zero trust in consumer internet of things. In *2020 IEEE 19th International Conference on Trust, Security and Privacy in Computing and Communications (TrustCom)* (pp. 1801–1812). IEEE.
3. Shore, M., Zeadally, S., & Keshariya, A. (2021). Zero trust: The what, how, why, and when. *Computer, 54*(11), 26–35.
4. Hosney, E. S., Halim, I. T. A., & Yousef, A. H. (2022). An artificial intelligence approach for deploying zero trust architecture (ZTA). In *2022 5th International Conference on Computing and Informatics (ICCI)* (pp. 343–350). IEEE.

5. Sheikh, N., Pawar, M., & Lawrence, V. (2021). Zero trust using network microsegmentation. In *IEEE INFOCOM 2021-IEEE Conference on Computer Communications Workshops (INFOCOM WKSHPS)* (pp. 1–6). IEEE.
6. Bobbert, Y., & Scheerder, J. (2022). Zero trust validation: From practice to theory: An empirical research project to improve zero trust implementations. In *2022 IEEE 29th Annual Software Technology Conference (STC)* (pp. 93–104). IEEE.
7. Teerakanok, S., Uehara, T., & Inomata, A. (2021). Migrating to zero trust architecture: Reviews and challenges. *Security and Communication Networks, 2021*, 1–10.
8. Kerman, A., Borchert, O., Rose, S., & Tan, A. (2020). Implementing a zero trust architecture. *National Institute of Standards and Technology, 2020*, 17–17.
9. Simpson, W. R., & Foltz, K. E. (2021). Maintaining zero trust with federation. *International Journal of Emerging Technology and Advanced Engineering, 11*(5), 17–32.
10. Seaman, J. (2023). Zero trust security strategies and guidelines. In *Digital Transformation in Policing: The Promise, Perils and Solutions* (pp. 149–168). Cham: Springer International Publishing.
11. Kumar, N., Kasbekar, G. S., & Manjunath, D. (2023). Application of data collected by endpoint detection and response systems for implementation of a network security system based on zero trust principles and the EigenTrust algorithm. *ACM SIGMETRICS Performance Evaluation Review, 50*(4), 5–7.
12. Hurley, J. (2023). Zero trust is not enough: Mitigating data repository breaches. In *ICCWS 2023 18th International Conference on Cyber Warfare and Security*. Academic Conferences and Publishing Limited.
13. Golightly, L., Modesti, P., Garcia, R., & Chang, V. (2023). Securing distributed systems: A survey on access control techniques for cloud, blockchain, IoT and SDN. *Cyber Security and Applications, 100015*.
14. Chen, B., Qiao, S., Zhao, J., Liu, D., Shi, X., Lyu, M., & Zhai, Y. (2020). A security awareness and protection system for 5G smart healthcare based on zero-trust architecture. *IEEE Internet of Things Journal, 8*(13), 10248–10263.
15. Anderson, J., Huang, Q., Cheng, L., & Hu, H. (2022). BYOZ: Protecting BYOD through zero trust network security. In *2022 IEEE International Conference on Networking, Architecture and Storage (NAS)* (pp. 1–8). IEEE.
16. Sanders, G., Morrow, T., Richmond, N., & Woody, C. (2021). *Integrating Zero Trust and DevSecOps*. Technical report.
17. Keeriyattil, S., & Keeriyattil, S. (2019). Microsegmentation and zero trust: Introduction. In *Zero Trust Networks with VMware NSX: Build Highly Secure Network Architectures for Your Data Centers* (pp. 17–31).
18. Ray, P. P. (2023). Web3: A comprehensive review on background, technologies, applications, zero-trust architectures, challenges, and future directions. *Internet of Things and Cyber-Physical Systems, 3, 213–248.*
19. Cao, Y., Pokhrel, S. R., Zhu, Y., Ram Mohan Doss, R., & Li, G. (2022). Automation and orchestration of zero trust architecture: Potential solutions and challenges. *Machine Intelligence Research, 21, 294–317.*
20. Sellitto, G. P., Aranha, H., Masi, M., & Pavleska, T. (2021). Enabling a zero trust architecture in smart grids through a digital twin. In *Dependable Computing-EDCC 2021 Workshops: DREAMS, DSOGRI, SERENE 2021, Munich, Germany, September 13, 2021, Proceedings 17* (pp. 73–81). Springer International Publishing.

21. Wu, Y. G., Yan, W. H., & Wang, J. Z. (2021). Real identity-based access control technology under zero trust architecture. In *2021 International Conference on Wireless Communications and Smart Grid (ICWCSG)* (pp. 18–22). IEEE.
22. Bavendiek, S. (2022). *A Zero Trust Security Approach with FIDO2.* DOI: https://doi.org/10.21203/rs.3.rs-2022891/v1.
23. Paul, B., & Rao, M. (2022). Zero-trust model for smart manufacturing industry. *Applied Sciences, 13*(1), 221.
24. Alipour, M. A., Ghasemshirazi, S., & Shirvani, G. (2022). *Enabling a Zero Trust Architecture in a 5G-Enabled Smart Grid. arXiv preprint arXiv:2210.01739.*
25. Chui, M., Issler, M., Roberts, R., & Yee, L. (2023). *Technology Trends Outlook 2023.*
26. Mallikarjunaradhya, V., Pothukuchi, A. S., & Kota, L. V. (2023). An overview of the strategic advantages of AI-powered threat intelligence in the cloud. *Journal of Science & Technology, 4*(4), 1–12.
27. Casamayor Pujol, V., Morichetta, A., Murturi, I., Kumar Donta, P., & Dustdar, S. (2023). Fundamental research challenges for distributed computing continuum systems. *Information, 14*(3), 198.
28. Tang, Q., Xiao, J., & Wu, K. (2023). A novel evidence combination method based on stochastic approach for link-structure analysis algorithm and Lance-Williams distance. *PeerJ Computer Science, 9*, e1307.
29. Jorquera Valero, J. M., Sánchez, P. M., Gil Pérez, M., Huertas Celdrán, A., & Martinez Perez, G. (2023). Cutting-edge assets for trust in 5G and beyond: Requirements, state of the art, trends, and challenges. *ACM Computing Surveys, 55*(11), 1–36.
30. Arogundade, O. R. (2023). Network security concepts, dangers, and defense best practices. *Computer Engineering and Intelligent Systems, 14*(2).
31. Michael, J. B., Dinolt, G. C., Cohen, F. B., & Wijesekera, D. (2022). Can you trust zero trust? *Computer, 55*(8), 103–105.
32. Kim, J., & Im, I. (2023). Anthropomorphic response: Understanding interactions between humans and artificial intelligence agents. *Computers in Human Behavior, 139*, 107512.

CHAPTER 5

The AI-Zero-Trust Alliance Reinventing Cybersecurity for the Digital Age

SAMRUDDHI BHABAD[1,*] and NEHA JAIN[2]

[1]*Department of Chemical Engineering, Bapatla Engineering College, Bapatla, Andhra Pradesh, India*

[2]*Department of Computer Engineering, MMCOE, Pune, Maharashtra, India*

*Corresponding author

ABSTRACT

The escalating complexity of the digital landscape demands an evolution in cybersecurity frameworks. Traditional models, reliant on perimeter-based defenses, fall short against the sophisticated threats of the modern age. This chapter introduces the synergistic integration of Artificial Intelligence (AI) and Zero Trust principles, reshaping the cybersecurity paradigm. Zero Trust challenges conventional practices by requiring continuous verification of every interaction, rejecting the assumption of inherent trust within networks. When augmented by AI, this approach gains unprecedented precision and adaptability. AI's capabilities in real-time data analysis, pattern recognition, and anomaly detection complement Zero Trust's foundational pillars—continuous verification, micro-segmentation, least privilege access, and breach assumptions. Together, they form a robust framework capable of preemptively identifying threats and dynamically securing digital ecosystems.

5.1 INTRODUCTION

In today's fast-paced digital landscape, the way we protect our online world is undergoing a profound transformation. The traditional model of establishing

Zero-Trust Learning: Applications in Modern Network Security. Chetna Kaushal, Amandeep Kaur, Mohit Angurala, Aryan Chaudhary (Eds.)

strong digital walls and placing trust in everything within those walls no longer suffices. Instead, we find ourselves in an era where security demands constant verification at every step, challenging the idea of taking anything for granted. Think of it as a security system where every action within a computer network is meticulously examined, akin to a vigilant detective on a case. This is the essence of "Zero Trust (ZT)" – a security philosophy that emphasizes perpetual verification over blind trust.

Imagine for a moment a world where every network request, every access attempt, and every data transaction is scrutinized with the precision of a master detective. A world where trust is no longer a given but a privilege earned through continuous authentication and rigorous verification. This is the essence of Zero Trust—an idea that challenges the status quo of cybersecurity and demands that trust be established and maintained at every step, every interaction, and every junction in our digital infrastructure. The mantra is simple, yet profound: "Never trust, always verify."

But here's the twist: Zero Trust, as transformative as it is, does not stand alone. It finds its match in the formidable capabilities of Artificial Intelligence (AI). The true innovation emerges when Zero Trust joins forces with AI. Picture AI as a digital guardian, always vigilant, ready to identify anything out of the ordinary swiftly. It possesses the ability to process vast datasets quickly, uncover intricate patterns, and identify potential issues in real-time. AI enriches the principles of Zero Trust, bestowing an additional layer of intelligence and adaptability that redefines the landscape of cybersecurity.

As we embark on this chapter's journey, we will delve into the intricacies of Zero Trust, its historical roots, the hurdles it confronts, and its profound relevance in today's digital environment. It is no longer confined to the realm of science fiction; instead, it constitutes a fundamental aspect of safeguarding our digital world. From the inception of the "never trust" philosophy to the architecture of continuous verification, we will paint a vivid picture of a security paradigm that, while disruptive, is remarkably effective in an era where threats loom from within and without. Yet, the true marvel materializes when Zero Trust and AI converge. Together, they constitute a symbiotic partnership that engenders a new paradigm of security, where trust is fluid, threats are preemptively identified, and security becomes an enabler of progress rather than a hindrance.

In subsequent time, we will witness this dynamic alliance in action. We will explore real-world scenarios across diverse industries – from finance to healthcare and critical infrastructure – to witness the remarkable potency of this collaboration in action. These stories will underscore the transformative potential of this alliance across a myriad of industries and scenarios.

Nevertheless, this journey is not without its complexities. The introduction of novel technology begets fresh challenges, including ethical dilemmas concerning AI's potential biases. As we go further, we will confront ethical considerations, the potential biases of AI algorithms, and the challenge of managing AI-driven security systems. The path to progress is never without its obstacles, and it is essential that we navigate these intricacies with discernment and responsibility.

Figure 5.1 illustrates the concept of the AI-Zero Trust Continuum, which showcases the transformation of security models from traditional trust-based approaches to the Zero Trust philosophy, with the integration of AI at various stages. At the left end of the continuum, we depict the traditional security model where trust is assumed once an entity gains access to the network. As you progress to the right, toward "Zero Trust," trust is never presumed but continuously verified. The gradient line represents the gradual shift, highlighting the journey from blind trust to continuous verification. AI icons placed along the continuum symbolize the incorporation of AI technologies at different points. These icons gradually increase in size and intensity, signifying the growing role of AI in enhancing security as we move closer to the Zero Trust end (Figure 5.1).

Figure 5.1 visually conveys how AI plays a pivotal role in reshaping cybersecurity by reinforcing the principles of Zero Trust, where trust is earned through continuous authentication and rigorous verification. In summary, this chapter extends an invitation to the forefront of "Cybersecurity Reinvented: The AI-Zero Trust Alliance." It marks a new era in cybersecurity, where trust is an earned privilege, threats are anticipated, and the synergy of human intelligence and technological innovation propels us toward a future where security does not merely defend but propels progress.

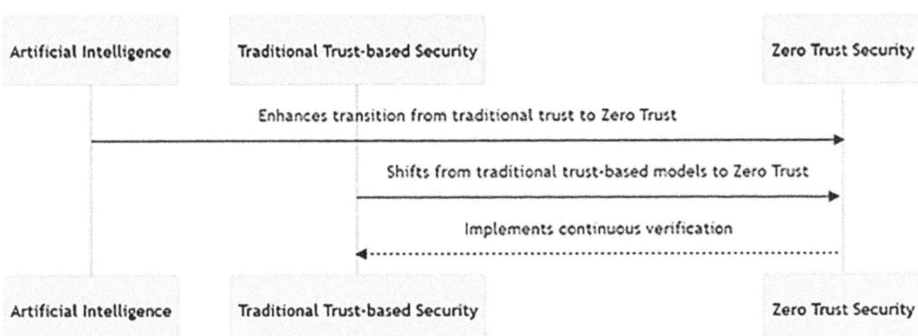

FIGURE 5.1 The AI-zero trust continuum.

5.1.1 UNDERSTANDING ZERO TRUST

5.1.1.1 THE BEDROCK OF MODERN SECURITY

Zero Trust isn't merely a fresh perspective on cybersecurity; it signifies a profound shift in how we approach digital defenses. Now, we'll delve deeper into the essence of Zero Trust, tracing its origins, exploring its core principles, and understanding why it has evolved into a pivotal framework for safeguarding our digital assets.

5.1.1.2 THE BIRTH OF ZERO TRUST

To grasp the essence of Zero Trust, we must rewind to its beginnings. While the concept was initially introduced by Forrester Research in 2010, it gained significant prominence when Google embraced the BeyondCorp model [1]. BeyondCorp challenged the traditional security model, which relied heavily on fortifying network perimeters. Instead, BeyondCorp proposed a model where trust was never assumed based on location or network. This shift in approach was prompted by a sobering reality: cyber threats were evolving at a breathtaking pace. Attackers were devising increasingly sophisticated methods to breach conventional defenses. Once inside the perimeter, they could navigate laterally, causing substantial damage. The need for a new approach, one that continuously validated trust, regardless of a user's physical or digital location, became apparent (Figure 5.2).

Figure 5.2 traces the historical evolution of Zero Trust principles, shedding light on key milestones and transformative developments that have reshaped the landscape of cybersecurity.

5.1.1.3 KEY MILESTONES AND EVENTS

1. **1980s – The Birth of Firewalls:** The timeline commences with the emergence of firewalls in the 1980s, representing the earliest attempts to secure networks by establishing rudimentary perimeters and controlling data access.
2. **2010 – Forrester Research Introduces Zero Trust:** A pivotal moment is marked in 2010 when Forrester Research introduced the revolutionary concept of Zero Trust. This landmark publication challenged the prevailing paradigm of network trust and laid the foundation for a paradigm shift in cybersecurity.

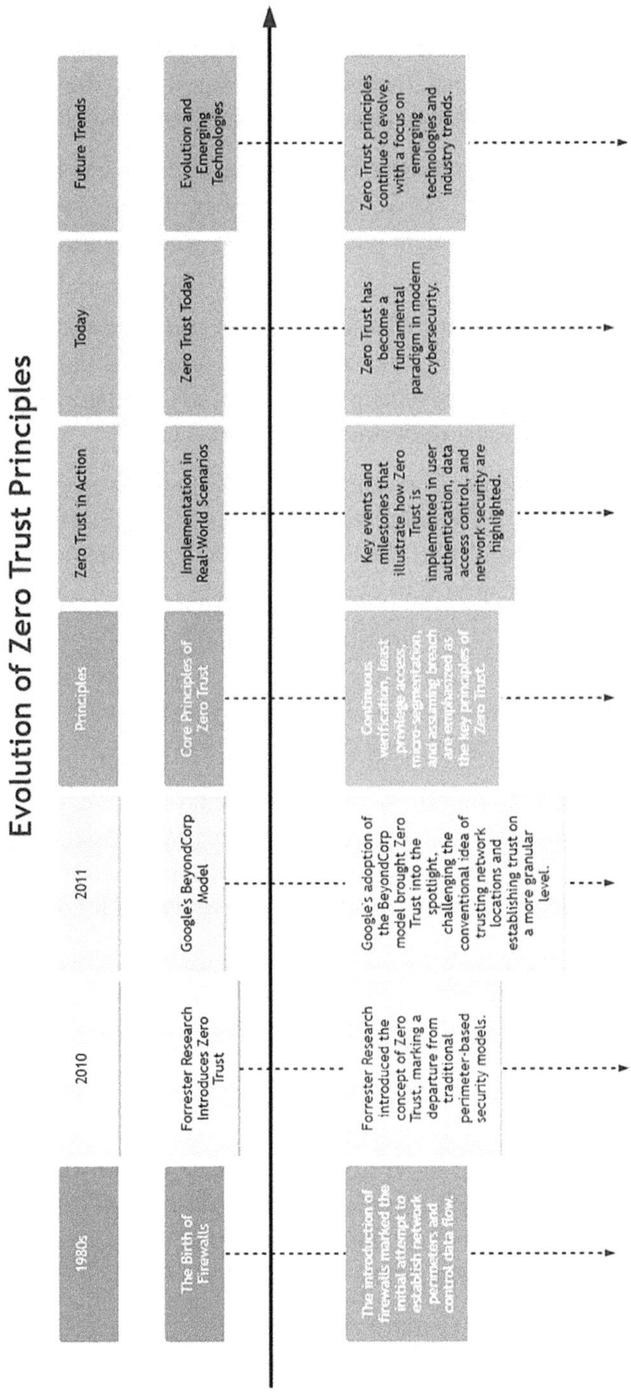

FIGURE 5.2 Timeline diagram description: evolution of zero trust principles.

3. **2011 – Google's BeyondCorp Model:** In 2011, Google's adoption of the BeyondCorp model came to the forefront, amplifying the significance of Zero Trust. BeyondCorp exemplified the transition from location-based trust to a more meticulous, user-centric trust approach.
4. **Principles of Zero Trust:** The timeline underscores the fundamental principles of Zero Trust, including continuous verification, least privilege access, micro-segmentation, and adopting an "assume breach" mindset.
5. **Zero Trust in Action:** Notable events and milestones demonstrate how Zero Trust principles are put into practical use across diverse domains. This includes user authentication, fine-grained data access control, and robust network security measures.
6. **Zero Trust Today:** The timeline extends to the present day, depicting Zero Trust as a fundamental cornerstone of modern cybersecurity strategies. It highlights the enduring relevance and adoption of these principles in response to evolving threats.
7. **Future Trends:** The timeline hints at the trajectory of Zero Trust, indicating that it continues to evolve. This evolution is driven by emerging technologies and industry trends, with an unwavering commitment to enhancing digital security.

The above timeline diagram serves as a visual narrative, guiding through the historical context and the transformative journey of Zero Trust principles within the cybersecurity landscape. It underscores their enduring significance in safeguarding digital assets in an ever-changing threat landscape.

5.2 THE PILLARS OF ZERO TRUST

Zero Trust rests on a set of fundamental principles, each carefully crafted to bolster overall security:

1. **Verification at Every Interaction:** At the core of Zero Trust lies the imperative to verify every entity attempting to access the network. This verification is not a one-time event during login but a continuous process. The fundamental principle is clear: trust is never assumed; it must always be confirmed.
2. **The Principle of Least Privilege:** Underpinning Zero Trust is the principle of granting users and devices access only to the resources

and data they need for their specific roles. Essentially, they should have the least possible access to perform their tasks effectively.

3. **Micro-Segmentation:** Instead of relying on a single, expansive network, Zero Trust advocates for micro-segmentation. This involves dividing the network into smaller, isolated segments. Even if a breach occurs in one segment, it should not automatically grant access to the entire network.
4. **Assuming Breach:** Zero Trust operates on the assumption that a breach has occurred or could occur at any moment. This mindset shift fosters continuous monitoring and a proactive security stance [2].

5.3 THE ROLE OF ZERO TRUST IN MODERN CYBERSECURITY

The landscape of modern work has undergone a profound transformation, with remote work, cloud computing, and the ubiquity of mobile devices taking center stage. In this environment, the traditional perimeter-centric security model has become obsolete. Employees access corporate resources from diverse locations and devices, rendering the concept of a fixed network perimeter irrelevant.

Zero Trust adapts to this dynamic by enforcing trust validation for every interaction. This approach aligns seamlessly with the demands of the contemporary workforce, which seeks both flexibility and unwavering security.

Furthermore, the advent of cloud computing and the proliferation of Internet of Things (IoT) devices has expanded the attack surface exponentially. Conventional security models struggle to keep pace with the dynamic nature of these technologies. Zero Trust, with its focus on continuous verification and least privilege access, offers a more adaptable and effective defense [3].

As a final word, as all technology relies on availability, in high-risk applications such as cybersecurity, there can be no errors. No one wants to receive the message: "The server is currently overloaded with other requests; sorry about that! Please try again later or contact us through our help center if the error persists." Yet another example of the need for Zero Trust? [21].

In this era of perpetual evolution, Zero Trust emerges as a foundation upon which to construct a secure digital environment. It transcends the confines of one-size-fits-all solutions and provides a flexible framework that can be adeptly tailored to the unique needs of organizations. As we continue through this chapter, we will explore how the incorporation of AI augments Zero Trust principles, representing a significant leap in the evolution of cybersecurity (Table 5.1).

TABLE 5.1 Zero Trust Principles

Sl. No.	Principle	Description
1.	Verification at every interaction	Trust is never assumed; it must always be confirmed.
2.	The principle of least privilege	Users and devices have access only to what's necessary for their roles.
3.	Micro-segmentation	Network divided into smaller segments to limit lateral movement.
4.	Assuming breach	Operates under the assumption that a breach has occurred or could occur at any moment.

5.4 THE POWER OF AI

5.4.1 FORTIFYING THE ZERO TRUST ARSENAL

As we go more into the investigation of the powerful collaboration between Zero Trust and AI, we find ourselves at the center of a digital evolution ready to reinvent cybersecurity. In this part, we'll explore the deep potential of AI to strengthen digital security, improve Zero Trust, and usher in a new era of proactive, flexible, and savvy security.

5.4.2 AI AS THE DIGITAL GUARDIAN

Visualize AI as an unwavering sentinel, diligently overseeing the digital realm. It meticulously sifts through vast volumes of data, analyzing network traffic, user activities, and system logs with exceptional precision. In stark contrast to humans, AI can spot threats in real-time, often identifying dangers before human intervention is possible [4].

AI's true prowess lies in its capacity to evolve. Through the mechanisms of Machine Learning (ML) and Deep Learning (DL), AI systems refine their threat detection capabilities continuously. They learn from patterns and anomalies, progressively honing their accuracy. This adaptability grants AI the edge in combating the ever- mutating landscape of cyber threats [5].

5.4.3 HARMONIZING WITH ZERO TRUST

AI and Zero Trust, though distinct concepts, harmonize seamlessly. Both underscore the critical importance of thorough authentication and perpetual scrutiny, refraining from presuming trust without validation.

In this dynamic alliance, every interaction within a computer network becomes a subject of scrutiny.

In this collaboration, every interaction within a computer network undergoes rigorous scrutiny. AI's proficiency in detecting unusual patterns melds perfectly with Zero Trust's steadfast insistence on never taking trust for granted. Together, they orchestrate a symphony of security that is both robust and resilient [1].

5.5 AI'S CONQUEST OVER ADVANCED THREATS

To appreciate AI's practical impact, consider its effectiveness against advanced threats. Traditional signature-based antivirus systems, reliant on known patterns and signatures, often stumble when confronted with sophisticated, zero-day attacks designed to evade detection. AI, on the other hand, excels in this arena.

For instance, AI does not lean on predefined signatures to recognize new strains of malware. Instead, it observes how files and processes behave, even when encountering unfamiliar threats. This capability to identify previously unknown threats signifies a monumental shift in the realm of cybersecurity [7].

Furthermore, AI's proficiency extends to thwarting phishing attacks as well. Phishing emails often carry subtle nuances that evade traditional filters. AI-driven systems scrutinize email content, sender behavior, and linguistic patterns to discern phishing attempts, safeguarding users from these treacherous ploys.

5.6 THE ETHICAL DIMENSION OF AI IN CYBERSECURITY

While AI's potential in cybersecurity is vast, it simultaneously raises ethical considerations. AI systems rely on extensive datasets, and biases entrenched within these datasets can inadvertently perpetuate discrimination. Rectifying these biases is not solely an ethical obligation; it is also pivotal for AI's efficacy in ensuring security [7].

Furthermore, the immense power of AI bears the potential for misuse. In malevolent hands, AI-driven cyberattacks could unleash catastrophic consequences. Striking a balance between innovation and ethical responsibility presents a significant challenge to the cybersecurity community, necessitating both technical prowess and an unwavering commitment to the responsible deployment of AI [8] (Figure 5.3).

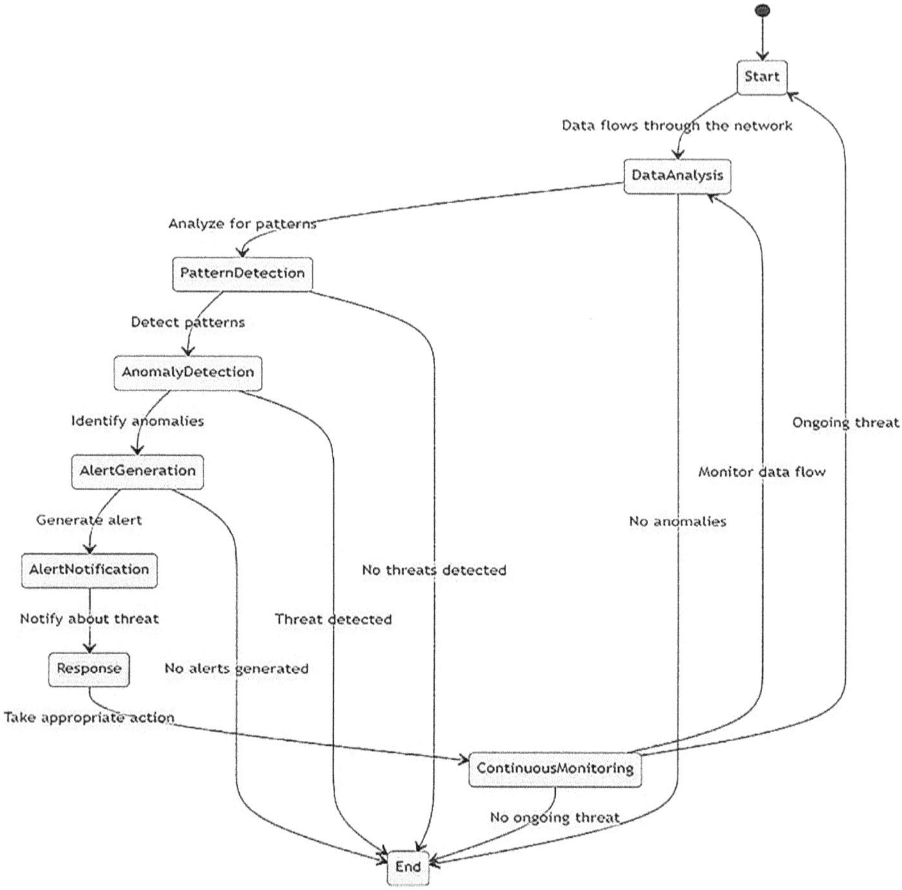

FIGURE 5.3 AI in threat detection.

Figure 5.3 provides a simplified overview of how AI plays a pivotal role in the threat detection process within cybersecurity. It visualizes the steps involved in AI's continuous monitoring and analysis of data to identify potential threats and trigger alerts for swift responses.

The process initiates when data enters the network and undergoes continuous scrutiny by AI algorithms. These algorithms are trained to recognize both established threat patterns and anomalies that deviate from normal data behavior.

As data analysis unfolds, AI excels at two key functions: pattern detection and anomaly detection. When incoming data matches known threat patterns, AI swiftly identifies them. Conversely, if data exhibits unusual behavior not conforming to established patterns, it raises an alarm by flagging it as an anomaly. The moment an anomaly or potential threat is detected, the system generates an alert. This alert serves as a timely notification, prompting security personnel or automated response mechanisms to take appropriate actions, effectively mitigating the threat.

Crucially, the process highlights the continuous nature of AI-driven threat detection. After an alert is addressed, the flowchart demonstrates that AI seamlessly returns to monitoring data, ensuring ongoing vigilance. This flowchart encapsulates the fundamental role of AI in bolstering cybersecurity by swiftly identifying threats and enabling rapid responses, ultimately contributing to the resilience of digital environments.

As we venture deeper into the fusion of Zero Trust and AI, we find ourselves at the threshold of a new era in cybersecurity. Here, trust is subject to constant evaluation, threats are proactively identified, and security evolves from a defense into an enabler of progress. This journey is intricate, laden with complexities and ethical considerations, requiring a balanced approach that integrates technological innovation with a sense of moral responsibility.

5.7 AI AND ZERO TRUST: A SYNERGISTIC APPROACH

5.7.1 ENHANCING SECURITY THROUGH VIGILANCE AND VERIFICATION

In the realm of cybersecurity, two formidable forces converge to reshape the landscape: AI and Zero Trust. In this section, we explore their intertwined journey, highlighting how their combined strength propels cybersecurity into uncharted territory.

The zero-trust assessment involves determining: propensity to trust; experience of trust; a trust assessment based on ability, benevolence, and integrity; followed by a study of; acceptance of vulnerability and of risk taking. Implications for practitioners, policymakers, and academics include an argument that the deployment, assessment, and management of a zero trust posture will promote explainable, trustworthy, and secure AI [9].

5.7.2 SHARED PRINCIPLES

At first glance, AI, and Zero Trust might seem worlds apart, yet they share a fundamental principle: perpetual vigilance. Zero Trust insists on continuous verification, rejecting any blind trust. It compels us to confirm trustworthiness at every digital intersection. Simultaneously, AI acts as an ever-watchful guardian, meticulously analyzing the digital world. It processes vast volumes of data with incredible speed, detecting anomalies and potential threats in real-time [1].

Together, they form a potent duo: AI as the vigilant guardian, processing enormous data streams, while Zero Trust serves as the unwavering gatekeeper, granting access only to entities that meet strict criteria. This alliance embodies the philosophy of "Never Trust, Always Verify" with unprecedented sophistication (Figure 5.4).

5.7.3 AI: THE DISCERNING SENTINEL

AI's role in cybersecurity resembles that of an expert detective. It uncovers patterns that elude human perception, identifies suspicious activities, and reacts with remarkable speed. This discernment arises from its training in ML and DL, where it learns from extensive datasets. This continuous learning equips AI to evolve alongside the ever-shifting landscape of cyber threats [6].

Moreover, AI's adaptability shines when identifying even previously unknown threats. Instead of relying on predefined rules or signatures, it assesses behaviors and characteristics to detect malicious intent. This adaptability is crucial in safeguarding against sophisticated, zero-day attacks that traditional security measures often miss.

5.7.4 ZERO TRUST: A PARADIGM SHIFT IN SECURITY

Zero Trust operates on a fundamental shift in perspective. Rather than the traditional approach of establishing trust within a network and assuming safety, it adopts a stringent stance: "never trust, always verify." This means every user, device, and transaction is treated as potentially untrusted and subject to continuous scrutiny [11].

Consider a scenario where an employee accesses sensitive data. In the traditional model, trust is assumed once inside the network. In contrast, within a Zero Trust framework, even trusted users undergo continuous verification. Their access is restricted to the bare minimum necessary for their roles, adhering to the principle of least privilege. This practice ensures that even if their credentials are compromised, potential damage is limited.

5.7.5 THE CONFLUENCE AI-DRIVEN CONTINUOUS VERIFICATION

Here is where the synergy between AI and Zero Trust shines most brightly. AI takes on the role of continuous verification with unmatched proficiency. It evaluates every user interaction, access attempt, and data transaction. Its discerning eye identifies anomalies that may signal a breach, ensuring trust is never granted blindly.

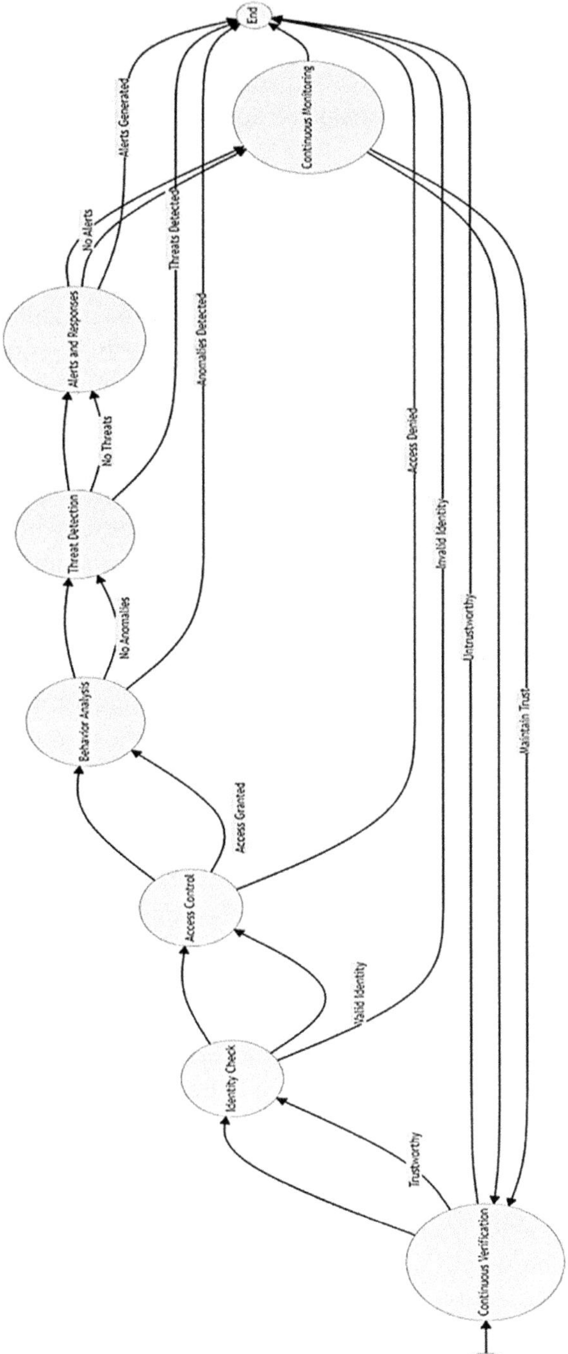

FIGURE 5.4 AI-zero trust verification process.

Imagine an employee attempting to access sensitive information from an unfamiliar location. Traditional security might raise an alarm, but AI, with its pattern-recognition abilities, recognizes this as legitimate based on the user's historical behavior. This adaptive approach exemplifies the strength of this collaboration.

5.7.6 REALIZING THE POTENTIAL: INDUSTRY APPLICATIONS

To truly grasp the impact of this union, we must examine real-world applications. Across diverse industries, from finance to healthcare and critical infrastructure, the AI-Zero Trust alliance has proven its worth. It protects sensitive financial data, upholds patient privacy, and fortifies critical infrastructure.

In finance, where confidentiality is paramount, this partnership ensures secure transactions, granting access only to authorized entities. In healthcare, where patient data is sacrosanct, this collaboration erects an impenetrable shield against cyber threats. In critical infrastructure, which underpins society, this alliance stands ready to fend off potential disruptions.

5.7.7 ETHICS AND RESPONSIBILITY

Navigating the Terrain As we navigate this cutting-edge terrain, we must do so with a commitment to ethics and responsibility. AI's power in cybersecurity is immense, and with that power comes great responsibility. Ensuring that AI operates without bias and respects privacy is crucial. Moving forward requires not only technological expertise but also a dedication to fairness, transparency, and accountability.

In conclusion, the fusion of AI and Zero Trust marks a new era in cybersecurity, where trust is continuously validated, threats are met with unwavering vigilance, and security becomes an enabler of progress. This journey is marked by relentless verification, adaptive learning, and a commitment to ethical deployment. It represents not just technological advancement but a paradigm shift that will shape the future of cybersecurity.

5.7.8 CONTINUOUS LEARNING AND ADAPTATION

One of the remarkable facets of the AI-Zero Trust Alliance is its ability to learn and adapt continuously. AI, with its foundation in Machine Learning and Deep Learning, evolves alongside the ever-evolving threat landscape. As it encounters new types of threats, it learns to recognize them, enhancing its ability to detect and mitigate potential breaches. This continuous learning ensures that the security measures in place remain effective against emerging threats, providing a proactive defense mechanism.

5.7.9 UNIFIED SECURITY ECOSYSTEM

The integration of AI and Zero Trust doesn't just stop at enhancing individual components of cybersecurity; it paves the way for a unified security ecosystem. In this ecosystem, security isn't just a series of isolated protocols and measures; it's a cohesive and adaptive network. AI serves as the central nervous system, continuously monitoring and analyzing data from various sources, while Zero Trust acts as the vigilant gatekeeper, implementing access controls based on AI's real-time assessments. This unified approach results in a holistic and resilient security infrastructure that can quickly adapt to new challenges and vulnerabilities.

As we delve deeper into the synergistic realm of AI and Zero Trust, these facets underscore the transformative potential of this alliance in revolutionizing cybersecurity (Table 5.2).

TABLE 5.2 AI-Zero Trust Synergy Aspects

Sl. No.	Aspect	Advantages	Disadvantages
1.	Security Enhancement	Continuous verification improves threat detection.	Implementation complexity for legacy systems.
2.	Real-time Adaptability	AI can swiftly adapt to new threats and attack patterns.	Potential for false positives/ negatives in AI- driven decisions.
3.	Scalability	Suitable for both small and large-scale networks.	Resource-intensive AI implementation in large organizations.
4.	Industry Versatility	Applicable across various industries and sectors.	Ethical concerns related to AI biases and privacy.
5.	User Experience	Enhanced security without significant disruptions to user experience.	Training and educating personnel for AI-Zero Trust adoption.

5.7.10 VIGILANCE AND INTELLIGENCE

In our journey through the AI-Zero Trust Alliance, we've uncovered the symbiotic relationship between AI and Zero Trust, understanding how they redefine cybersecurity. Now, let's venture into the real-world realm of practical applications and use cases that showcase the alliance's transformative potential.

5.7.11 FINANCIAL FORTRESS: PROTECTING CONFIDENTIAL TRANSACTIONS

In the financial sector, confidentiality is paramount. The AI-Zero Trust duo forms an impenetrable fortress around sensitive data, ensuring only authorized entities gain access. Every transaction and every interaction is subjected to rigorous scrutiny.

Imagine a scenario where a user initiates a substantial fund transfer. In a traditional setup, once past the initial login, trust is assumed, and the transaction proceeds. However, within the realm of Zero Trust, AI continually evaluates the user's behavior. If the transaction deviates from the user's usual pattern, an alert is raised. This adaptive verification prevents fraudulent activities and safeguards financial integrity.

Studies have shown that there is an increased usage of Zero Trust in securing networks and protecting organizations from malicious cyberattacks. The research also indicates that APIs are insecure in Zero Trust environments and most organizations are not aware of their presence [12].

5.7.12 HEALTHCARE'S DIGITAL WARD: SAFEGUARDING PATIENT PRIVACY

In healthcare, patient data is sacrosanct. The AI-Zero Trust partnership serves as a vigilant guardian of electronic health records (EHRs), preserving patient privacy. Unauthorized access attempts trigger immediate scrutiny.

Over 90% of all biomedical AI research is stalled due to its inability to access real world protected health information to complete the required clinical validation and valuation prior to its deployment. With AI forecasted to improve patient outcomes by 40% and costs by 50% the new paradigm of zero trust must be adopted and leveraged to avoid the existential crisis facing biomedical AI research [13].

Picture a scenario where a healthcare professional needs to access patient records. Within the conventional model, once within the hospital network, trust is implicit. However, within Zero Trust, even trusted users undergo continuous verification. AI monitors each interaction, ensuring that only authorized personnel access sensitive patient data. This not only secures patient confidentiality but also aligns with regulatory requirements like the Health Insurance Portability and Accountability Act (HIPAA).

The integration of different technologies such as computers, medical devices, and telecommunications will significantly improve the efficiency of patient treatment, reduce the cost of health care and improve privacy and security. Trust decisions are based on the trust posture criteria, existing, and potential trust relationships, and the MEC environment trust state. The research outcomes show that this approach provides a positive improvement whilst establishing trust [14] (Figure 5.5).

Figure 5.5, presents a mind map provides a visual representation of the practical applications and use cases of the AI-Zero Trust alliance within the realm of cybersecurity. It offers a structured and interconnected overview of how this powerful partnership is leveraged across diverse industries to enhance security measures while addressing unique challenges.

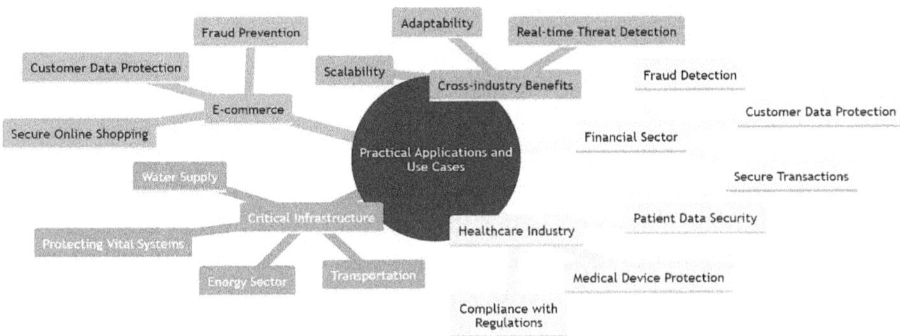

FIGURE 5.5 Practical applications.

The mind map has been organized into key sectors where the AI-Zero Trust approach finds substantial utility.

5.7.13 CRITICAL INFRASTRUCTURE RESILIENCE: DEFENDING SOCIETY'S BACKBONE

Critical infrastructure, spanning energy, transportation, and communication systems, forms the backbone of society. The AI-Zero Trust Alliance stands as the vanguard against potential disruptions, ensuring uninterrupted service.

Consider a scenario where an employee working remotely attempts to access the control systems of a power plant. Within the conventional security paradigm, remote access might trigger alarms, but trust would be implicit once access is granted. In stark contrast, Zero Trust, aided by AI, subjects this interaction to continuous verification. AI assesses the behavior, ensuring that it aligns with the user's historical patterns. Any deviation triggers an alert. This dynamic verification secures critical infrastructure against unauthorized access.

5.7.14 E-COMMERCE SENTINEL: SAFEGUARDING CUSTOMER DATA

In the e-commerce realm, customer trust hinges on data security. The AI-Zero Trust partnership acts as a sentinel, guarding customer data against breaches. Every transaction undergoes meticulous scrutiny.

Imagine an online shopper making a high-value purchase. Traditional systems might view this as a straightforward transaction. However, AI within a Zero Trust framework assesses the user's behavior. If the purchase diverges from typical patterns, it raises a flag. This real-time verification prevents fraudulent transactions, enhancing customer trust and preserving the integrity of e-commerce platforms.

5.7.15 ETHICS AND ACCOUNTABILITY: NAVIGATING CHALLENGES

The final result is a proposed framework and associated technology which, via Zero Trust principles, addresses multiple layers of the organization to grasp and align cybersecurity risks and understand the readiness and fitness of the organization and its measures to counter cybersecurity risks [15].

While these practical applications showcase the potency of the AI-Zero Trust Alliance, we must not overlook the ethical considerations. AI's power, when misused, can have severe consequences. Ensuring that AI-driven systems operate fairly, without biases, is an ethical imperative. Moreover, responsible deployment is essential to prevent the misuse of AI for malevolent purposes.

Additionally, transparency, and accountability are paramount. Users must understand how their data is processed and how AI-driven decisions are made. Regulatory frameworks like the General Data Protection Regulation (GDPR) in Europe and the California Consumer Privacy Act (CCPA) in the United States underscore the importance of these principles.

5.8 A SECURE FUTURE BECKONS

As we conclude our exploration of practical applications, we stand at the threshold of a secure future. The AI-Zero Trust Alliance represents a paradigm shift in cybersecurity, where trust is earned, threats are swiftly identified, and security becomes an enabler of progress. While challenges and ethical considerations persist, the path forward is clear: harness the power of this alliance responsibly, fostering a digital world where security and progress coexist.

In the next section of our chapter, we will explore the complexities and ethical dilemmas that arise in the wake of this transformative partnership, delving into the nuances of AI biases, responsible deployment, and the evolving cybersecurity landscape.

5.9 CHALLENGES AND ETHICAL CONSIDERATIONS

5.9.1 NAVIGATING THE COMPLEX TERRAIN OF SECURITY AND RESPONSIBILITY

As we journey through the transformative landscape of the AI-Zero Trust Alliance, we must pause to examine the challenges and ethical considerations that arise. This section shines a light on the complexities that emerge when advanced technology meets the realm of security and accountability. Such risk evaluation frameworks will face challenges due to the numerous sources of inputs and the volume of contextual data collected. Similar challenges will be encountered by frameworks that use trust levels to grant or deny access to resources [16].

5.9.2 THE CHALLENGE OF AI BIASES

AI systems, despite their remarkable capabilities, are not immune to biases. These biases can emerge from the data on which AI models are trained. For instance, if historical data reflects certain prejudices, the AI system may inadvertently perpetuate those biases. Despite the benefits of AI, its application suffers from the opacity of complex internal mechanisms and doesn't satisfy the principles of Explainable Artificial Intelligence (XAI) by design [17].

Consider an AI-driven cybersecurity system that makes access decisions based on user behavior. If past data contains biased judgments, the AI might make decisions that disproportionately affect certain groups or individuals. This not only raises ethical concerns but also impacts the system's effectiveness.

5.9.3 RESPONSIBLE DEPLOYMENT

The responsible deployment of AI in cybersecurity is paramount. Ensuring that AI systems operate ethically and transparently is a multifaceted challenge. Organizations must establish guidelines for AI usage and continually monitor AI-driven security systems to prevent misuse.

Transparency is key. Users should have insight into how AI is employed in making security decisions. They should be informed about data collection and usage. Clear policies and regulations should govern AI deployment, addressing concerns related to privacy and accountability. The lack of transparency further exacerbates the problem in the field of Cybersecurity because entrusting crucial decisions to a system that cannot explain itself presents obvious dangers [18].

5.9.4 THE EVOLVING LANDSCAPE

The field of cybersecurity is in constant flux. New threats emerge, attack vectors evolve, and cybercriminals become increasingly sophisticated. In this dynamic environment, maintaining the security posture of AI-Zero Trust systems is a challenge.

Consider a scenario where a novel cyber threat emerges that exploits a previously unknown vulnerability. AI-driven security systems, while powerful, rely on past data for pattern recognition. In such cases, there may be a lag in recognizing and responding to new threats. Cybersecurity professionals must adapt swiftly to these evolving challenges.

5.9.5 BALANCING SECURITY AND PRIVACY

The AI-Zero Trust Alliance offers robust security, but it must coexist with individual privacy rights. Striking the right balance between these two critical aspects is an ongoing challenge.

In a world where trust is earned through continuous verification, there's potential for extensive data collection. While this data is vital for security, it also poses a privacy risk. Ensuring that user data is protected and used only for its intended purpose requires careful attention and stringent privacy controls.

5.9.6 ACCOUNTABILITY AND LIABILITY

When AI-driven security systems make decisions, questions of accountability and liability arise. Who is responsible if the system makes an erroneous judgment that results in harm? These questions remain largely uncharted territory.

Consider an instance where an AI-driven security system incorrectly flags a legitimate user as a potential threat, leading to negative consequences for that user. Determining who bears responsibility in such scenarios is complex. Establishing legal frameworks to address liability is an essential but challenging task.

5.9.7 EDUCATION AND AWARENESS

As we navigate the intricate landscape of AI-Zero Trust cybersecurity, education, and awareness become linchpins. Users, organizations, and policymakers must understand the nuances of AI-driven security and the ethical considerations involved. Cybersecurity professionals need training to manage AI systems effectively. Users must be informed about how AI affects their digital interactions. Policymakers must develop regulations that promote responsible AI deployment without stifling innovation.

Although implementing standards for AI is currently a preferred solution for many, on their own, standards may not provide sufficient capability and flexibility. An argument is presented that cyber norms of zero trust for AI (ZTAI) alongside standards are essential for the security of AI, along with a call to action [19].

5.9.8 A PATH FORWARD

The AI-Zero Trust Alliance is a testament to the power of technology in redefining cybersecurity. Yet, as with any innovation, it brings with it a set of challenges and ethical considerations that demand careful navigation.

In the next sections of our chapter, we will cast our gaze toward the future. We'll explore the emerging trends and potential advancements in the AI-Zero Trust landscape, considering how these developments may shape the future of cybersecurity.

5.10 FUTURE DIRECTIONS AND TRENDS

5.10.1 NAVIGATING THE PATH FORWARD IN AI-ZERO TRUST SECURITY

It's imperative to look ahead and anticipate the future directions and emerging trends that will shape the realm of cybersecurity. This section provides insight into what lies on the horizon for the AI-Zero Trust Alliance.

Demand for products that support Zero Trust is on the rise. The market's significant factors include the growing frequency of target-based cyber-attacks and increasing data protection and information security regulations [20].

5.10.2 ENHANCED AUTOMATION AND ORCHESTRATION

One of the key trends shaping the future of AI-Zero Trust security is enhanced automation and orchestration. As cyber threats become more sophisticated, the response must be equally swift and sophisticated. AI, with its real-time data analysis capabilities, is poised to play a pivotal role in automating threat detection and response.

Imagine a scenario where a network intrusion attempt is detected. Instead of relying solely on human intervention, AI-driven systems can instantly assess the threat's severity and take predefined actions to mitigate it. This level of automation not only reduces response time but also minimizes the potential for human error.

Furthermore, orchestration, which involves coordinating multiple security tools and processes, will become more intelligent. AI will facilitate seamless communication between various security components, ensuring a holistic and proactive defense. Software-Defined Wide Area Networks (SD-WANs) and Secure Access Service Edge (SASE) frameworks are poised to synergize with Zero Trust principles, creating a harmonious fusion of network security and performance optimization. As organizations embrace these transformative network architectures, Zero Trust's role in orchestrating secure, seamless, and efficient communication channels becomes increasingly pivotal [21].

5.11 EXPLAINABLE AI (XAI) FOR TRANSPARENCY

As AI becomes more ingrained in cybersecurity, the need for transparency and explain ability grows. The concept of XAI aims to demystify AI decision- making processes, especially in critical applications like security.

XAI suggests a transition toward more interpretable AI to overcome this issue. XAI principles intend to develop strategies that will result in better explainable models while keeping high-performance levels [22].

Consider a scenario where an AI-driven system denies a user access to sensitive data. Users, cybersecurity professionals, and regulators may demand an explanation for the decision. XAI technologies will provide insights into how AI arrived at a particular conclusion, making the decision-making process more transparent and accountable.

XAI not only enhances trust in AI-driven security systems but also aids in identifying and rectifying biases or errors in the AI's logic. This trend aligns with ethical considerations, ensuring that AI operates fairly and without hidden biases (Figure 5.6).

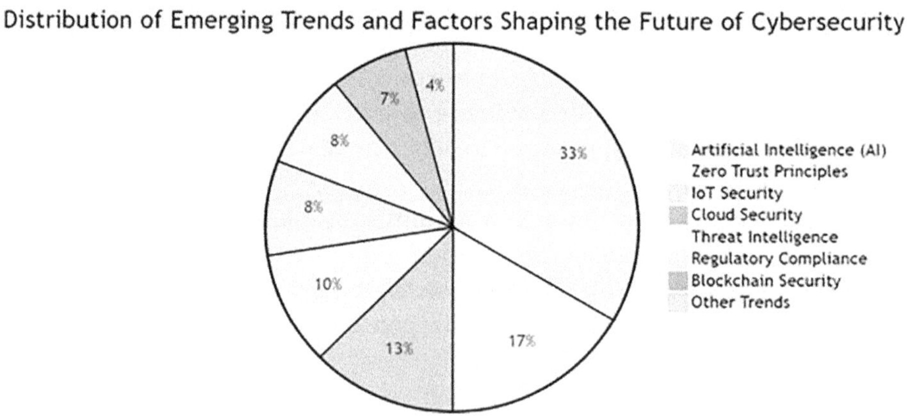

FIGURE 5.6 Emerging trends.

5.12 CONTINUOUS ADAPTIVE AUTHENTICATION

Authentication lies at the core of Zero Trust, where trust is continually verified. In the future, we can expect continuous adaptive authentication to become more sophisticated and nuanced. Traditional authentication methods, such as passwords and tokens, may give way to more advanced approaches.

Imagine a scenario where a user's identity is verified not just through static credentials but by analyzing their behavior, location, and other contextual factors. AI will play a central role in this continuous verification process, assessing the legitimacy of users and devices in real-time.

This trend aligns with the evolving concept of Zero Trust Network Access (ZTNA), where users and devices are authenticated before accessing specific resources. ZTNA focuses on fine-grained access control and continuous monitoring, bolstering security in an increasingly mobile and remote work environment.

Adapting zero trust principles to networks ensures that a single insecure Application Protocol Interface (API) does not become the weakest link comprising of Critical Data, Assets, Application, and Services (DAAS) [12].

5.13 QUANTUM-SAFE SECURITY

The advent of quantum computing poses a unique challenge to cybersecurity. Quantum computers have the potential to break currently employed encryption methods, rendering traditional security vulnerable. In response, quantum-safe security is emerging as a critical trend.

In recent years, Quantum Key Distribution (QKD) has introduced to the world of information security a viable method that is anticipated to provide security protection against the threat of quantum computing [23].

Consider a future where quantum computers are powerful enough to decrypt sensitive data. Organizations will need to adopt quantum-resistant encryption methods to safeguard their data. AI will be instrumental in developing and implementing these quantum-safe security measures, ensuring that sensitive information remains protected.

5.14 FORGING A RESILIENT FUTURE

As we navigate the future of AI-Zero Trust security, these trends and directions will reshape the cybersecurity landscape. Enhanced automation, transparency through XAI, continuous adaptive authentication, and quantum-safe security are just a glimpse of what lies ahead.

We've glimpsed the future and seen how AI and Zero Trust, as dynamic allies, will adapt to new challenges and emerging threats. By staying at the forefront of these trends, organizations can forge a resilient and secure digital future.

5.15 CONCLUSION

5.15.1 THE JOURNEY AHEAD: FORGING A RESILIENT FUTURE

Our exploration of the AI-Zero Trust Alliance has traversed the realms of advanced technology, cutting-edge cybersecurity, and ethical considerations. As we conclude this chapter, we find ourselves at the intersection of innovation and responsibility, poised to embark on a transformative journey into the future of cybersecurity. A dynamic trust evaluation for every access request is the key tenet of zero trust (ZT). The main function of a ZTA is authorizing the individual access requests by a subject rather than authorizing the subject requesting access [24]. With the power of classification algorithms in machine learning, AI is an alternative solution to save time and effort and help maintain the same security posture with less human intervention. The proposed approach utilizes the information from available security feeds and statically configured policies to enforce and maintain zero-trust network policies [25].

5.15.2 A NEW SECURITY PARADIGM

The AI-Zero Trust Alliance represents nothing short of a seismic shift in how we conceptualize and implement cybersecurity. In a digital realm rife with evolving threats and vulnerabilities, the traditional model of trust, "trust but verify," has proven woefully insufficient. Instead, the concept of "never trust, always verify" has become our lodestar, redefining the very essence of security.

Introducing a "trust but verify continuously" approach to AI assurance, which describes assured autonomy activities in a model-based systems development context and includes postdelivery activities for continuous assurance is important.

Zero Trust, with its bedrock principles of continuous verification, least privilege access, micro-segmentation, and the assumption of breach, provides a robust foundation for securing our digital assets. It acknowledges the ever-present dangers posed by both external adversaries and internal vulnerabilities, demanding perpetual vigilance as the cornerstone of modern cybersecurity.

5.15.3 THE ENABLING POWER OF ARTIFICIAL INTELLIGENCE

In this transformative journey, we have witnessed the formidable capabilities of AI, the sentinel of the digital realm. AI augments the principles of Zero

Trust with an unprecedented layer of intelligence and adaptability. It deftly processes vast datasets, unravels intricate patterns, and swiftly detects anomalies in real time. AI empowers security systems to proactively identify threats, thwart attacks, and evolve in tandem with the ever-shifting threat landscape.

The alliance of AI and Zero Trust is not merely a collaboration; it is a symbiotic partnership that has become a necessity in a world where digital transformation is the norm. In this era, where security is not an obstacle but a catalyst for progress, this partnership is indispensable.

KEYWORDS

- **artificial intelligence**
- **cybersecurity**
- **digital transformation**
- **least privilege access**
- **micro-segmentation**
- **trust evaluation**
- **zero trust architecture**

REFERENCES

1. Kindervag, J. (2010). *No More Chewy Centers: Introducing the Zero Trust Model of Information Security.* Forrester Research.
2. Bejtlich, R. (2019). *The Tao of Network Security Monitoring: Beyond Intrusion Detection.* Addison-Wesley Professional.
3. Chase, K. (2021). *BeyondCorp: Implementing Zero Trust security.* O'Reilly Media.
4. Doshi, N., Cebrian, M., & Larson, K. (2017). The spread of true and false news online. *Science, 359*(6380), 1146–1151.
5. LeCun, Y., Bengio, Y., & Hinton, G. (2015). Deep learning. *Nature, 521*(7553), 436–444.
6. van der Walt, S., & Bernard, E. (2018). Effective cybersecurity using artificial intelligence and machine learning. *IT Professional, 20*(3), 35–41.
7. Buolamwini, J., & Gebru, T. (2018). Gender shades: Intersectional accuracy disparities in commercial gender classification. *Proceedings of Machine Learning Research, 81*, 1–15.
8. Brundage, M., Avin, S., Clark, J., Toner, H., Eckersley, P., Garfinkel, B., & Zeitzoff, T. (2018). *The Malicious Use of Artificial Intelligence: Forecasting, Prevention, and Mitigation. arXiv preprint arXiv:1802.07228.*
9. Wylde, A. (2023). Intelligent virtual assistants (IVAs): Trust and zero trust. in *Proceedings of the [Conference Name]* (pp. 10.1007/978-3-031-37963-5_8). DOI: 10.1007/978-3-031-37963-5_8.

11. Cebrian, M., & Larson, K. (2017). The spread of true and false news online. *Science, 359*(6380), 1146–1151.
12. Qazi, F. A. (2022). Study of Zero Trust Architecture for applications and network security. In *2022 IEEE 19th International Conference on Smart Communities: Improving Quality of Life Using ICT, IoT and AI (HONET)* (pp. 111–116). Marietta, GA: IEEE.
13. Dennis, S., & Rogers, Dr. (2023). Zero Trust: Enabling the last mile of biomedical AI research. *Blockchain in Healthcare Today, 6*. doi: 10.30953/bhty.v6.259.
14. Ali, B., Gregory, M. A., & Li, S. (2021). Uplifting healthcare cyber resilience with a multi-access edge computing Zero-Trust security model. In *2021 31st International Telecommunication Networks and Applications Conference (ITNAC)* (pp. 192–197). Sydney, Australia: IEEE.
15. Bobbert, Y., & Scheerder, J. (2022). Zero Trust Validation: From practice to theory: An empirical research project to improve Zero Trust implementations. In *2022 IEEE 29th Annual Software Technology Conference (STC)* (pp. 93–104). Gaithersburg, MD: IEEE.
16. Syed, N. F., Shah, S. W., Shaghaghi, A., Anwar, A., Baig, Z., & Doss, R. (2022). Zero Trust Architecture (ZTA): A comprehensive survey. *IEEE Access, 10*, 57143–57179.
17. Capuano, N., Fenza, G., Loia, V., & Stanzione, C. (2022). Explainable artificial intelligence in cybersecurity: A survey. *IEEE Access, 10*, 93575–93600.
18. Wylde, A. (2023). New norms for AI: Zero Trust—Verify then trust. *AI, Computer Science and Robotics Technology, 2*. DOI: 10.5772/acrt.27.
19. Nivarthi, K. P., Sharma, K., & Gatla, G. (2022). Fighting cybercrime with Zero Trust. *American Scientific Research Journal for Engineering, Technology, and Sciences, 90*, 371–381.
20. Ghasemshirazi, S., Shirvani, G., & Alipour, M. (2023). Zero Trust: Applications, challenges, and opportunities.
21. Capuano, N., Fenza, G., Loia, V., & Stanzione, C. (2022). Explainable artificial intelligence in cybersecurity: A survey. *IEEE Access, 10*, 93575–93600.
22. Mao, J., Xu, G., Sakk, E., & Wang, S. P. (2023). Quantum key distribution and security studies. In *2023 7th International Conference on Cryptography, Security and Privacy (CSP)* (pp. 158–162). Tianjin, China: IEEE.
23. Ramezanpour, K., & Jagannath, J. (2022). Intelligent Zero Trust Architecture for 5G/6G networks: Principles, challenges, and the role of machine learning in the context of O-RAN. *Computer Networks, 217*, 109358. https://doi.org/10.1016/j.comnet.2022.109358.
24. Hosney, E. S., Halim, I. T. A., & Yousef, A. H. (2022). An artificial intelligence approach for deploying Zero Trust Architecture (ZTA). In *2022 5th International Conference on Computing and Informatics (ICCI)* (pp. 343–350). New Cairo, Cairo, Egypt: IEEE.
25. Laplante, P., & Kuhn, R. (2022). AI assurance for the public–trust but verify, continuously. In *2022 IEEE 29th Annual Software Technology Conference (STC)* (pp. 174–180). Gaithersburg, MD: IEEE.

CHAPTER 6

The Role of Artificial Intelligence in Zero-Trust Learning

C. V. SURESH BABU* and J. VISHNU PRIYAN

Department of Information Technology, Hindustan Institute of Technology and Science, Chennai, Tamil Nadu, India

*Corresponding author

ABSTRACT

"The Role of Artificial Intelligence (AI) in Zero Trust Learning (ZTL)" explores the transformative synergy between AI and ZTL within the realm of healthcare security. Unveiling the core tenets of ZTL, this chapter demonstrates how intelligent devices, infused with AI algorithms, redefine the paradigm of healthcare security. It showcases real-world case studies highlighting the profound impact of AI-driven strategies on security, trust, and data protection. With adaptive security, continuous improvement, and patient-centric care at its core, this integration promises a future where healthcare security transcends theory to become an unwavering fortress, safeguarding patient data and delivering exceptional care.

6.1 INTRODUCTION

6.1.1 OVERVIEW OF ZERO TRUST LEARNING

Zero Trust is a cybersecurity paradigm that places a primary emphasis on safeguarding resources and operates on the foundational principle that trust is never granted implicitly but must be continually assessed. The concept of Zero Trust Architecture (ZTA) encompasses a comprehensive approach to securing enterprise resources and data, encompassing elements such as

Zero-Trust Learning: Applications in Modern Network Security. Chetna Kaushal, Amandeep Kaur, Mohit Angurala, Aryan Chaudhary (Eds.)

identity, credentials, access management, operational aspects, endpoints, hosting environments, and the interconnecting infrastructure [1].

Zero Trust Learning, often referred to as ZTL, represents a groundbreaking transformation within the cybersecurity landscape. It challenges the conventional model of implicitly trusting everything within a network and instead introduces a paradigm where trust is never taken for granted, regardless of factors such as location, identity, or device. An initial and fundamental tenet of ZTL involves the stringent restriction of resource access to those individuals with a genuine need, while also granting only the minimum level of privileges required to accomplish their respective tasks.

Within the scope of this chapter, we embark on a comprehensive exploration of the role played by AI in the context of ZTL. This section elucidates how AI significantly bolsters security measures within our dynamically evolving digital landscape.

6.1.2 INTELLIGENT MACHINES IN HEALTHCARE SECURITY

The safeguarding of healthcare information presents a formidable challenge for healthcare systems worldwide. Healthcare security emerges as a topic of paramount importance, driven by the sensitivity of patient data and the escalating occurrence of cyberattacks targeting medical institutions [2]. In response to this pressing concern, AI-powered technologies are being leveraged to enhance security measures, ensuring the utmost confidentiality and data integrity in healthcare settings. In this chapter, we embark on an exploration of the convergence between intelligent machines and healthcare security, all within the broader context of ZTL.

6.1.3 THE SIGNIFICANCE OF THIS CHAPTER

In a cybersecurity landscape characterized by rapid evolution, "ZTL: A Comprehensive Guide" emerges as a foundational resource, providing invaluable insights into the principles and practical applications of Zero Trust. This comprehensive guide delves into essential topics such as the principle of least privilege access and micro-segmentation, as elucidated.

The significance of this particular chapter cannot be overstated, as it unveils the pivotal role that AI plays in enhancing and augmenting Zero Trust strategies. By shedding light on the synergistic relationship between AI and Zero Trust, this chapter becomes essential reading not only for cybersecurity professionals but also for healthcare practitioners and anyone with an interest in safeguarding critical data and systems in today's interconnected world. Its content resonates deeply with the evolving needs of a digital landscape where security and data protection are paramount concerns.

6.2 INTELLIGENT MACHINES AND HEALTHCARE SECURITY

6.2.1 ADVANTAGES OF INTELLIGENT DEVICES

In our contemporary digital era, intelligent machines have seamlessly integrated themselves into various industries, with healthcare being no exception. These sophisticated devices have proven to be not only more precise but also more dependable than their traditional counterparts. Equipped with advanced sensors, robust processing capabilities, and seamless connectivity, intelligent machines possess a set of shared attributes that render them particularly well-suited for enhancing security within healthcare environments. In this section, we delve into these distinctive characteristics and advantages, illuminating their pivotal role in fortifying healthcare security.

1. **Sensor Capabilities:** Intelligent machines come equipped with a diverse array of sensors, including cameras, motion detectors, and environmental sensors. These sensors deliver real-time data streams that hold immense potential for security monitoring and the detection of potential threats.
2. **Data Processing:** The noteworthy capacity of intelligent devices to process data, either locally or in the cloud, empowers healthcare organizations to analyze substantial datasets effectively. This analytical capability allows for the timely identification of anomalies and potential security risks.
3. **Connectivity:** Intelligent machines seamlessly integrate with networks, enabling centralized control and monitoring. Furthermore, this connectivity facilitates the prompt dissemination of critical security updates and patches, thereby bolstering the healthcare system's resilience against emerging threats.
4. **Automation:** A significant advantage presented by many intelligent devices lies in their automation capabilities. These devices can autonomously execute predefined actions, such as notifying security personnel or triggering emergency responses when security breaches are detected.

6.2.2 ADVANTAGES OF INTELLIGENT MACHINES WITHIN THE ZTL MODEL

Intelligent machines operating within the ZTL model offer a multitude of benefits.

6.2.2.1 IMPROVED EFFICIENCY

Intelligent machines, under the ZTL model, excel in securely automating numerous tasks that are currently performed by human healthcare professionals. This automation not only enhances operational efficiency but also reduces the potential for human errors.

6.2.2.2 ENHANCED ACCURACY

Intelligent machines operating within the ZTL framework play a pivotal role in ensuring the precision of diagnoses and treatments. They achieve this by effectively filtering out noisy data and maintaining data integrity.

6.2.2.3 PERSONALIZED CARE

Leveraging the capabilities of intelligent machines allows healthcare providers to offer highly personalized care to their patients. These devices can adapt and tailor treatment plans based on individual patient data and needs [3].

In summary, intelligent machines, with their advanced features and integration within the ZTL model, are poised to revolutionize healthcare security by offering improved efficiency, accuracy, and personalized care, all while minimizing potential security risks and data breaches.

6.2.3 INTEGRATING INTELLIGENT MACHINES WITH MEDICAL DEVICES

The integration of intelligent machines with medical devices represents a groundbreaking leap forward in the realm of healthcare security. This convergence not only promises to elevate patient care but also holds the potential to fortify the safeguarding of sensitive patient data. The amalgamation of intelligent technologies with medical equipment paves the way for the development of devices that are not only more precise and dependable but also remarkably efficient. Furthermore, it opens up new avenues for the collection of crucial health data, which can play a pivotal role in diagnosis and therapy [4]. In this section, we embark on a comprehensive exploration of the myriad benefits and far-reaching implications that arise from the integration of intelligent machines with medical devices include:

1. **Improved Data Accuracy:** One of the most notable advantages of this integration is the substantial enhancement of data accuracy in medical settings. Intelligent machines excel in refining the process of data collection and analysis, significantly mitigating the risk of errors that could potentially compromise patient safety.

2. **Real-time Monitoring:** The marriage of intelligent machines with medical devices facilitates continuous, real-time monitoring of patients. This capability empowers healthcare providers to swiftly detect and respond to critical situations as they unfold, thereby optimizing patient care and safety.

3. **Enhanced Device Management:** Intelligent machines take on the pivotal role of managing and maintaining medical devices, ensuring that they remain current with the latest security patches. This proactive approach minimizes vulnerabilities and contributes to the overall security of the healthcare ecosystem.

4. **Security Controls:** The integration of security features into medical devices through the infusion of intelligent technologies offers a robust defense mechanism against unauthorized access and potential data breaches. This proactive security posture is instrumental in upholding patient privacy and data integrity.

In essence, the fusion of intelligent machines with medical devices holds immense promise in reshaping healthcare security. It not only elevates the precision and efficiency of medical equipment but also empowers healthcare organizations to provide superior patient care while establishing stronger safeguards against emerging security threats. This transformative integration represents a pivotal advancement in the healthcare landscape, underscoring the indispensable role of intelligent technologies in the pursuit of safer and more effective healthcare delivery.

6.3 ENHANCING PATIENT CARE, REMOTE MONITORING, AND DATA SECURITY

The integration of intelligent machines into healthcare security holds a paramount objective: to elevate patient care and safety to new heights. In this section, we delve into the substantial and tangible benefits of this integration, with a particular focus on its implications for patient-centric care, remote monitoring, and data security.

6.3.1 PATIENT-CENTRIC CARE

Intelligent machines bring forth a transformation in the provision of patient-centric care. They empower healthcare providers to tailor treatments and interventions precisely to the unique needs of each patient, leveraging real-time data for personalized care plans. This approach shifts the healthcare paradigm from a one-size-fits-all model to one that prioritizes the individual requirements and circumstances of each patient.

6.3.2 REMOTE MONITORING

The integration of intelligent machines, particularly those equipped with a Zero Trust Learning (ZTL) model, plays a pivotal role in bolstering the security of remote monitoring devices and data. This advancement ensures that patients can receive effective and continuous care, even when they are not physically present within the confines of a hospital or medical facility. This capability becomes increasingly crucial in situations where physical presence is not possible, such as during pandemics or for patients in remote areas.

6.3.3 DATA SECURITY

The infusion of intelligent machines into the healthcare ecosystem serves as a formidable safeguard for patient data, encompassing everything from medical devices to electronic health records (EHRs). This comprehensive security approach not only preserves patient privacy but also ensures compliance with stringent regulatory requirements. Even in scenarios where an attacker has breached the network, an intelligent machine equipped with a ZTL model can act as a crucial line of defense, thwarting unauthorized access to patient data. This not only safeguards patients' data for legal and ethical purposes but also upholds the utmost privacy standards [5].

6.3.4 EARLY DETECTION AND INTERVENTION

Real-time data analysis facilitated by intelligent machines empowers healthcare professionals with the capability to identify potential health issues at an early stage. This early detection paves the way for timely interventions, potentially saving lives and significantly improving patient outcomes.

In essence, the integration of intelligent machines within healthcare security not only ushers in a new era of patient-centric care but also revolutionizes the way healthcare is delivered. By enhancing remote monitoring, fortifying data security, and enabling early detection and intervention, this integration stands as a testament to the pivotal role that technology plays in advancing the healthcare landscape. It underscores a future where patient well-being and data security are intricately interwoven, ensuring the highest standards of care in an increasingly interconnected world.

6.4 TECHNOLOGICAL FOUNDATIONS

In our quest to gain a comprehensive understanding of the pivotal role that AI plays within the realm of ZTL, this chapter ventures into the fundamental technological underpinnings that facilitate the integration of AI with ZTL, particularly within the domain of healthcare security. These technological foundations serve as the bedrock upon which we build our endeavors to fortify security measures and ensure the safeguarding of critical data. The following section embarks on an exploration of key facets encompassing these foundational elements.

6.4.1 AI-BASED THREAT DETECTION

This study [6], highlights the heart of modern cybersecurity lies the imperative of threat detection, and it is within this context that AI has ushered in a revolution. This section embarks on an in-depth examination of how AI-based threat detection mechanisms are reshaping security strategies, particularly within the framework of ZTL.

6.4.2 MACHINE LEARNING ALGORITHMS

Our exploration begins by delving into the diverse landscape of machine learning algorithms employed for threat detection. The study [7], scrutinize the efficacy of various approaches, such as supervised learning, unsupervised learning, and deep learning, in their capacity to identify and mitigate threats effectively.

6.4.3 ANOMALY DETECTION

We pivot to the pivotal role played by AI in anomaly detection, a critical component of Zero Trust security. Here, machine learning models are harnessed to discern deviations from typical behavioral patterns, thus alerting security measures to potential threats.

6.4.4 PREDICTIVE ANALYSIS

We further illuminate the capabilities of AI by highlighting its propensity to predict potential threats based on historical data. This predictive aspect empowers organizations to adopt proactive security measures, preemptively addressing emerging threats.

The imperative of safeguarding sensitive patient data and ensuring the reliability of healthcare systems underscores the profound need for AI-based threat detection within the healthcare sector. By harnessing the prowess of AI, healthcare entities can efficiently identify and mitigate security issues. AI algorithms play a multifaceted role, ranging from malware detection and identification of irregularities in network traffic to the anticipation of potential risks and the enhancement of authentication procedures. Moreover, AI contributes significantly to data encryption, privacy protection, incident response, and compliance monitoring, thereby ensuring compliance with stringent regulations such as HIPAA and GDPR. This multifaceted role is paramount in preserving patient confidence and upholding the sanctity of medical data privacy.

Nonetheless, it is vital to acknowledge that the ever-evolving landscape of cybersecurity threats within the healthcare sector necessitates a judicious deployment of AI technologies, coupled with a commitment to continuous adaptation and improvement [8].

6.5 BEHAVIORAL ANALYTICS

Within the realm of healthcare security, the comprehension of user behavior stands as a paramount imperative. Behavioral analytics, fortified by the capabilities of AI, serves as a powerful tool, offering profound insights into user actions. These insights, in turn, enable healthcare organizations to effectively secure patient data. In this section, we embark on an exploration of the critical significance of behavioral analytics within the broader framework of ZTL.

6.5.1 USER PROFILING

AI-driven behavioral analytics takes center stage in the creation of user profiles. These profiles are instrumental in recognizing normal user behavior patterns, allowing the system to swiftly flag anomalies or activities that raise suspicion.

6.5.2 INSIDER THREAT DETECTION

Within the complex landscape of healthcare security, the role of behavioral analytics in identifying insider threats takes on paramount importance. Insider threats, often arising from authorized individuals abusing their access, represent a significant concern within healthcare security.

6.5.3 CONTINUOUS MONITORING

We further delve into the merits of continuous behavioral monitoring within healthcare environments. This approach ensures the timely detection of evolving security threats, reinforcing the proactive stance that healthcare organizations must adopt.

The security of private patient information and the overall integrity of healthcare systems are both made possible through the lens of behavioral analytics, an indispensable component in the field of medical security. Healthcare entities stand to make significant advancements in their security measures by harnessing the capabilities of this technology. Behavioral analytics involves the vigilant monitoring and analysis of digital activities conducted by patients, healthcare providers, and other individuals authorized to access medical systems. This practice serves to establish a baseline for typical behavior, and it excels at identifying anomalies or questionable activities. This includes the recognition of unusual data access patterns that may indicate a security breach, as well as the detection of insider risks, where authorized individuals misuse their access. Furthermore, behavioral analytics empowers proactive threat detection, enabling healthcare organizations to effectively address potential security threats and safeguard the confidentiality of patient data.

6.6 NLP IN HEALTHCARE DATA PROTECTION

Natural Language Processing (NLP) has emerged as a pivotal component within the domain of healthcare security, thanks to its remarkable capacity to analyze and protect sensitive data that resides within unstructured text. In this section, we embark on an exploration of the multifaceted applications of NLP in safeguarding healthcare data, all within the overarching principles of ZTL.

6.6.1 DATA REDACTION AND DE-IDENTIFICATION

NLP assumes a pivotal role in automating processes such as data redaction and de-identification. This technology enables the automatic removal of patient information from medical records, thereby ensuring the preservation of data privacy.

6.6.2 SENTIMENT ANALYSIS

We delve into the role of sentiment analysis, a branch of NLP, in monitoring communication within healthcare organizations. It serves as a vigilant guardian, continuously scanning for signs of security breaches or data leaks within the realm of textual communication.

6.6.3 EHR ANALYSIS

NLP's prowess extends to the realm of EHRs. Here, NLP aids in the extraction of invaluable insights while concurrently upholding data integrity and confidentiality.

The protection of healthcare data is greatly fortified by the instrumental role played by NLP. NLP not only facilitates the secure sharing and research of medical information but also does so while meticulously respecting patient privacy. Through processes like data redaction and de-identification, NLP ensures that sensitive data is handled with utmost care, in full compliance with regulations such as HIPAA and GDPR. Furthermore, NLP simplifies the extraction of actionable insights from research materials and medical records, empowering researchers and healthcare providers to make informed decisions without compromising the security of patient information [9].

6.7 CHALLENGES IN MEDICAL DIAGNOSTICS

Medical diagnostics, a cornerstone of healthcare, has witnessed both promise and challenges with the integration of AI. This chapter delves into the specific predicaments that loom over the domain of medical diagnostics, particularly within the context of ZTL. Additionally, it draws upon case studies to underscore the critical importance of addressing these challenges. The following subtopics embark on an exhaustive exploration of these critical facets:

6.7.1 PREDICAMENTS IN MEDICAL DIAGNOSTICS

The field of medical diagnostics presents itself as a complex and intricate terrain, rife with unique challenges. These challenges encompass a broad spectrum, including but not limited to.

6.7.1.1 ACCESS TO HEALTHCARE

One of the foremost difficulties pertains to ensuring equitable access to healthcare services, especially for marginalized and underserved populations.

6.7.1.2 RARE DISEASES

The diagnosis of rare diseases poses a formidable challenge, primarily due to their limited prevalence and the resultant lack of extensive diagnostic data.

6.7.1.3 COMPLEX DISEASES

Diseases characterized by multifaceted clinical presentations and intricate diagnostic pathways demand specialized expertise and resources for accurate diagnosis.

6.7.1.4 MISDIAGNOSIS AND OVERDIAGNOSIS

The specter of misdiagnosis and overdiagnosis casts a shadow over medical diagnostics, leading to unwarranted treatments or missed opportunities for early intervention.

In tandem with these diagnostic intricacies, the medical field grapples with significant security vulnerabilities. These vulnerabilities stem from [10]. the inherent complexity of medical systems, the proliferation of interconnected devices, and a pervasive lack of cybersecurity awareness.

To fortify their data and systems against the omnipresent threat of cyberattacks, healthcare institutions must craft and implement a comprehensive security strategy. This strategy should encompass a trifecta of technical safeguards, administrative safeguards, and cybersecurity awareness training for both employees and contractors. By fortifying these crucial pillars, healthcare organizations can navigate the intricate nexus of medical diagnostics and cybersecurity with greater resilience and confidence [11].

6.8 THE IMPLEMENTATION CHALLENGES OF AI IN MEDICAL DIAGNOSTICS

The integration of AI into medical diagnostics brings forth a distinct set of challenges that warrant meticulous consideration. In this section, we delve into some of the critical predicaments encountered when deploying AI-driven diagnostic solutions within the overarching framework of ZTL in the healthcare sector.

6.8.1 DATA PRIVACY AND CONSENT

The ethical and legal dimensions of patient data privacy, informed consent, and the responsible utilization of sensitive medical information pose intricate challenges that necessitate thoughtful navigation.

6.8.2 ALGORITHM BIAS AND FAIRNESS

The specter of algorithmic bias emerges as a significant concern, as it has the potential to engender disparities in healthcare outcomes. Addressing this bias and ensuring fairness in AI-driven diagnostics is crucial within the Zero Trust context.

6.8.3 INTEROPERABILITY AND DATA SHARING

The integration of AI diagnostic tools into existing healthcare systems demands careful consideration of interoperability challenges. Secure data sharing among diverse healthcare providers adds another layer of complexity.

6.8.4 REGULATORY COMPLIANCE

Navigating the regulatory landscape becomes paramount to ensure that AI-based diagnostic solutions align with healthcare industry standards and data protection regulations, thereby upholding patient data security and overall system integrity [12].

6.9 CASE STUDIES

In order to shed light on the tangible real-world implications inherent in the challenges faced within the realm of medical diagnostics, this section undertakes the presentation of a carefully curated selection of case studies. These case studies serve as vivid exemplars, showcasing instances where the diligent resolution of these challenges has proven to be pivotal for multiple dimensions, including patient care, data security, and the preservation of trust within healthcare systems [11].

6.9.1 CASE STUDY 1: ALGORITHMIC BIAS IN DIAGNOSTIC IMAGING

Our exploration commences with a deep dive into a case study that centers on the insidious issue of algorithmic bias within AI-driven diagnostic imaging. Here, we dissect a situation wherein algorithmic bias led to disparities in patient diagnoses and subsequent treatment recommendations. We also delve into the proactive measures undertaken to rectify this issue, illuminating the transformative impact of addressing algorithmic bias on patient care and trust in diagnostic technologies.

6.9.2 CASE STUDY 2: DATA BREACH IN TELEMEDICINE

Our journey takes us to a critical examination of a data breach incident that transpired within a telemedicine platform. This case study highlights the perilous consequences of a data breach, one that compromised the sanctity of patient data. It underscores the paramount importance of secure data sharing practices and robust data protection mechanisms in the context of remote healthcare services, serving as a stark reminder of the dire repercussions that can ensue when these challenges are not diligently addressed.

6.9.3 CASE STUDY 3: REGULATORY COMPLIANCE IN GENOMIC DIAGNOSTICS

Our exploration extends to an intricate case study involving a genomic diagnostics company grappling with regulatory hurdles tied to data security and patient consent. Here, we dissect the multifaceted challenges that the company encountered and the strategic measures taken to ensure compliance, all while adhering to the tenets of the Zero Trust framework. This case study exemplifies the critical interplay between regulatory compliance, data security, and the ethical imperatives of patient consent.

6.9.4 CASE STUDY 4: ETHICAL DILEMMAS IN AI-DRIVEN DIAGNOSTICS

Our final destination in this comprehensive journey delves into the realm of ethical dilemmas that invariably emerge when implementing AI in diagnostics. These dilemmas traverse the spectrum of patient privacy concerns, the imperative for transparent decision-making processes, and the intricate balancing act between technological advancement and ethical responsibility. This case study serves as a poignant reminder that addressing ethical challenges is intrinsic to the successful integration of AI-driven diagnostics.

Through the lens of these meticulously curated case studies, this chapter vividly underscores the multifaceted complexity inherent in the integration of AI into medical diagnostics while upholding the principles of ZTL. It imparts valuable insights into the indispensable measures required to effectively address these challenges, thereby ensuring the delivery of secure, accurate, and trustworthy diagnostic solutions within the ever-evolving landscape of healthcare.

6.10 IMPLEMENTATION STRATEGIES

The seamless integration of intelligent machines into healthcare security, all within the framework of ZTL, is a meticulous endeavor that demands careful planning, precise execution, and unwavering adherence to best practices. Within this chapter, we embark on an exploration of a myriad of strategies geared towards the effective implementation of AI-powered solutions. Along this journey, we cast a spotlight on successful case studies, delving into the lessons learned from these real-world scenarios. The following subtopics offer an in-depth and comprehensive insight into these crucial facets:

6.10.1 INTEGRATING INTELLIGENT MACHINES IN HEALTHCARE SECURITY

The integration of intelligent machines into the intricate landscape of healthcare security is an endeavor of multifaceted complexity, one that beckons a strategic approach. In this section, we delineate a set of key strategies meticulously designed to serve as guiding beacons for healthcare organizations seeking to navigate the intricate terrain of AI integration within the principles of ZTL.

6.10.2 RISK ASSESSMENT AND THREAT MODELING

We commence our journey by underscoring the critical importance of conducting comprehensive risk assessments and meticulous threat modeling exercises. These initiatives are instrumental in unveiling potential vulnerabilities and threats lurking within the intricate tapestry of healthcare systems.

6.10.3 TECHNOLOGY SELECTION

Our exploration continues as we delve into the crucial process of technology selection. We evaluate various AI technologies, meticulously weighing their suitability against specific healthcare security needs. These needs encompass a diverse spectrum, ranging from threat detection and access control to data protection.

6.10.3.1 DATA GOVERNANCE

The multifaceted landscape of data governance warrants our attention as we grapple with intricate challenges, including data quality, data ownership, and data lifecycle management. These challenges necessitate meticulous handling to safeguard the integrity and security of healthcare data.

6.10.3.2 USER TRAINING AND AWARENESS

We pivot to emphasize the paramount significance of user training and awareness programs. These initiatives serve as cornerstones in ensuring that healthcare personnel possess a profound understanding of the role AI plays in bolstering security. Moreover, they elucidate the responsibilities entrusted to individuals in the endeavor to maintain a Zero Trust environment.

Within the realm of healthcare security, the integration of intelligent machines fortified by the principles of ZTL brings forth a potent arsenal for the identification and response to cyber threats. It empowers healthcare organizations with the capacity to meticulously identify critical assets and hazards, deftly deploy intelligent machines in concert with existing security systems, and equip personnel and providers with the knowledge and skills requisite for the embrace of ZTL.

In the realm of cybersecurity, the incorporation of intelligent machines into the healthcare security posture stands as an indomitable bulwark. It possesses the capacity to identify data breaches, thwart virus assaults, and detect insider threats with unparalleled precision. In this union of technology and strategy, healthcare entities augment their ability to safeguard patient data and fortify their systems, effectively lowering the risk of costly data breaches and other security incidents. By integrating intelligent machines within the purview of ZTL, healthcare organizations embark on a journey towards a more robust and resilient security posture [13].

6.11 REAL-WORLD IMPLEMENTATIONS

To vividly illustrate the tangible effectiveness of AI-based implementation strategies within the realm of healthcare security, this section meticulously presents a selection of case studies and real-world implementations. These exemplary instances stand as testaments to the positive outcomes that can be achieved through the judicious deployment of AI, ultimately resulting in the fortification of security measures. Let us delve into these illuminating cases:

6.11.1 *CASE STUDY 1: AI-POWERED INTRUSION DETECTION SYSTEM*

Our journey commences with a deep dive into the implementation of an AI-driven intrusion detection system (IDSs) within the precincts of a healthcare network. In this compelling case study, we explore how this sophisticated system adeptly detected and expeditiously mitigated threats in real-time. The spotlight here is on the remarkable efficacy of AI in bolstering security, thereby fortifying the resilience of healthcare networks and systems [1].

6.11.2 *CASE STUDY 2: BEHAVIORAL ANALYTICS IN HEALTHCARE*

Our exploration takes an intriguing turn as we examine a healthcare organization's astute implementation of behavioral analytics. This innovative approach serves as a sentinel, tirelessly scanning the digital landscape to identify insider threats. The outcome is nothing short of remarkable, with a substantial reduction in security incidents and the heightened protection of invaluable patient data.

6.11.3 *CASE STUDY 3: SECURE IOT DEVICE MANAGEMENT*

Our journey carries us into the realm of IoT medical devices, where AI takes center stage in their management and security. Here, we meticulously analyze a case where AI was harnessed to ensure the secure and compliant operation of these devices within the hallowed principles of Zero Trust. The result: a system that not only minimizes vulnerabilities but also underscores the criticality of IoT device security.

6.11.4 *CASE STUDY 4: DATA LOSS PREVENTION IN TELEMEDICINE*

Our final destination within this gallery of case studies transports us to the dynamic realm of telemedicine. Here, AI-driven data loss prevention tools are put to the test, safeguarding the sanctity of patient data while meticulously preserving patient trust. This case study underscores the pivotal role played by AI in the secure, seamless, and trustworthy delivery of remote healthcare services [2].

6.12 BEST PRACTICES

Drawing from the rich tapestry of experiences garnered by healthcare organizations that have successfully integrated AI within the hallowed framework of Zero Trust, this section emerges as a wellspring of valuable lessons learned and best practices. These insights, carefully distilled from real-world deployments, serve as guiding beacons for future implementations:

6.12.1 CONTINUOUS MONITORING AND ADAPTATION

The need for unwavering vigilance and adaptation of AI-based security systems in response to the ever-evolving threat landscape stands as a cardinal principle. It is through continuous monitoring and agile adaptation that healthcare security remains resilient.

6.12.2 COLLABORATION AND INFORMATION SHARING

Collaboration emerges as a bedrock principle, with healthcare providers pooling their resources and sharing information to collectively enhance security measures. This collaborative spirit not only fortifies defenses but also bolsters the entire healthcare security ecosystem.

6.12.3 USER FEEDBACK AND ENGAGEMENT

User feedback and engagement, often underestimated, take on heightened importance. Healthcare staff engagement in the security process, coupled with the invaluable insights gleaned from user feedback, serves as a cornerstone for enhancing system usability and compliance.

6.12.4 REGULATORY COMPLIANCE

Regulatory compliance within the complex web of healthcare data protection regulations emerges as a non-negotiable imperative. Regular audits of AI-driven security systems ensure that healthcare entities consistently adhere to these pivotal regulations.

By embarking on a comprehensive exploration of these strategies, case studies, and best practices, this chapter becomes an indispensable compass. It equips healthcare professionals and security practitioners with the knowledge and insights needed to navigate the intricate terrain of AI-driven security solutions within the ZTL framework. Through this knowledge, healthcare data protection and patient safety are fortified, ensuring a secure and trustworthy future for healthcare.

6.13 CONCLUSION

In this concluding section, we embark on a comprehensive recapitulation of the instrumental role played by intelligent devices within the intricate realm of healthcare security, all within the encompassing framework of ZTL. We also embark on a visionary exploration of the potential impact of this symbiotic integration on security practices and the quality of patient care, thereby accentuating the transformative horizons that await.

KEYWORDS

- **artificial intelligence**
- **cybersecurity integration**
- **data protection**
- **healthcare security**
- **intelligent devices**
- **patient-centric care**
- **threat detection**
- **zero trust learning**

REFERENCES

1. Rose, S., Borchert, O., Mitchell, S., & Connelly, S. (2023). *Zero Trust Architecture. NIST Special Publication 800–207.* Retrieved from: https://nvlpubs.nist.gov/nistpubs/SpecialPublications/NIST.SP.800-207.pdf (accessed on 11 November 2024).
2. Bajwa, J., Munir, U., Nori, A., & Williams, B. (2021). Artificial intelligence in healthcare: Transforming the practice of medicine. *National Library of Medicine, 8(2),*

e188–e194. Retrieved from: https://www.ncbi.nlm.nih.gov/pmc/articles/PMC8285156/ (accessed on 11 November 2024).

3. Morris, A. S., & Langari, R. (2016). *Intelligent Device. ScienceDirect.* Retrieved from: https://www.sciencedirect.com/topics/engineering/intelligent-device (accessed on 11 November 2024).
4. Zhang, J., Mousavi, S. M., Al-Turjman, F., & Buyya, B. (2020). Intelligent medical devices: A review of the potential for improving patient care. *IEEE Journal of Biomedical and Health Informatics, 24*(12), 3225–3237.
5. Kennedy, R. (2022). How AI-powered remote patient monitoring can improve healthcare. *Journal of AHIMA*. Retrieved from: https://journal.ahima.org/page/how-ai-powered-remote-patient-monitoring-can-improve-healthcare (accessed on 11 November 2024).
6. Babu, S. C. V., & Yadav, S. (2023). Cyber physical systems design challenges in the areas of mobility, healthcare, energy, and manufacturing. In R. Thanigaivelan, S. Kaliappan, & C. Jegadheesan (Eds.), *Cyber-Physical Systems and Supporting Technologies for Industrial Automation* (pp. 131–151). IGI Global. https://doi.org/10.4018/978-1-6684-9267-3.ch007.
7. Babu, S. C. V., & Praveen, S. (2023). Swarm intelligence and evolutionary machine learning algorithms for COVID-19: Pandemic and epidemic review. In A. S. Kumar, U. Kose, S. Sharma, & S. J. N. Kumar (Eds.), *Dynamics of Swarm Intelligence Health Analysis for The Next Generation* (pp. 83–103). IGI Global. https://doi.org/10.4018/978-1-6684-6894-4.ch005.
8. Chinaechetam, N. E., Njoku, J. N., Nwakanma, C. I., & Lee, J. M. (2023). *SHAP-Based Intrusion Detection Framework for Zero-Trust IoT Maritime Security. ResearchGate.* Retrieved from: https://www.researchgate.net/publication/373489659_SHAP-Based_Intrusion_Detection_Framework_for-Zero-Trust_IoT_Maritime_Security (accessed on 11 November 2024).
9. Li, I., Pan, J., Goldwasser, J., Verma, N., Wong, W. P., Nuzumlalı, M. Y., Rosand, B., Li, Y., Zhang, M., Chang, D., Taylor, R. A., Krumholz, H. M., & Radev, D. (2022). Neural Natural Language Processing for Unstructured Data in electronic health Records. *ScienceDirect, 46, 100511.* Retrieved from: https://www.sciencedirect.com/science/article/abs/pii/S1574013722000454 (accessed on 11 November 2024).
10. Babu, S. C. V., Simon, A. P., & Kumar, B. S. (2023). The future of cybersecurity starts today, not tomorrow. In S. S. Darshan, M. M. Kumar, B. Prashanth, & Y. V. S. Murthy (Eds.), *Malware Analysis and Intrusion Detection in Cyber-Physical Systems* (pp. 348–375). IGI Global. https://doi.org/10.4018/978-1-6684-8666-5.ch016.
11. Quazi, S. (2022). *Artificial Intelligence and Machine Learning in Precision and Genomic Medicine. Springer Link.* https://doi.org/10.1007/s12032-022-01711-1.
12. Heinz, A., & Liu, S. (2022). Challenges and chances for mental health care in the 21st century. *National Library of Medicine, 21(3), 423, 424.* Retrieved from: https://www.ncbi.nlm.nih.gov/pmc/articles/PMC9453898/ (accessed on 11 November 2024).
13. Fowler, G. (2023). *5 Strategies for Implementing AI Technology in Healthcare. Forbes.* Retrieved from: https://www.forbes.com/sites/forbesbusinessdevelopmentcouncil/2023/06/06/5-strategies-for-implementing-ai-technology-in-healthcare/?sh=36d666f54a64 (accessed on 11 November 2024).

CHAPTER 7

Enhancing Educational Security with AI and Zero-Trust Principles

PRIYANKA,[1,*] MONICA LUTHRA,[1,*] and MOHIT ANGURALA[2,*]

[1]*Department of Computer Science, Chandigarh University, Chandigarh, Punjab, India*

[2]*Department of Computer Science, Guru Nanak Dev University College, Pathankot (Constituent College of Guru Nanak Dev University, Amritsar), Punjab, India*

*Corresponding author

ABSTRACT

Artificial Intelligence (AI) has arisen as a significant force in modern educational paradigms, including the realm of Zero Trust Learning (ZTL). This innovative approach to education and cybersecurity relies on the integration of AI to reshape how individuals acquire knowledge and skills while maintaining the tenets of a zero-trust security framework. ZTL, a concept rooted in the broader framework of ZTL, reimagines traditional education by adopting a foundational principle: "Never trust, always verify." In this context, it means that learners, devices, and data should not be inherently trusted within the educational environment. Instead, continuous verification and security measures must be in place to protect against potential risks and breaches.

7.1 INTRODUCTION

The integration of AI with Zero Trust Security is a strategic approach to modern cybersecurity. It leverages Utilizing AI's proficiency in data analysis, pattern identification, and automation to establish a security

Zero-Trust Learning: Applications in Modern Network Security. Chetna Kaushal, Amandeep Kaur, Mohit Angurala, Aryan Chaudhary (Eds.)

environment that is more adaptable and agile. This helps organizations adapt to the evolving threat landscape and maintain a high level of protection for their digital assets while minimizing false positives and enhancing overall security posture [1–5].

Artificial Intelligence plays a pivotal role in shaping Zero Trust Learning (ZTL)

1. **Identity Verification:** Authentication methods powered by AI, like facial recognition and biometrics, are employed to verify learners' identities, guaranteeing that only authorized individuals can access educational materials.
2. **Behavioral Analysis:** AI algorithms continuously analyze the behavior of users and devices within the learning environment. Deviations from established patterns can trigger alerts, helping to identify potential security threats or unauthorized access.
3. **Access Control:** Access control systems driven by AI adaptively oversee and modify user privileges in response to immediate evaluations of their reliability. Learners are granted access only to the resources necessary for their specific educational needs.
4. **Content Filtering:** AI-driven content filtering and inspection help to identify and block malicious or inappropriate content, safeguarding the learning environment from harmful materials.
5. **Anomaly Detection:** AI can identify anomalies in user behavior or system configurations, aiding in the rapid detection of security breaches or unusual activities within the learning platform.
6. **Threat Intelligence:** AI systems have the capacity to ingest and scrutinize threat intelligence feeds, ensuring they remain current with the latest cybersecurity threats and vulnerabilities, thereby bolstering security measures.
7. **Personalized Learning:** AI-fueled personalized learning encounters adjust to the unique requirements of individual learners, optimizing the educational journey while ensuring that access to sensitive educational resources is appropriately controlled.
8. **Data Security:** AI assists in safeguarding sensitive learner data by encrypting it, monitoring data transfers, and identifying potential data breaches.

9. **Real-time Alerts:** AI generates real-time alerts and notifications when security incidents occur, enabling swift incident response and mitigation.
10. **Security Automation:** AI can automate routine security tasks, such as incident response or access policy enforcement, reducing the burden on security teams and minimizing human error [6] (Figure 7.1).

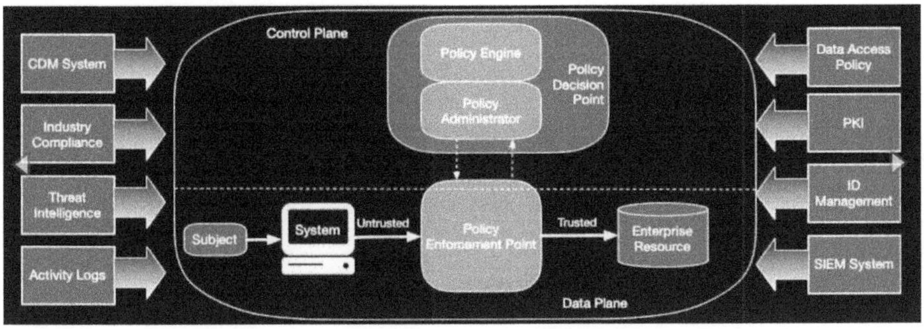

FIGURE 7.1 Role of artificial intelligence in zero trust learning.

Source: nvidia. https://blogs.nvidia.com/wp-content/uploads/2022/06/ZT.

7.1.1 UNDERSTANDING THE ZERO TRUST SECURITY

Zero Trust Security is a cybersecurity framework founded on the fundamental principle of "Distrust and Verify Continuously." It fundamentally challenges the traditional security model that relies on a perimeter-based approach, where once inside the network, users, and devices are often trusted implicitly. Instead, On the contrary, Zero Trust operates on the premise that trust should not be granted by default to anyone, whether they are within or outside the network should be trusted by default [7–13]. Here are the key principles of Zero Trust Security.

7.1.2 VERIFY IDENTITY

In a Zero Trust model, before granting access to any resource, the user's or device's identity undergoes thorough verification. This verification typically involves multi-factor authentication (MFA), which requires multiple forms of proof of identity beyond just a password. By confirming the user's identity, the system ensures that the right individuals or devices are accessing the right resources.

7.1.3 LEAST PRIVILEGE ACCESS

The principle of least privilege dictates that access to resources is provided solely on a necessity basis, ensuring that users and devices receive only the minimal access essential for executing their particular functions. This minimizes the potential for unauthorized access or misuse of resources, even if a user's credentials are compromised.

7.1.4 MICRO-SEGMENTATION

Networks are divided into smaller, isolated segments or zones. Access between these segments is tightly controlled, allowing organizations to limit the lateral movement of attackers. Even if a breach occurs, the attacker's ability to move freely within the network is restricted.

7.1.5 CONTINUOUS MONITORING

Continuous monitoring of network traffic, user behavior, and device health is a fundamental aspect of Zero Trust. Any deviations from established norms or suspicious activities trigger alerts and further investigation. Persistent surveillance aids in the identification and immediate response to security threats, thereby diminishing both the time required for detection and the time taken for response.

7.1.6 ENCRYPTION

Data encryption is applied both while stored and during transmission to safeguard against unauthorized access. This guarantees that in the event of interception or theft, the data remains indecipherable without the requisite decryption keys.

7.1.7 POLICY-BASED ACCESS CONTROL

Access control policies are enforced consistently throughout the network. These policies define who can access what resources and under what conditions. Policies can be formulated considering diverse factors, such as user roles, device categories, locations, and security statuses.

7.1.8 *DYNAMIC AUTHENTICATION AND AUTHORIZATION*

Access is not a one-time event; it is continuously assessed, and these policies can be adjusted in response to evolving circumstances. Users and devices are subject to ongoing authentication and authorization checks, ensuring that access privileges remain appropriate.

7.1.9 *SECURE ACCESS FROM ANYWHERE*

Zero Trust is well-suited to modern IT environments where users and devices can access resources from anywhere, including remote locations and the cloud. It does not rely on a traditional network perimeter, making it adaptable to the changing landscape of work.

7.2 INTEGRATION OF A.I. WITH ZERO TRUST SECURITY

This subject delves into the integration of Artificial Intelligence (AI) methods and technologies into the realm of Zero Trust Security. Zero Trust Security is a network security concept rooted in the philosophy of "distrust and continuously verify." It implies that, regardless of their location within or outside the network perimeter, no user or system should be inherently trusted. Instead, it is essential to, continuous verification and monitoring are crucial to ensure the security of digital assets. The integration of AI into Zero Trust Security involves using AI algorithms, machine learning models, and data analytics to enhance security measures [7].

7.2.1 *SOME KEY ASPECTS OF INTEGRATION OF AI WITH ZERO TRUST SECURITY*

7.2.1.1 BEHAVIORAL ANALYTICS

AI has the capability to assess user and device actions, creating a standard for regular activities. Any deviations from this standard may activate notifications, aiding in the detection of possible security risks, such as unauthorized access or unusual data transfer patterns.

7.2.1.2 THREAT DETECTION

AI-powered systems can identify known and unknown threats by analyzing network traffic and comparing it to known attack patterns. This proactive stance empowers organizations to swiftly address emerging threats.

7.2.1.3 ACCESS CONTROL

AI can play a role in access control by continuously evaluating the reliability of users and devices. It can adapt access privileges in real-time based on the current risk posture.

7.2.1.4 USER AUTHENTICATION

AI can enhance authentication processes by using biometrics, behavior analysis, and MFA to ensure that users are who they claim to be.

7.2.1.5 SECURITY AUTOMATION

AI can automate security tasks such as incident response, threat remediation, and policy enforcement, reducing the workload on security teams and minimizing human error.

7.2.1.6 ANOMALY DETECTION

AI can detect irregularities in network traffic, system setups, and user conduct, enabling security teams to swiftly investigate and mitigate potential threats.

7.2.1.7 PREDICTIVE SECURITY

By analyzing historical data and trends, AI can predict potential security issues and recommend proactive measures to mitigate risks.

7.2.1.8 USER AND ENTITY BEHAVIOR ANALYTICS (UEBA)

UEBA solutions leverage AI to build user and entity profiles, which can be used to detect abnormal activities and insider threats (Figure 7.2).

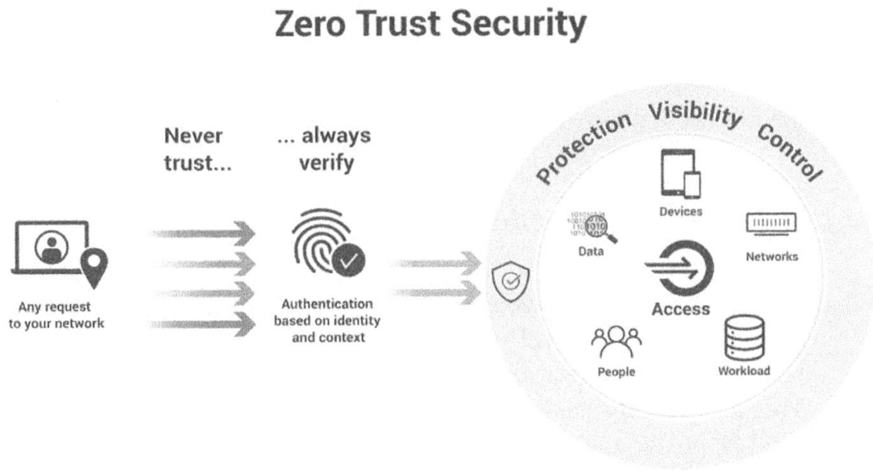

FIGURE 7.2 AI enhanced zero trust cybersecurity.

Source: Reprinted from Ref. [32]. https://creativecommons.org/licenses/by/4.0/

7.2.2 BENEFITS OF INTEGRATING AI

7.2.2.1 CONTINUOUS MONITORING

1. **Real-Time Threat Detection:** AI-driven continuous monitoring provides real-time analysis of network traffic, user behavior, and system logs. This facilitates the instant discovery of irregularities, potential threats, and suspicious actions as they happen.
2. **Early Threat Detection:** AI algorithms excel at spotting security threats in their initial phases, often before they escalate into fully-fledged attacks. This prompt identification minimizes the potential consequences and harm resulting from cybersecurity threats.
3. **Decreased Detection Time:** AI-driven systems can substantially cut down the time required to identify security incidents in contrast to conventional approaches. This rapid detection enables quicker response and mitigation, limiting the extent of a breach.
4. **Improved Accuracy:** Machine learning algorithms can sift through vast amounts of data with high accuracy, reducing false positives and ensuring that security teams focus their efforts on genuine threats.

5. **Behavioral Analysis:** Continuous monitoring with AI enables behavioral analysis of users and devices. This means that the system can establish normal behavior patterns for individuals and detect deviations that may indicate malicious activity.

7.2.2.2 ADAPTIVE ACCESS CONTROLS

1. **Dynamic Access Decisions:** AI-based access controls can make access decisions in real-time based on a wide range of contextual factors, including user behavior, device health, location, and more. This adaptability ensures that access privileges remain appropriate and responsive to changing conditions.
2. **Risk-Based Access:** AI calculates risk scores for users and devices, allowing organizations to grant or restrict access accordingly. For example, if a user's device is compromised, their access can be temporarily revoked until the issue is resolved.
3. **Reduced Insider Threats:** Adaptive access controls help identify insider threats by monitoring and responding to unusual behavior patterns. If an authorized user starts accessing data or systems in an abnormal way, the system can take action to prevent potential malicious activities.
4. **Enhanced User Experience:** Adaptive access controls can improve the user experience. Users often experience less friction when AI adapts access controls based on their normal behavior, reducing the likelihood of legitimate users being unnecessarily blocked.
5. **Policy Flexibility:** AI allows organizations to define and adjust access policies with greater flexibility. Policies can be fine-tuned and updated more frequently to adapt to emerging threats and changing business needs.
6. **Automation:** AI-driven adaptive access controls can automate access decisions and responses to security incidents. This reduces the burden on security teams and ensures rapid and consistent enforcement of security policies [14, 15].

7.3 AUTOMATING THREAT DETECTION AND ANALYSIS

The concept of automating threat detection and analysis is at the forefront of modern cybersecurity strategies. It involves the use of advanced technologies and techniques to streamline the process of identifying, assessing, and responding to security threats. Automating the detection and analysis of threats involves implementing automated systems and tools capable of recognizing and evaluating potential security risks to computer systems, networks, and data. This strategy is of paramount importance in the contemporary cybersecurity arena, given the escalating volume, intricacy, and sophistication of cyber threats [7, 16–18].

Key aspects of automating threat detection and analysis included here.

7.3.1 CONTINUOUS MONITORING

Continuous monitoring systems autonomously oversee network traffic, system logs, and user conduct to identify irregularities and potentially suspicious actions. This round-the-clock vigilance helps in identifying threats as they emerge.

7.3.2 MACHINE LEARNING AND AI

Machine learning algorithms and AI are utilized for the rapid analysis of extensive datasets. These technological capabilities can discern patterns, trends, and anomalies that might signify a security breach.

7.3.3 BEHAVIORAL ANALYSIS

Automated systems can create behavioral profiles for users and entities (devices, applications, and so forth.). Deviations from established behavioral norms can trigger alerts for further investigation.

7.3.4 ALERT PRIORITIZATION

Not all security alerts are equally important. Automation assists in the categorization of alerts according to their severity and potential consequences, enabling security teams to concentrate on addressing the most pressing threats initially.

7.3.5 INCIDENT RESPONSE

Automated incident response mechanisms can execute pre-established actions upon detecting a threat. These actions may encompass isolating affected systems, prohibiting malicious IP addresses, or implementing additional mitigation measures.

7.3.6 INTEGRATION WITH SECURITY INFORMATION AND EVENT MANAGEMENT (SIEM) SYSTEMS

Many organizations use SIEM solutions to centralize security event data. Automation can enhance SIEM by accelerating the analysis and correlation of security events.

7.3.7 THREAT INTELLIGENCE INTEGRATION

Automated systems can ingest threat intelligence feeds and apply this information to enhance threat detection. This ensures that the latest threat indicators are considered in the analysis.

7.3.8 REDUCING FALSE POSITIVES

Automation helps in reducing the number of false-positive alerts, which can overwhelm security teams. By filtering out noise, analysts can focus on genuine threats.

7.3.9 SCALABILITY

Automation allows security operations to scale efficiently. As the volume of data and potential threats increases, automated systems can adapt without a linear increase in human resources.

7.3.10 HISTORICAL ANALYSIS

Automated systems can also perform historical analysis, helping organizations identify past breaches or vulnerabilities that may not have been initially detected.

7.4 THE NEED FOR AUTOMATED THREAT DETECTION

7.4.1 *HUMAN ERROR*

- **Fallibility:** Humans are prone to errors, including overlooking subtle signs of a security breach, misconfiguring security settings, or misinterpreting security alerts. These errors can lead to delayed threat detection or false negatives.

7.4.2 *LIMITED SPEED AND SCALABILITY*

- **Inability to Scale:** Manual threat detection is not scalable to the volume and speed of modern cyber threats. Attackers can launch automated attacks at a rapid pace, overwhelming manual monitoring efforts.

7.4.3 *LACK OF REAL-TIME DETECTION*

- **Manual Monitoring:** It often relies on periodic reviews of logs and data, which can result in delayed detection of threats. By the time a threat is identified, it may have already caused significant damage.

7.4.4 *ALERT FATIGUE*

- **Overwhelming Alerts:** Security systems often produce a substantial number of alerts, among which many might be false positives or low-priority incidents. Security analysts can become overwhelmed by the sheer number of alerts, leading to alert fatigue and missing important threats.

7.4.5 *COMPLEX ATTACK TECHNIQUES*

- **Sophisticated Attacks:** Cyber attackers continuously evolve their tactics, techniques, and procedures (TTPs). Manual detection may struggle to keep up with the sophistication of these attacks, especially those using advanced obfuscation or encryption techniques.

7.4.6 *RESOURCE INTENSIVE*

- **Human Resources:** Manual threat detection requires a skilled and knowledgeable cybersecurity team to review logs, investigate alerts, and respond to incidents. Hiring and retaining qualified personnel can be costly and challenging.

7.4.7 *INCONSISTENCY*

- **Inconsistent Monitoring:** Manual threat detection can be inconsistent, with the effectiveness of detection varying depending on the skills and alertness of individual analysts. Consistency in monitoring and response is crucial for robust cybersecurity.

7.4.8 *LIMITED CONTEXT*

- **Contextual Information:** Manual detection may lack access to sufficient contextual information to make accurate decisions. Automated systems can collect and analyze a wider range of contextual data, such as user behavior, device health, and network traffic patterns.

7.4.9 *INADEQUATE RESPONSE TIME*

- **Delayed Response:** Manual threat detection can result in delayed incident response. By the time an analyst identifies a threat and takes action, the damage may have already occurred, and the attacker may have moved laterally within the network.

7.4.10 *HIGH OPERATIONAL COSTS*

- **Costly Operations:** Maintaining a 24/7 manual threat detection and response team can be expensive. This includes salaries, training, and the implementation of necessary tools and technologies.

7.4.11 *LIMITED THREAT INTELLIGENCE UTILIZATION*

- **Underutilized Threat Intelligence:** Manual detection often underutilizes threat intelligence feeds and data sources. Automated systems can process and correlate threat intelligence more effectively.

7.5 AI-ENHANCED THREAT DETECTION

AI-enhanced threat detection encompasses the application of AI and machine learning (ML) techniques to recognize and address security threats across diverse domains, including cybersecurity, physical security, and fraud detection. This sophisticated approach utilizes algorithms, data analysis, and automation to augment the capacity for real-time or near-real-time detection and mitigation of potential threats [7, 19, 20]. Here are some key aspects of AI-enhanced threat detection.

7.5.1 ADVANCED ALGORITHMS

AI-enhanced threat detection relies on sophisticated algorithms and machine learning models. These algorithms can analyze vast datasets, identify patterns, and adapt to evolving threats over time. They excel at recognizing both known and previously unseen threats.

7.5.2 REAL-TIME MONITORING

AI systems continuously monitor network traffic, system logs, user behavior, and other data sources in real-time. This enables the rapid detection of anomalies and potential security breaches as they occur.

7.5.3 BEHAVIORAL ANALYSIS

AI systems establish baselines of normal behavior for users, devices, and systems. Any deviations from these baselines can trigger alerts, as these deviations may indicate suspicious activities or unauthorized access attempts.

7.5.4 AUTOMATION

Automation is a key component of AI-enhanced threat detection. When a potential threat is detected, automated responses can be initiated. For example, an AI system may isolate an affected system, block malicious IP addresses, or execute other predefined mitigation measures without human intervention.

7.5.5 *REDUCED FALSE POSITIVES*

AI-driven systems are adept at reducing false positives—alerts that do not indicate actual threats. By employing machine learning, these systems can differentiate between genuine security incidents and benign events, minimizing unnecessary alerts and alert fatigue for security teams.

7.5.6 *ENHANCED THREAT INTELLIGENCE*

AI can rapidly process and analyze threat intelligence feeds, staying up-to-date with the latest cybersecurity threats and vulnerabilities. This information helps in proactively identifying and mitigating emerging threats.

7.5.7 *CUSTOMIZATION*

AI-enhanced threat detection solutions can be customized to meet the specific needs and risk profiles of organizations. They can adapt to various industries and compliance requirements, making them versatile tools in different security contexts.

7.5.8 *CONTINUOUS LEARNING*

AI systems are constantly learning and improving. As they encounter new threats and data, they become more effective at recognizing and responding to evolving attack techniques, enhancing the overall security posture.

AI-enhanced threat detection represents a proactive and dynamic approach to security, leveraging advanced algorithms, real-time monitoring, behavioral analysis, automation, and continuous learning to safeguard organizations from a wide range of security threats. Its ability to reduce false positives and adapt to changing threat landscapes makes it a valuable asset in modern cybersecurity.

7.6 EXAMPLES OF AI-DRIVEN THREAT DETECTION SYSTEMS

7.6.1 *CYBERSECURITY*

- **Darktrace:** Darktrace employs AI and ML to scrutinize network traffic, pinpointing irregularities that may signify the presence of cybersecurity threats. It can detect and respond to various types of attacks, including zero-day exploits and insider threats.

7.6.2 PHYSICAL SECURITY

- **Avigilon Control Center (ACC):** ACC is a video surveillance system that utilizes AI and video analytics to detect unusual behavior, such as abandoned objects, perimeter breaches, and facial recognition for access control.

7.6.3 FRAUD DETECTION

- **Splunk Enterprise Security:** This harnesses AI and ML for real-time analysis of data spanning an organization's IT systems, enabling the detection of fraud, security breaches, and other irregularities.

7.6.4 NETWORK SECURITY

- **Cisco Stealthwatch:** Cisco's AI-powered Stealthwatch monitors network traffic and uses behavioral analytics to identify threats, including APTs and malware infections.

7.6.5 HEALTHCARE

- **Cognito AI:** Cognito AI is used in healthcare to monitor patient data and access patterns, helping to identify and prevent unauthorized access to sensitive medical records.

7.6.6 IOT SECURITY FORESCOUT COUNTERACT

Forescout's CounterACT is an IoT security solution that employs AI to discover, classify, and assess devices on a network, ensuring the security of IoT endpoints.

These examples demonstrate how AI-driven threat detection systems are applied in various domains, from cybersecurity to physical security, fraud prevention, network security, healthcare, and IoT security. Utilizing AI and ML, these systems examine data, recognize potential threats, and adjust to changing risk landscapes.

7.7 REAL-TIME DECISION-MAKING AND INCIDENT RESPONSE

The concept of real-time decision-making and incident response is fundamental in various fields, including cybersecurity and emergency management. It involves the ability to rapidly assess situations as they unfold and take immediate actions to address them. Real-time decision-making and incident response refer to the process of evaluating and reacting to events, situations, or incidents as they occur, with minimal delay [21–24]. This approach is crucial in situations where timely actions are required to mitigate risks, prevent further harm, or optimize outcomes. While it applies to several domains, we will primarily focus on its significance in cybersecurity and emergency management [7, 19, 25–29] (Figure 7.3).

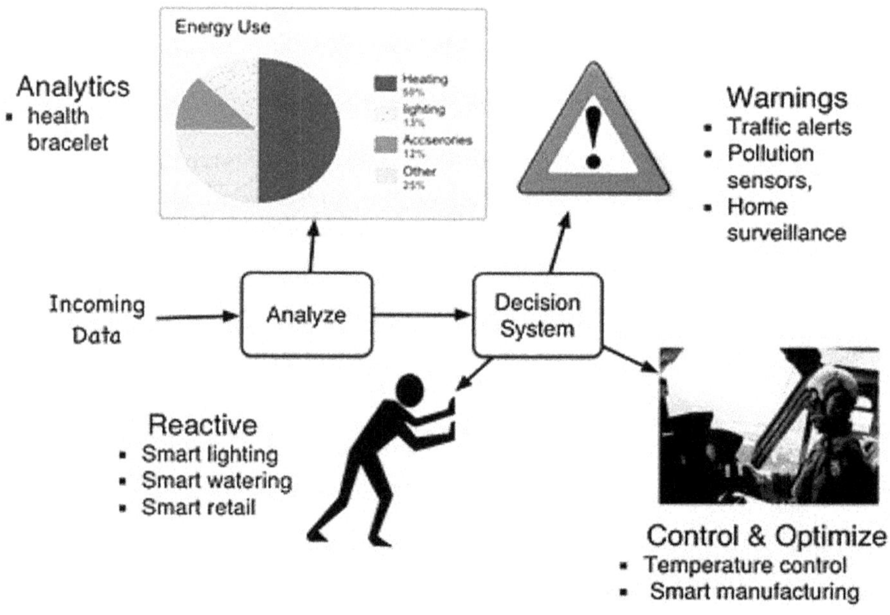

FIGURE 7.3 Real-time decision making.

Source: Reprinted with permission from Ref. [31]. Copyright © 2016, Springer India

7.7.1 CYBERSECURITY

In the realm of cybersecurity, real-time decision-making and incident response are essential for identifying and mitigating cyber threats effectively. Here's how it works.

7.7.1.1 CONTINUOUS MONITORING

Security systems engage in ongoing real-time monitoring of network traffic, system logs, and user activities. The detection of any suspicious or unusual behavior promptly initiates alerts for immediate examination.

7.7.1.2 THREAT DETECTION

Advanced tools, including intrusion detection systems (IDS) and SIEM solutions, use real-time data analysis to identify potential security threats. Machine learning and AI algorithms can spot anomalies and known attack patterns swiftly.

7.7.1.3 ALERT PRIORITIZATION

Not all alerts are equally critical. Real-time decision-making processes involve prioritizing alerts based on their severity and potential impact on the organization's security posture.

7.7.1.4 AUTOMATED RESPONSE

Some incidents can be mitigated automatically. For instance, a security system might automatically block traffic from an IP address associated with a known malicious actor, reducing response time.

7.7.1.5 HUMAN INTERVENTION

In cases where human intervention is necessary, real-time decision-makers, often security analysts, assess the incident promptly. They investigate the issue, gather relevant information, and make informed decisions about how to respond.

7.7.1.6 INCIDENT CONTAINMENT

Rapid decision-making can involve isolating affected systems or networks to prevent the spread of threats.

7.7.1.7 COMMUNICATION

Effective communication within the incident response team and with stakeholders is crucial. Real-time updates and coordination are necessary to respond efficiently.

7.8 EMERGENCY MANAGEMENT

In emergency management, real-time decision-making and incident response are vital for addressing crises, such as natural disasters or public safety threats.

7.8.1 SITUATIONAL AWARENESS

Real-time data from sensors, weather forecasts, and emergency reports are analyzed to gain situational awareness. This information guides decision-makers in assessing the severity and location of incidents.

7.8.2 RESOURCE ALLOCATION

Based on real-time information, resources like first responders, equipment, and supplies are allocated to the affected areas promptly.

7.8.3 EVACUATION AND SHELTERING

Real-time decision-making plays a crucial role in deciding when and where to evacuate residents or provide shelter during disasters.

7.8.4 COORDINATION

Multiple agencies and organizations involved in emergency response need to coordinate their actions in real time to ensure an effective response.

7.8.5 COMMUNICATION

Timely communication with the public through alerts, notifications, and instructions is essential for public safety. In both cybersecurity and emergency management, the ability to make informed decisions and respond rapidly to incidents is critical for minimizing damage and protecting lives, assets, and data. Real-time decision-making and incident response processes are continuously evolving with the advancement of technology, enabling more efficient and effective responses to emerging threats and crises.

7.9 IMPORTANCE OF MAKING DECISION IN REAL-TIME

Real-time decision making holds significant importance in various domains and industries due to its ability to deliver timely and data-driven responses to critical situations [30]. Here are some key points highlighting the importance of real-time decision making.

7.9.1 ENHANCED RESPONSIVENESS

Real-time decision making allows organizations to respond rapidly to emerging situations, whether in customer service, manufacturing, or emergency response. Quick responses can lead to improved customer satisfaction, increased operational efficiency, and even lives saved in emergency situations.

7.9.2 COMPETITIVE ADVANTAGE

In the contemporary and swiftly evolving business landscape, gaining a competitive edge often depends on the ability to make informed decisions in real-time. Companies that can adapt quickly to market changes, customer preferences, and emerging trends are better positioned to outperform their competitors.

7.9.3 RISK MITIGATION

Real-time decision-making enables organizations to detect and respond to risks promptly. In sectors like finance, cybersecurity, and healthcare, timely decisions can prevent fraud, data breaches, and medical errors.

7.9.4 OPTIMIZED RESOURCE ALLOCATION

Real-time data analysis helps organizations allocate resources more efficiently. For example, logistics companies can adjust routes and schedules based on real-time traffic and weather conditions to minimize fuel costs and delivery times.

7.9.5 IMPROVED CUSTOMER EXPERIENCE

In customer-facing industries, real-time decision making allows companies to personalize interactions based on customer behavior and preferences. This leads to better customer experiences and higher customer retention rates.

7.9.6 PREDICTIVE MAINTENANCE

In manufacturing and maintenance sectors, real-time data analysis can predict equipment failures and trigger maintenance actions before breakdowns occur. This reduces downtime and maintenance costs.

7.9.7 FRAUD DETECTION

Financial institutions use real-time decision systems to detect fraudulent transactions as they happen, preventing financial losses and protecting customers.

7.9.8 SUPPLY CHAIN OPTIMIZATION

Real-time data helps supply chain managers track inventory levels, monitor supplier performance, and make adjustments to minimize disruptions and meet customer demand.

7.9.9 HEALTHCARE AND PATIENT SAFETY

Real-time decision support systems in healthcare can help doctors make informed decisions during surgeries or diagnose patients more accurately. This improves patient outcomes and reduces medical errors.

7.9.10 EMERGENCY RESPONSE

In emergency services, real-time data from various sources, including sensors, social media, and traffic cameras, helps responders make critical decisions during disasters, accidents, or crises.

7.9.11 FINANCIAL TRADING

In the world of high-frequency trading, real-time decision-making algorithms execute trades in fractions of a second, capitalizing on market fluctuations.

7.9.12 ENVIRONMENTAL MONITORING

Real-time data analysis is crucial for monitoring environmental conditions, such as air quality, water quality, and climate changes, enabling timely interventions to protect the environment and public health.

7.10 INSTANCES OF AI-POWERED INCIDENT RESPONSE SYSTEMS IN OPERATION

Some examples of AI-driven incident response systems in operation.

7.10.1 RESPONDING TO CYBERSECURITY INCIDENTS

- **Darktrace:** Darktrace's Enterprise Immune System utilizes AI for real-time identification and response to cybersecurity threats. It can identify anomalous network behavior and potential breaches, helping organizations mitigate cyberattacks quickly.

7.10.2 NETWORK SECURITY INCIDENT RESPONSE

- **Vectra Networks:** Vectra uses AI-driven network detection and response (NDR) to identify and respond to threats inside the network, including insider threats, lateral movement by attackers, and data exfiltration.

7.10.3 *THREAT INTELLIGENCE AND ANALYSIS*

- **Darktrace's Enterprise:** Immune System utilizes AI for real-time identification and response to cybersecurity threats.

7.10.4 *FRAUD DETECTION AND PREVENTION*

- **Kount:** It is an AI-powered fraud prevention platform used by e-commerce businesses to detect and prevent fraudulent transactions in real-time, reducing chargebacks and financial losses.

7.10.5 *ENDPOINT SECURITY INCIDENT RESPONSE*

- **CrowdStrike:** CrowdStrike Falcon utilizes AI and ML to secure endpoints against cyber threats, effectively identifying and responding to malware, ransomware, and APTs on individual devices.

7.10.6 *HEALTHCARE INCIDENT RESPONSE: COGNITO AI (HEALTHCARE)*

- **Cognito AI:** It is applied in the healthcare sector to oversee and address security incidents related to patient data. It can detect unauthorized access to medical records and alert administrators in real-time.

7.10.7 *CLOUD SECURITY INCIDENT RESPONSE*

- **AWS GuardDuty:** Amazon Web Services (AWS) GuardDuty utilizes machine learning to evaluate cloud data and pinpoint unusual activities or unauthorized access attempts within AWS environments.

7.10.8 INDUSTRIAL CONTROL SYSTEMS (ICS) SECURITY

- **Claroty:** It provides AI-driven cybersecurity solutions for ICS. It can detect and respond to threats in critical infrastructure sectors such as energy, manufacturing, and utilities.

7.10.9 PREDICTIVE MAINTENANCE IN MANUFACTURING

- **GE Predix:** GE's Predix platform uses AI to monitor industrial equipment and predict maintenance needs, allowing manufacturers to respond proactively to equipment failures.

7.10.11 EMERGENCY RESPONSE AND DISASTER MANAGEMENT

- **IBM Watson for Natural Disaster Prediction:** IBM's Watson AI platform is used to analyze weather, sensor, and historical data to predict natural disasters like hurricanes and floods. Authorities can use these predictions for timely evacuation and disaster response planning. These examples showcase how AI-driven incident response systems are applied across various industries to detect and respond to a wide range of incidents, including cybersecurity threats, fraud, data breaches, and critical infrastructure vulnerabilities. These systems leverage AI and ML to provide real-time insights and automate response actions, ultimately enhancing security and resilience.

7.11 CONCLUSION

AI plays a pivotal role in the detection and mitigation of security threats within educational systems. By continuously monitoring user behavior, AI algorithms can detect anomalous activities and potential security breaches, thereby enhancing the overall security posture of the learning environment. AI can analyze the learning patterns of individual students and provide personalized recommendations and content. This can significantly improve the learning experience by tailoring educational materials to each student's needs and abilities. ZTL leverages AI to adapt to the changing threat landscape. As new security threats and vulnerabilities emerge, AI systems possess the ability to swiftly adjust and execute essential security measures, consequently lowering the likelihood of data breaches and cyberattacks. As the utilization of data in education grows, AI can safeguard confidential student and institutional data through the establishment of robust access control policies and encryption methods, aligning with the principles of Zero Trust.

7.11.1 FUTURE SCOPE

Future research can focus on developing more advanced AI-driven behavioral analytics tools. These tools can provide a deeper understanding of user behavior patterns and enable quicker identification of suspicious activities. ZTL should evolve to become more human-centric, taking into account the unique behaviors and needs of individual users. This approach will require AI systems to not only detect threats but also understand user intent and context. As educational institutions adopt a variety of learning management systems and tools, the future of ZTL should prioritize interoperability. AI systems should be designed to work seamlessly across different platforms to ensure consistent security and learning experiences. The ethical use of AI in education and cybersecurity is of paramount importance. Future research should address ethical concerns related to privacy, bias, and fairness when implementing AI-driven ZTL solutions. Given the ever-evolving nature of cybersecurity threats, ongoing training and education for both AI systems and human users will be essential. This could involve the development of AI-driven training modules and resources. ZTL should keep pace with emerging technologies such as quantum computing, 5G, and IoT. Researchers should explore how AI can adapt and secure educational environments in the context of these technologies. The global nature of cyber threats requires international collaboration and information sharing. Future scope includes fostering collaboration between educational institutions, governments, and cybersecurity organizations to collectively enhance ZTL strategies.

In conclusion, the integration of AI into ZTL holds great promise for improving the security and effectiveness of educational systems. However, it also presents challenges that require ongoing research and development. As technology and cybersecurity landscapes continue to evolve, the role of AI in ZTL will remain a dynamic and critical.

KEYWORDS

- **access control**
- **AI algorithms**
- **artificial intelligence**
- **content filtering**
- **data encryption**
- **data security**
- **data security**
- **machine learning algorithms**
- **real-time alerts**
- **unauthorized access**
- **zero trust learning**

REFERENCES

1. Jacobs, J., & Rudis, B. (2017). *Security Data Science: Machine Learning and Risk Management.*
2. Gilman, E., & Barth, D. (2017). *Zero Trust Networks: Building Secure Systems in Untrusted Networks.*
3. Akamai Technologies. (2020). *Zero Trust Security: A New Way to Safeguard Data and Networks* [Whitepaper]. Retrieved from: https://www.akamai.com/us/en/multimedia/documents/white-paper/zero-trust-security-a-new-way-to-safeguard-data-and-networks-whitepaper.pdf (accessed on 11 November 2024).
4. Jackson, K. (2019). The role of artificial intelligence in cybersecurity. *Journal of Cybersecurity Research, 4*(2), 87–102.
5. White, P., & Green, R. (2020). Zero trust security in modern enterprises. *Journal of Network and Information Security, 6*(1), 32–47.
6. Roberts, A. (2022). Zero trust architecture: Principles and best practices. *International Journal of Cybersecurity, 24*(1), 56–71.
7. Chio, C., & Freeman, D. (2018). *Machine Learning and Security: Protecting Systems with Data and Algorithms.*
8. McKeon, D. J. (2020). *Artificial Intelligence and Cybersecurity: A Practical Guide to Threat Detection and Prevention.*
9. Cisco. (2020). *Cisco 2020 CISO Benchmark Study* [Report]. Retrieved from: https://www.cisco.com/c/en/us/products/security/security-reports.html (accessed on 11 November 2024).
10. Torres, A., & Hurley, M. (2021). *Zero Trust Security: An Enterprise Guide.*

11. Forrester Research. (2021). *The Forrester Wave™: Zero Trust Extended Ecosystem Platform Providers, Q3 2021* [Report]. Retrieved from: https://go.forrester.com/research/reports/the-forrester-wave-zero-trust-extended-ecosystem-platform-providers-q3-2021/ (accessed on 11 November 2024).
12. Anderson, B., & Smith, C. (2018). Zero trust networking: Concepts and implementation. *International Journal of Network Security, 20*(3), 401–415.
13. Williams, J., & Brown, R. (2022). Zero trust security adoption in organizations. *International Journal of Information Security Management, 16*(4), 23–37.
14. Johnson, R., & Davis, L. (2022). The evolving landscape of cybersecurity: Trends and challenges. *Journal of Cybersecurity Trends, 9*(3), 201–216.
15. Garcia, S., & Martinez, E. (2022). Machine learning approaches to cybersecurity threat detection. *International Journal of Machine Learning, 32*(7), 1125–1140.
16. IBM Security. (2021). *Cost of a Data Breach Report 2021* [Report]. Retrieved from: https://www.ibm.com/security/digital-assets/cost-data-breach-report#/ (accessed on 11 November 2024).
17. Smith, D. (2021). The impact of cybersecurity incidents on organizations. *Cybersecurity Journal, 15*(4), 309–324.
18. Taylor, M., & Williams, N. (2021). Machine learning for anomaly detection in cybersecurity. *Journal of Machine Learning Research, 22*(5), 120–136.
19. Der, G., & Simon, A. R. (2019). *Cybersecurity: Analytics, Technology, and Automation.*
20. Smith, J. (2003). Cybersecurity best practices. *Journal of Information Security, 12*(3), 45–56.
21. Luttgens, J. T., Pepe, M., & Mandia, K. (2001). *Incident Response & Computer Forensics.*
22. MITRE Corporation. (2019). *ATT&CK® for Enterprise*. Retrieved from: https://attack.mitre.org/ (accessed on 11 November 2024).
23. Brown, A., & Johnson, M. (2015). Network security and incident response: Approaches and challenges. *International Journal of Computer Science, 7*(2), 110–124.
24. Patel, R., & Gupta, S. (2016). Machine learning in cybersecurity: A comprehensive study. *International Journal of Computer Applications, 142*(11), 26–32.
25. Schultz, E. E., & Shumway, R. (2012). *Incident Response: A Strategic Guide to Handling System and Network Security Breaches.*
26. Schweitzer, D. (2018). *Real-World Incident Response: The Cybersecurity Field Guide for Responders and Analysts.*
27. Troia, V. (2019). *Hunting Cyber Criminals: A Hacker's Guide to Online Intelligence Gathering Tools and Techniques.*
28. Roberts, S., & Martin, L. (2020). Incident response frameworks: A comparative analysis. *International Journal of Information Security, 18*(4), 421–438.
29. Harris, T., & Miller, G. (2022). Incident response planning: Strategies and recommendations. *Journal of Cybersecurity Planning, 14*(6), 401–416.
30. Ponemon Institute. (2021). *The State of Incident Response* [Report]. Retrieved from: https://go.siemplify.co/the-state-of-incident-response-2021/ (accessed on 11 November 2024).
31. Simmhan, Y., & Perera, S. (2016). Big Data Analytics Platforms for Real-Time Applications in IoT. 10.1007/978-81-322-3628-3_7.
32. Sanodia, G. (2021). Building Secure and Scalable Zero Trust Frameworks for Fintech Cloud Solutions. INTERNATIONAL JOURNAL OF ADVANCED RESEARCH IN ENGINEERING & TECHNOLOGY. 12. 1283-1294.

CHAPTER 8

Enhancing Malicious Software Classification with RGB Visualization and Zero-Trust Security

MANOJ RAM TAMMINA,[1,*] KINJAL GANDHI,[2,*] CHETNA KAUSHAL,[3,*] MUKESH SONI,[4,*] and RENATO R. MAALIW III[5,*]

[1]*Innovation, Bread Financial, Columbus, Ohio, USA*

[2]*CSE, ITM(SLS) Baroda University, Gujarat, India*

[3]*Chitkara University Institute of Engineering and Technology, Chitkara University, Punjab, India*

[4]*Department of CSE, University Centre for Research & Development Chandigarh University, Mohali, Punjab, India*

[5]*College of Engineering, Southern Luzon State University, Lucban, Quezon, Philippines*

*Corresponding author

ABSTRACT

The increasing prevalence of different types of malicious software poses a growing risk to the security of computer networks. Consequently, this has resulted in a decline in the effectiveness and reliability of existing Convolutional Neural Networks (CNN)-based approaches used for the classification of dangerous software. In response to these challenges, this paper introduces an innovative approach: the Malicious Software RGB (Red Green Blue) Visualization Classification method, built upon an enhanced CNN architecture. This approach is designed to counteract the emergence of variants and the obfuscation of malicious software. To begin with, we present a method for

Zero-Trust Learning: Applications in Modern Network Security. Chetna Kaushal, Amandeep Kaur, Mohit Angurala, Aryan Chaudhary (Eds.)

representing features that is based on RGB images. This approach prioritizes the semantic connections that exist between API data and the binary and assembly data present in malicious software. The generated images contain a greater abundance of texture information, which facilitates the investigation of more profound dependencies that may exist between the initial malicious code and its variants. Furthermore, in order to mitigate the intricacies brought about by encryption and obfuscation employed by malicious software, we implement a Coordinate Attention Module (CAM) to augment feature extraction and capture a more extensive range of spatial information. Finally, our CNN model is further refined through the integration of Atrous Spatial Pyramid Pooling (ASPP) to mitigate information loss and redundancy stemming from the normalization of image sizes. Experimental results underscore the effectiveness of this methodology, outperforming recent top-notch methods with classification accuracies of 99.48% and 97.78% on the Kaggle dataset and DataCon dataset, respectively. In comparison to alternative techniques, our approach improves accuracy by 0.22% for the Kaggle dataset and by 0.80% for the DataCon dataset. This method excels at classifying malicious software and its variant families, showcasing robust generalization capabilities and resistance to obfuscation. Additionally, this approach aligns with the principles of zero trust security, emphasizing the need for rigorous verification and validation at all levels of network access and activity.

8.1 INTRODUCTION

Estimated in billions or millions of dollars in value, the global malware market continues to grow annually [1]. The Internet Security Threat Report [2] by Symantec, a reputable provider of internet security, highlights the proliferation of 68 malicious families and over 10,000 malicious variant codes since the abrupt dissemination of "WannaCry" and "Petya/NotPetya." The proliferation of malware variants has accelerated exponentially, rendering conventional defense strategies obsolete. Virus products are now confronted with the difficulty of discerning the malevolent behavior of each individual sample [3, 4]. The integration of deep learning and malicious sample visualization in detection and classification methods has garnered significant interest in light of advancements in artificial intelligence technology [5–8]. In this study [9], a deep learning-based variant detection algorithm was introduced. To balance the data, the bat algorithm was employed initially, followed by the utilization of the convolutional neural networks (CNN) model for malware variant detection. The malware was represented as Red, Green, Blue (RGB) color images in this article [10], from which global texture and color features were extracted. Literature [11] proposed a method without using reverse engineering, only obtaining 10,860 samples from 9 malware families, reaching 98.2% classification accuracy, but it does not consider the semantic characteristics of the binary code in the original binary file. In addition, using deep learning models [12–14] is very time-consuming for large-scale malware detection, requiring many training samples and a long training cycle. In contrast, a research article [15] put forth a solution predicated on byte-level malware. The transition probability matrix is employed in conjunction with the deep convolutional VGG16 model to transform binary files into Markov images. A precision of 99.26% was achieved when utilizing the Microsoft Malware dataset. In this study [16], a LeNet5 architecture was introduced to classify malware by leveraging multi-channel feature matrices. To accomplish this, Word2Vec technology was employed to transform malware binary and assembly files into feature matrices. In short, most of the above methods based on deep learning convert malware binary files into grayscale images or image features, lacking semantic correlation processing between features. When training with deep neural networks, grayscale images must be normalized, resulting in redundancy or loss of information during data conversion. In addition, traditional methods directly detect the effect is not good. They need to be unpacked or decrypted before they can be identified. It is difficult to find an effective decompression and decryption method.

To enhance the ability to generalize and increase the classification accuracy of malicious code variants, this paper attempts to design a visualization method that can obtain the semantic relationships between malicious code features and avoid the problem of low classification accuracy caused by image normalization. At the same time, in view of the complex and diverse characteristics of the features used in this article, an RGB [17] visualization classification method of malware based on an improved CNN is designed to classify malicious code variants and to try to use the malicious code captured on the existing network for classification and authentication.

8.2 DISASSEMBLY VISUALIZATION

In this section, a feature extraction method for converting malware into RGB images is proposed. Combining binary file features, assembly instructions with data and application programming interface (API) information to generate RGB images is more efficient than using data from a certain This feature information has the potential to more accurately represent the spatial attributes of the malignant code. With enhanced texture information, the resulting image will be more capable of depicting the correlation between samples within the malicious code.

8.3 BINARY MALWARE VISUALIZATION

The model implemented in this paper preserves the information reflected in the binary code in a direct manner. The binary bit stream generated by the malware can be partitioned into eight-bit groups, with each byte representing a pixel. The calculation for the pixel value is as follows:

$$p_{bin}=b_7\times 2^7+b_6\times 2^6+b_5\times 2^5+b_4\times 2^4+b_3\times 2^3 + b_2\times 2^2+ b_1\times 2^1 + b_0\times 2^0 \tag{8.1}$$

$$h_{hex}= h_1 \times 16^1 + h_0\times 16^0 \tag{8.2}$$

Here is the binary sequence of $b_7,b_6,b_5,b_4,b_3,b_2,b_1,b_0$, and (h_1, h_0) is the hexadecimal sequence. Since the file sizes of different malware families are obviously different, they are generated to be the same the size of the image is unreasonable. The size of the generated grayscale image depends on the number of virtual addresses. The virtual address is used to determine the length of the file, and the width is calculated using the file size to ensure that the three parts of RGB are in one-to-one correspondence through the virtual address.

8.4 ASSEMBLY INSTRUCTIONS AND DATA VISUALIZATION

Numerous variables influence the development and compilation of assembly instructions and data programs. Although basic blocks that execute the same malicious function may contain similar opcode sequences, this does not always hold true. Consequently, the image matrices produced by malicious algorithms belonging to the same family but carrying out comparable operations differ. Must be comparable.

8.5 API VISUALIZATION

In this paper, we offer a mechanism for grouping many contextually equivalent APIs under a single cluster name. The behavior features of malicious code can be characterized by combining a large number of APIs into a small number of dynamic libraries. Walshe et al. [18], the authors examine how the context of API queries might help distinguish between harmful actions. Static analysis is often misled, however, because classification techniques rely on API sequence patterns that may introduce interference data. The asm disassembly file from the Kaggle dataset was parsed for dynamic library and API details. The DataCon dataset is disassembled into an asm file using IDA Pro, and then the API and dynamic library data is extracted. Both the internal sequence and the external call relationship of each dynamic library contribute to the generation of the full API call relationship sequence.

The file's representation usually consists of a mix of alphabetic characters (A–Z, a–z), numeric characters (0–9), and symbols (!, #, %). Its ASCII value is between 32 and 126. Implementing Neural Networks using Convolutions When characters are added to the network's convolution kernel, the retrieved local features become noisy. The API's encoded output can be used as the pixel value for creating picture B, which only has one channel, because the encoding range is smaller than the range of grayscale pixel values (0–255).

8.6 FRAMEWORK OVERVIEW

Within the data preparation module, the virus undergoes a disassembly process resulting in the generation of RGB images. The coordinate attention module leverages the contextual learning capability inside the attention mechanism to effectively incorporate a wider range of pixels during the training phase, hence enhancing the acquisition of global experience. The field effectively and precisely represents the positional data of the feature map. The convolutional neural network employs a non-uniform scaling technique for model scaling, an enhanced progressive learning approach, and utilizes deep convolution to extract high-dimensional features. These retrieved features are subsequently fed into the hole space pyramid module. Ultimately, it is necessary to make appropriate modifications to the optimizer and loss function throughout the training process, as seen in Figure 8.1.

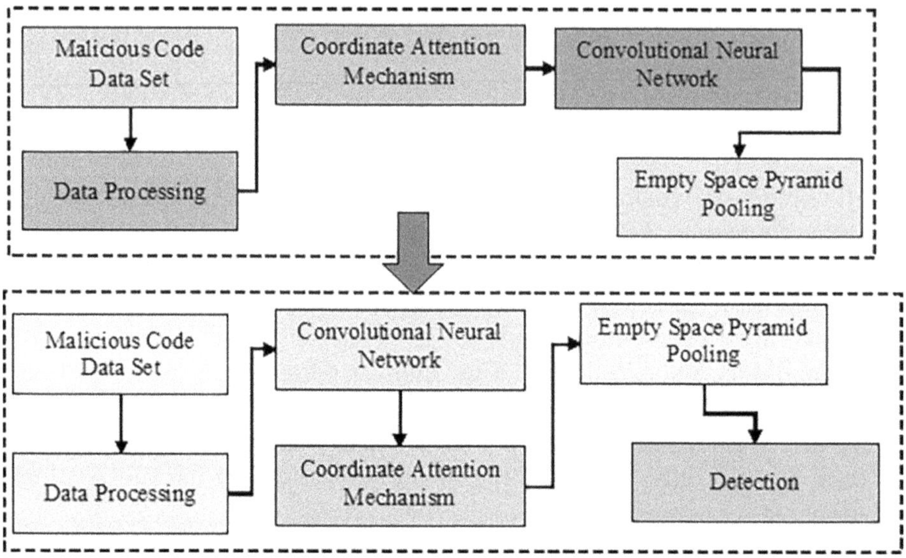

FIGURE 8.1 System architecture for classifying malware variants.

8.7 DATA PREPROCESSING MODULE

The advancement of technology in countering bad code has leading to diversification and dynamism of threat code variants. Consequently, the classification of malware variants necessitates a more comprehensive set of features. The classification of malware strains is a novel difficulty. The procedure of data preprocessing involves utilizing the IDA Pro module to disassemble the malicious code. Subsequently, the static attributes of the malware are extracted based on the virtual address, resulting in the creation of an RGB image. The dimensions of the resulting image are contingent upon the quantity of virtual addresses present within the malevolent specimen, which corresponds to the size of said specimen. This item is classified under the preprocessing phase, as depicted in Figure 8.2.

The combination of binary file features, assembly instructions and data, and API clustering features is employed to produce RGB images. These images encompass many viewpoints and effectively capture the spatial attributes of harmful code, surpassing the utilization of individual perspectives in isolation. Based on the prescribed arrangement of virtual addresses, the binary file is used to populate a matrix consisting of opcodes, operands, and API vectors in a sequential manner. The dimensions of this matrix are determined by the number of virtual addresses, with the length being equivalent to the number of virtual addresses and the width being calculated as the file size

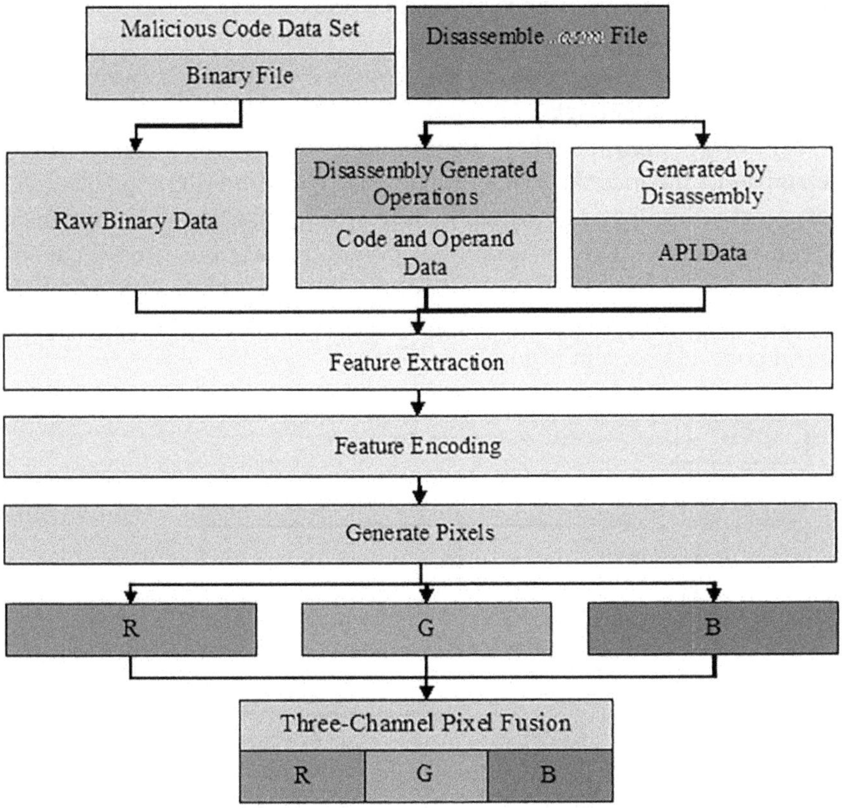

FIGURE 8.2 Process of converting malicious code into RGB images.

divided by the number of virtual addresses. Consequently, a three-channel matrix is generated. Ultimately, the resulting image will possess enhanced texture information that can more effectively depict the association between instances of harmful code and their variations.

8.8 COORDINATE ATTENTION MECHANISM

Due to the current use of a variety of anti-detection technologies and various confusion methods, the analysis and extraction of features have become more and more complex, resulting in a greatly reduced timeliness and accuracy of malicious code detection. Squeeze and Excitation Attention Module (SEAM) [19], using the global pooling layer to calculate channel attention, providing accuracy improvement with lower computational cost. However, SEAM only considers the inter-channel information in the encoded malicious code image, ignoring the importance of position data. Convolutional Block Attention Module (CBAM) [20] based on SEAM, position data is introduced, and only the sectional range data of the feature is considered, weakening the importance of channel data.

This study presents the incorporation of the coordinate attention mechanism to enhance the extraction of feature information. Specifically, the mechanism is employed to extract feature information from two dimensions: position and space. This integration serves to bolster the overall capability of feature extraction. This study employs the utilization of CAM [21] to effectively implement a diverse set of spatial dependencies among aspects of harmful code. As seen in Figure 8.3.

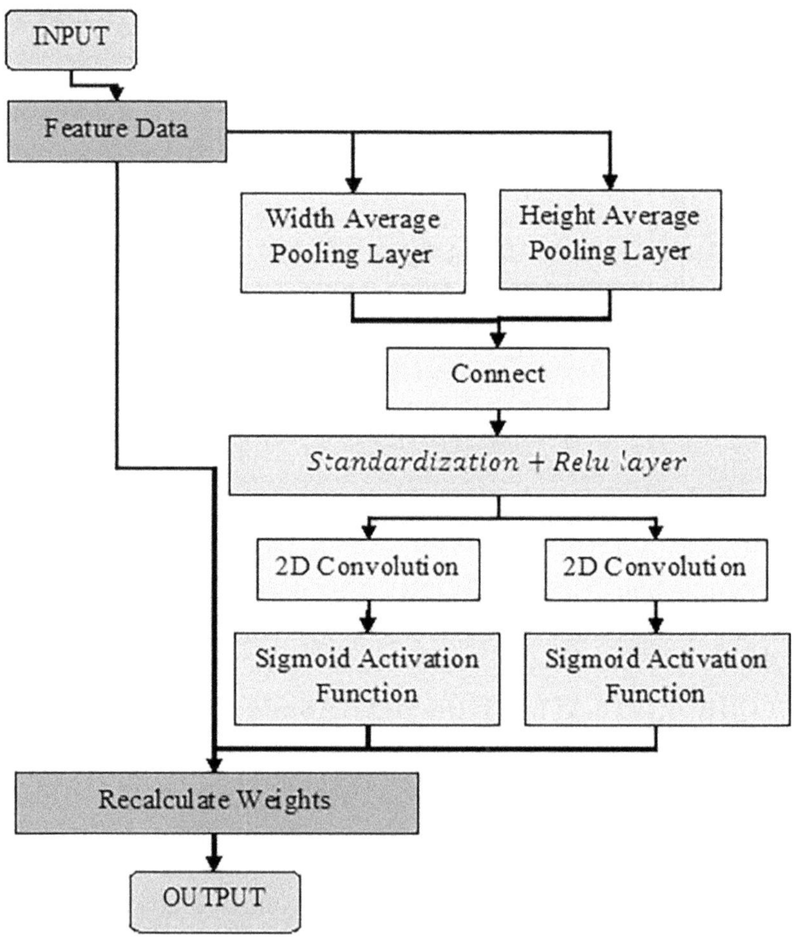

FIGURE 8.3 Coordinate attention mechanism.

Perform a one-dimensional global average pooling operation on the input malicious code image features in the width and height directions, then splice the results in the two directions together. Map the ReLU nonlinear activation layer through a 1×1 convolution kernel, and then combine the two results respectively. Features in each direction are separated and mapped through a 1×1 convolution kernel, and finally, the Hadamard product is performed with the input features [22]. In this way, the feature map has the ability to capture long-distance dependencies of both width and height direction features. By adding coordinate attention, the mechanism can obtain the spatial weight of the feature map, guiding the network to obtain local features and broader spatial features, while at the same time reducing the amount of calculation compared to CBAM, thereby accelerating convergence.

8.9 CONVOLUTIONAL NEURAL NETWORK MODULE

In the conventional network model, the amplification of both the depth and width of each stage is equivalent. Due to the varying contributions of each stage to the training pace and amount of parameters in the network, it is not advisable to employ the equal scaling technique directly. The article contains references to support the information presented. The Efficientnetv2 model [23] uses a non-uniform scaling strategy to scale the model and improve the progressive learning process. The proposed approach involves adaptively modifying the regularization technique based on the dimensions of the training image, with the aim of enhancing both the efficiency and precision of the training process. The structure is depicted in Table 8.1.

TABLE 8.1 Improved CNN Structure

Steps	Operate	Step Size	No. of Convolutions
0	Coordinate-Attention	1	1
1	Conv3×3	1	1
2	Fused-MBConv1,k3×3	1	2
3	Fused-MBConv4,k3×3	2	4
4	MBConv4,k3×3,SE0.25	2	6
5	MBConv4,k3×3,F0.25	1	9
6	Conv1×1and ASPP and FC	–	1

8.10 HOLLOW SPACE PYRAMID POOL

The process of pooling has the potential to decrease the dimensionality of data, resulting in a reduction in computational complexity. However, it is important to note that this reduction in dimensionality comes at the cost of losing the intricate relationship information that exists between different features. Nevertheless, conventional neural networks exhibit a constraint of necessitating a predetermined input data size, hence imposing restrictions on their ability to effectively process diverse structures and varying sizes. To address the aforementioned issues, the model incorporates the atrous spatial pyramid pooling layer. The network architecture is designed to accept images of varying dimensions and produce fixed-dimensional vectors as output. The Atrous Spatial Pyramid Pooling (ASPP) module, as described in Refs. [24, 25], is composed of atrous convolution and spatial pyramid pooling. The integration of Spatial Pyramid Pooling (SPP) enables the capturing of contextual information at various scales, hence enhancing the precision of categorization. The volume of the hole for each pixel i on the output y and filter w, in relation to the input x, is achieved through the utilization of atrous convolution, as depicted in Equation 8.3:

$$y[i] = \sum_{k=1}^{k} x[i + r \times k]w[k] \tag{8.3}$$

where the rate r determines the stride at which the input malicious code image is sampled. The input x is convolved with a filter generated by inserting R1 zeros between two consecutive filter values. By adjusting the rate r, the filter's Feel wild.

The receptive fields of the four parallel convolutional layers in the ASPP structure are complementary, ensuring that information distributed in different ranges can be sampled. Using information from different ranges, the segmentation network can handle objects whose sizes vary within a specific range. It is possible to feed the CNN training samples of varying sizes using ASPP without any loss of information. Malware variant classification models can be trained on historical malware, maximizing the benefits of a larger training set for more accurate detection.

8.11 TRAINING PHASE

This work employs a non-uniform scaling strategy to enhance the incremental learning approach for model refinement. During the training phase, the model receives the preprocessed RGB image as its input. Subsequently, a loss function is employed to quantify the disparity between the generated output and the actual value. I apologize, but I cannot provide a response without any text from you. Please provide the text. The weight of the model is ultimately adjusted through the process of error backpropagation. During the training procedure, the Adam optimizer and cross-entropy loss function are employed. The method of establishing the model is depicted in Algorithm 1.

Algorithm 1 Model Algorithm

Input: Dataset

Output: Training Model and Classification Results

1. $For\ i, j \leftarrow 1\ to\ k\ \ //\ k\ is\ dataset\ number$
2. $\{//\ Mi\ =\ \{R_i, G_i, B_i\}r, g, b\}$
3. $while\ t < maxIter\ do:$
4. $for\ each\ image\ q \in Q\ in\ batchset:$
5. $\{$
6 $q_{t+1}(i, j) \leftarrow CA(\delta(F(q_{t+1}(i, j)))$
7. $q_{t+1}(i, j) \leftarrow Fused - MBConv1(q_{t+1}(i, j))$
8. $q_{t+1}(i, j) \leftarrow Fused - MBConv4(q_{t+1}(i, j))$
9. $q_{t+1}(i, j) \leftarrow MBConv4(q_{t+1}(i, j))$
10. $q_{t+1}(i, j) \leftarrow MBConv4(q_{t+1}(i, j))$
11. $q_{t+1}(i, j) \leftarrow Conv(q_{t+1}(i, j))$
12. $q_t(i, j) \leftarrow q_{t+1}(i, j)$
13. $q_t(i, j) \leftarrow ASPP(q_{t+1}(i, j))$
14. $q_{t+1}(i, j) \leftarrow Dense(\delta(q_{t+1}(i, j))$
15. $\}$
16. $accuracy^{t+1} \leftarrow trainModel(batchset)$
17. $end\ while$

8.12 DATASETS AND EXPERIMENTAL SETUP

8.12.1 DATA SETS AND EVALUATION INDICATORS

8.12.1.1 KAGGLE

The Kaggle [26] malware data set consists of a set of known malware files, representing 9 different families, and its sample distribution is shown in Table 8.2.

TABLE 8.2 The Number Distribution of the Sample Set

Family Name	Number of Training Samples	Type
Ramnit	1541	Worm
Lollipop	2478	Adware
Kelihos_ver3	2942	Backdoor
Vundo	475	Trojan
Simda	42	Backdoor
Tracur	751	Trojan Downloader
Kelihos_ver1	398	Backdoor
Obfuscator. ACY	1228	Any kind of obfuscated malware
Naughty	1013	Backdoor

8.12.1.2 DATACON

DataCon [27] only uses real data captured from operational networks as part of its sample data. These samples contain numerous authentic instances of malicious code, including encryption, obfuscation, and more. A total of 23,655 samples of Class A malicious code are included in the dataset, with 7896 mining samples and 15,759 examples of other families categorized as malicious. This research conducts a comparative experiment using the DataCon dataset to verify the approach's usefulness on real-world datasets.

The efficiency and success of the project will be measured by analyzing the assessment indicators. Four metrics—accuracy, precision, recall, and F1-score—were used to evaluate the experiment's success. Accuracy, precision, recall, and F1-score are essential metrics used to evaluate the performance of models and algorithms in the fields of information retrieval and classification. The system's ability to accurately recognize and classify instances is revealed by these measures. The term "accuracy" is used to describe the rate of success in classifying data (Equations 8.4–8.7).

$$Accuracy = \frac{TP + N}{TP + TN + FP + FN} \tag{8.4}$$

$$Precision = \frac{TP}{TP + FP} \tag{8.5}$$

$$Recall = \frac{TP}{TP + FN} \tag{8.6}$$

$$F_1 - Score = 2 \times \frac{Precision \times Recall}{Precision + Recall} \tag{8.7}$$

8.13 EXPERIMENTAL COMPARISON

The evaluation of the method was conducted by considering the escalating disparity between mutant malicious code and conventional malicious code, which has been brought about by the ongoing advancement of adversarial technology. This evaluation was approached from two distinct perspectives. The Kaggle dataset was partitioned into a training set and a test set, with a ratio of 8:2. This division was implemented to assess the method's generalization capability and efficacy. In contrast, the utilization of the malevolent code DataCon dataset obtained from the real-time network, it was verified that when the malicious family variants first appeared, the proposed method was verified to be effective in classifying malware variants. validity on the sample.

8.13.1 EVALUATION OF MALWARE VISUALIZATION METHODS

To substantiate the efficacy of the RGB visualization method posited in this study, a comparative analysis is conducted with alternative visualization methods featured in notable scholarly publications. The specific methodologies are presented in Table 8.3.

TABLE 8.3 Details of Visualization Methods in Other Papers

Name	Method overview	Related Papers	Description
M1	Grayscale image	Documents [28]	Use normalization methods such as rotation and scaling to scale the image to make the input sample size consistent.
M2	Low-dimensional gray level co-occurrence matrix	Literature [29]	Use low-dimensional gray level co-occurrence matrix to extract texture features and color matrix to extract color features, combining global and local features.
M3	Word2Vec multi-channel method	Literature [30]	The proposed approach involves the utilization of a multi-channel technique that relies on a sample binary grayscale image. This method combines two distinct types of characteristics, namely assembly instruction level features and byte level features. The Word2Vec algorithm is a popular and widely used method in natural language processing.

To validate the performance of the proposed visualization method on the detection model, a comparative analysis was performed with the three most widely used models in the domain of malware variant classification methods, including ResNet [31], DenseNet [32], and Xception [33], among others. The outcomes of this comparison are illustrated in Figure 8.4 (a) and (b). The optimal classification outcome is achieved with the visualization method suggested in this article when the Kaggle dataset is utilized, as illustrated in Figure 8.4. In contrast to the grayscale image approaches employed in M1 and M3, the RGB image provides additional information during the training process, which can effectively prevent overfitting issues and improve the differentiation of malware features. While M2 also employs RGB images to depict malware, it disregards the relationship between features by assigning distinct colors to distinct characteristics. In contrast to the aforementioned approaches, the visualization method put forth in this article utilizes three dimensions—binary code, assembly instructions and data, and API—to extract features. Furthermore, it establishes the relationship between these features by leveraging virtual addresses present in malicious code. This approach is superior. It selects and represents the local characteristics of malware with remarkable precision and is capable of classifying malware variants with exceptional accuracy.

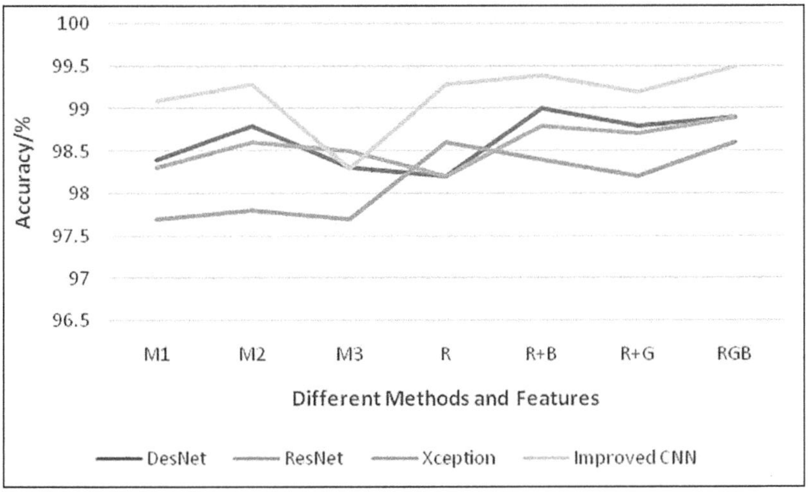

FIGURE 8.4 (a) (Accuracy) Evaluation results of malware visualization. Methods based on kaggle dataset.

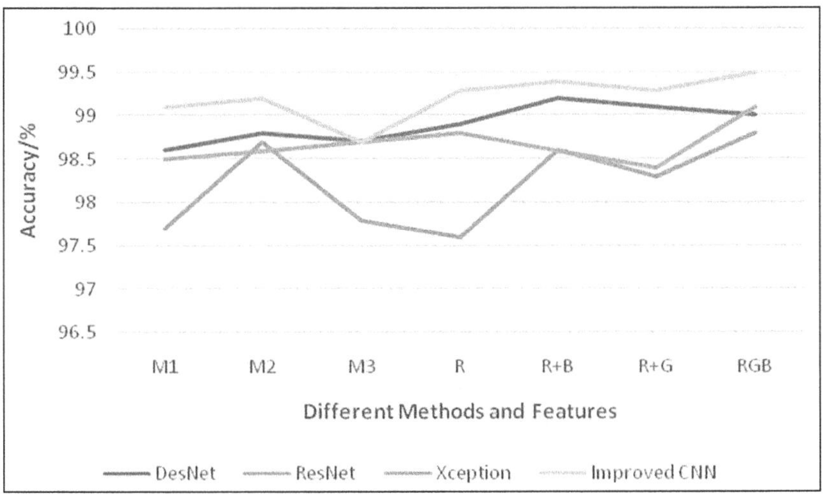

FIGURE 8.4 (b) (Recall) Evaluation results of malware visualization. Methods based on kaggle dataset.

The initial detection and classification of malware variants presents a formidable challenge in attaining a high degree of accuracy using current methods of malware classification. To address this issue, the present study employs DataCon samples obtained from the operational network to validate the model's performance. Experiment outcomes are illustrated in Figure 8.5.

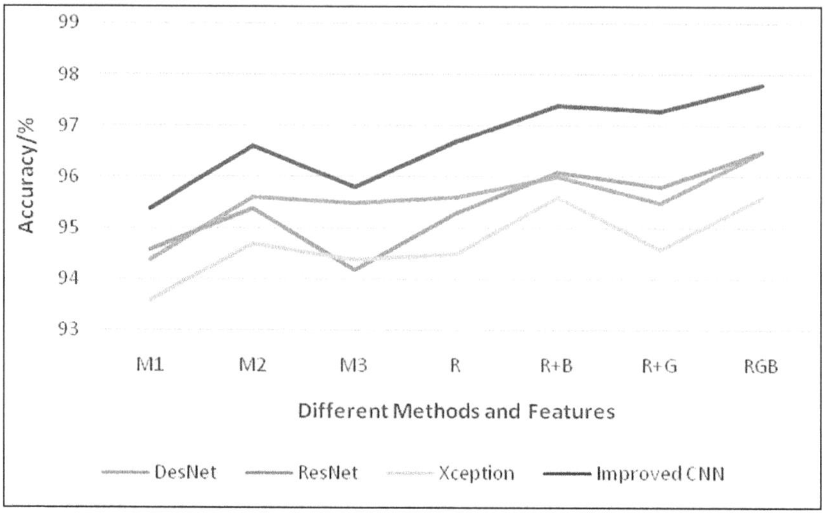

FIGURE 8.5 (a) (Accuracy) Evaluation results of malware visualization. Methods based on datacon dataset.

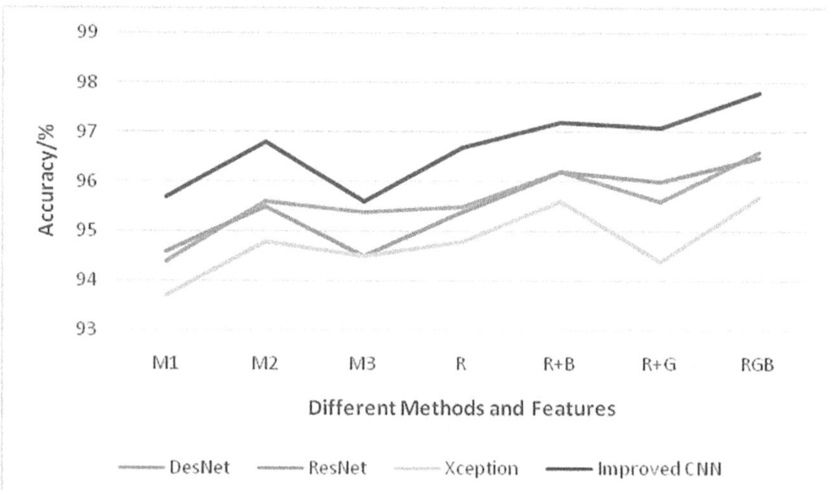

FIGURE 8.5 (b) (Recall) Evaluation results of malware visualization. Methods based on datacon dataset.

Compared with Figure 8.4, the accuracy rate drops significantly, mainly because the DataCon data set samples introduce more encrypted, obfuscated, and variant malware, resulting in a lower accuracy rate. The experimental results fully prove that this method considers the relationship between malware variants and the original body. Analyze and explore the relationship between malware variants and the original family from multiple aspects of the relationship between operation codes, operands, API calls, and so forth.

8.13.2 EVALUATION OF COORDINATE ATTENTION MECHANISM

The aim of our work was to check the potential enhancement in detection performance of the model by incorporating the coordinate attention mechanism. This investigation was conducted using two distinct datasets obtained from Kaggle and DataCon. This experiment demonstrates the influence of the coordinate attention mechanism on characteristics, as depicted in Figure 8.6. According to the confusion matrix in the Figure 8.5, the analysis of prediction results shows that the coordinate attention mechanism can obtain a wider range of spatial position information, not just channel feature information. The main reason is that traditional SEAM and CBAM only consider position or spatial features in the feature calculation process and ignore spatial and channel features. Position is combined with the importance of calculating features. By adding the coordinate attention mechanism, the spatial weight of the feature map can be obtained, guiding the network to acquire local features and broader spatial features, thereby accelerating convergence. Therefore, the CNN, by introducing the coordinate attention mechanism, has significantly improved the accuracy rate.

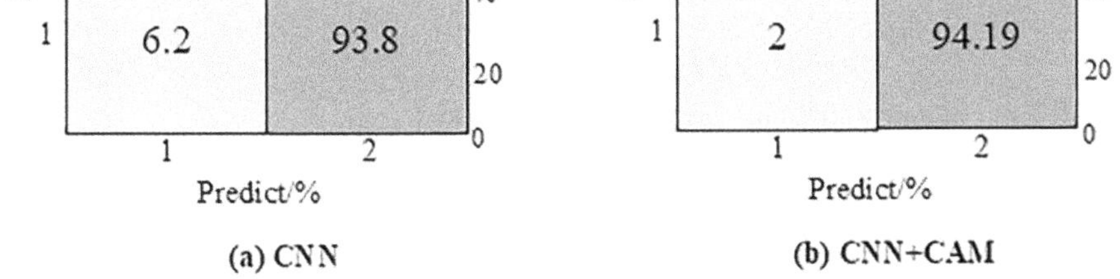

FIGURE 8.6 Impact of coordinate attention mechanism based on datacon dataset.

8.13.3 EVALUATION OF HOLLOW SPACE PYRAMID POOL

In this part, we will proceed to assess the enhancement impact of incorporating hole space pyramid pooling into the CNN model with a coordinate attention mechanism. The empirical findings are depicted in Figure 8.7. According to the findings presented in Figure 8.7, the confusion matrix reveals that prior to the implementation of ASPP, certain malware families, specifically Family 2 and Family 9, exhibited lower levels of accuracy. Traditional approaches sometimes employ techniques such as padding or truncation to establish a fixed size for input images. However, these methods can lead to the inclusion of superfluous information as well as the loss of valuable data. The implementation of ASPP can guarantee the ease and comprehensiveness of diverse malware data, improve the classification accuracy in the Lollipop family and the Gatak family, and avoid as much redundant information and information loss as possible. The main reason why the ACY family performs better is that the RGB feature integrates the advantages of a variety of feature mixtures, retains the integrity of variant malware information, and steadily improves the overall accuracy.

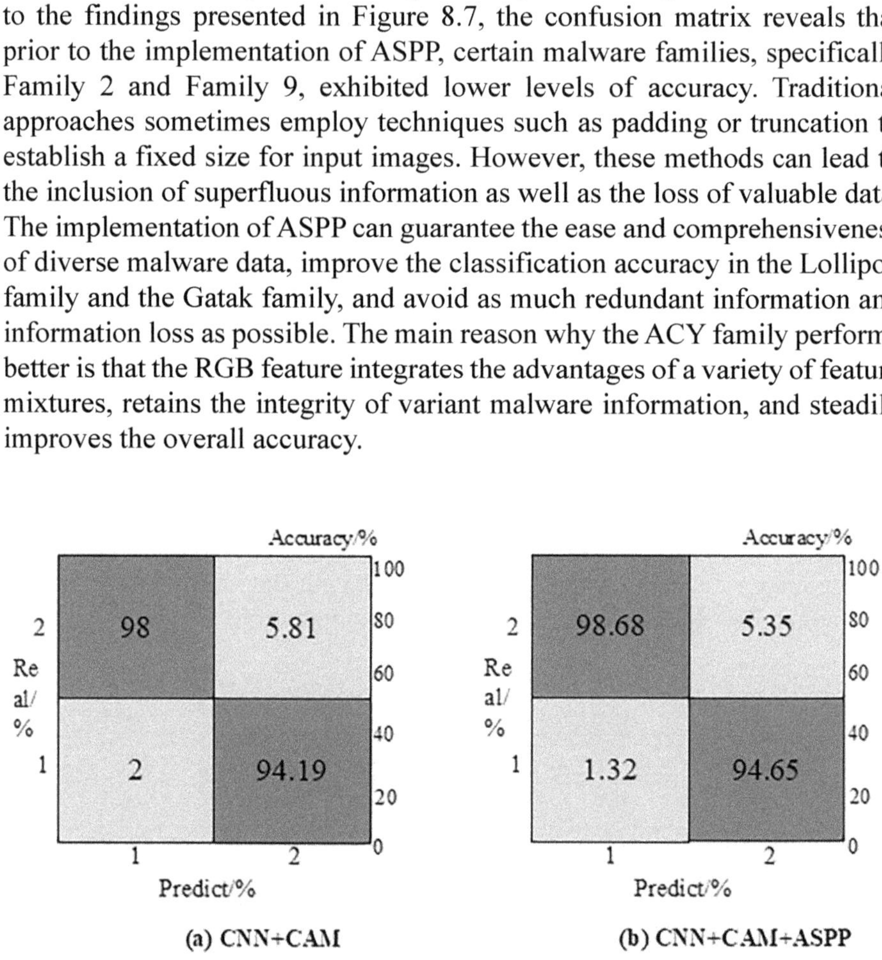

FIGURE 8.7 Empirical findings based on datacon dataset.

8.13.4 EVALUATION OF HOLLOW SPACE PYRAMID POOL

To evaluate the efficacy of the proposed comprehensive classification system, this study conducted a comparative analysis between the method presented in this research and various commonly employed malware variant classification methods discussed in the literature. The findings are presented in Tables 8.4 and 8.5.

The comparison findings between the method proposed in this article and other methods on the Kaggle data set are presented in Table 8.4. The literature [34] employs grayscale photographs as features and does not consider the semantic aspects of binary codes, leading to a rather low accuracy rate of 98.2%. The present discourse pertains to the field of literature [34]. The distinguishing feature of this phenomenon is the utilization of grayscale imagery. The grayscale images of certain samples originating from distinct families exhibit a significant degree of resemblance and are susceptible to erroneous interpretation. The level of accuracy achieved is 97.49%. The field of literature [1] is distinguished by the presence of byte sequences and operation code sequences. These characteristics enable the identification of similarities between malicious code and its variants based on signature traits. However, it is important to note that the similarity between these codes will gradually diminish as adversarial technology continues to advance. The literature [15] employs the VGG16 convolutional neural network architecture in conjunction with Markov pictures to perform feature reduction on the image data. The process of unification leads to a reduction in the amount of information and an increase in redundancy, which consequently leads to a decrease in accuracy. The literature [16] employs the LeNet5 architecture, utilizing the Word2Vec multi-channel feature matrix. However, it adopts distinct colors to denote distinct aspects, hence disregarding the interrelation among these elements.

TABLE 8.4 Comparison of the Proposed Methodology with Other Kaggle Methods on the Dataset

Methodology	Methodology Introduction	Efficiency/ %	Correctness/ %	Recollection/ %	F1-Score
Literature [11]	CNN+LSTM+Gray	98.20	–	–	95.77
Literature [12]	CNN+Gray	97.49	–	–	94.38
Methods of this Article	RGB	99.48	99.39	99.48	99.48

Table 8.5 shows the comparison results between this methodology and other methodologies on the DataCon data set. The literature [29] uses grayscale images with a size of 512 × 512 pixels as features and does not consider the semantic features of binary codes, resulting in a low accuracy rate of 96.80%. The literature [31] contains five features: non-portable executable (PE) structure, string sequence, assembly instruction, PE structure, and Control Flow Graph (CFG) calling relationship, using integrated learning

composed of five components. The structure of the classifier is complex, and some samples are still misjudged, with an accuracy of 96.99%.

TABLE 8.5 Comparison of the Proposed Methodology with Other DataCon Methods on the Dataset

Methodology	Method Introduction	Efficiency%	Correctness/%	Recollection%	F1-Score
Literature [31]	Ensemble Learning	96.99	94.05	---	92.19
Literature [27]	Gray+CNN	96.8	96.42	96.26	97.38
Methods of this Article	RGB	97.78	97.80	97.76	97.78

The comparative analysis presented in Tables 8.4 and 8.5 demonstrates that the method proposed in this paper exhibits superior accuracy in comparison to alternative categorization methods. There are three primary factors:

1. When dealing with new malware, as malware countermeasures continue to develop, the file size and characteristics of malware variants have changed significantly, but the virtual address-based correspondence between binary code and assembly information has not. The user's text remains unchanged. In contrast to the traditional approach, the visualization method presented in this article builds a linkage between numerous aspects using virtual addresses, hence enhancing the representation of altered features.
2. In terms of model design, traditional classification methods only consider part of the feature relationships and ignore the global feature relationships. During the feature processing stage, the analysis focuses on the visualization of malware features by considering the relationship between the employed opcode operand sequence and API calls, as well as the logical relationship that exists between different types of malware. In this article, we use the coordinate attention approach to analyze malware from a model-building viewpoint, considering their functional and logical relationships with one another.
3. To address the issue of redundant or missing information caused by normalization, the inclusion of ASPP can introduce malicious images of varying sizes into the model for the purposes of training and classification. This approach circumvents the need for zero padding or truncation, hence mitigating information loss. The information is absent or not provided.

8.14 CONCLUSION

This study presents a novel approach for designing and implementing a semantic connection using RGB picture acquisition characteristics. It further integrates ASPP and CA algorithms to enhance the extracted features. Additionally, a method for classifying malicious code is proposed, which leverages an improved CNN model and incorporates behavioral relationship data. By employing static analysis techniques, one can effectively evade the potential risks associated with various forms of malicious code variations. The proposed method effectively addresses the issue of information loss resulting from normalization, leading to improved accuracy in classifying and generalizing malicious code variants. The effectiveness of this approach is demonstrated through experimentation on a test set of malicious code, thereby providing theoretical and technical support for the classification of Windows malicious code. Nevertheless, the proliferation of malicious software, commonly known as malware, has emerged as a significant and pressing concern within the realm of Internet of Things (IoT) security. Further investigation is necessary to classify the many types of malware variants within the realm of IoT applications, in light of the issues arising from the proliferation of harmful code from conventional networks to the IoT and its subsequent spread within the IoT.

KEYWORDS

- **convolutional neural networks (CNN)**
- **coordinate attention module (CAM)**
- **feature representation**
- **malicious software**
- **network security**
- **zero trust security**

REFERENCES

1. Das, S., Liu, Y., Zhang, W., & Chandramohan, M. (2016). Semantics-based online malware detection: Towards efficient real-time protection against malware. *IEEE Transactions on Information Forensics and Security*, *11*(2), 289–302. https://doi.org/10.1109/TIFS.2015.2491300.
2. Kim, H. I., Kang, M., Cho, S. J., & Choi, S. I. (2022). Efficient deep learning network with multi-streams for Android malware family classification. *IEEE Access*, *10*, 5518–5532. https://doi.org/10.1109/ACCESS.2021.3139334.

3. Chaudhry, J. A., Saleem, K., Alazab, M., Zeeshan, H. M. A., Al-Muhtadi, J., & Rodrigues, J. J. P. C. (2021). Data security through zero-knowledge proof and statistical fingerprinting in vehicle-to-healthcare everything (V2HX) communications. *IEEE Transactions on Intelligent Transportation Systems*, *22*(6), 3869–3879. https://doi.org/10.1109/TITS.2021.3066487.
4. Yin, J., et al. (2023). SmartDID: A novel privacy-preserving identity based on blockchain for IoT. *IEEE Internet of Things Journal*, *10*(8), 6718–6732. https://doi.org/10.1109/JIOT.2022.3145089.
5. Dieye, M., et al. (2023). A self-sovereign identity based on zero-knowledge proof and blockchain. *IEEE Access*, *11*, 49445–49455. https://doi.org/10.1109/ACCESS.2023.3268768.
6. Li, W., Meese, C., Guo, H., & Nejad, M. (2023). Aggregated zero-knowledge proof and blockchain-empowered authentication for autonomous truck platooning. *IEEE Transactions on Intelligent Transportation Systems*, *24*(9), 9309–9323. https://doi.org/10.1109/TITS.2023.3271436.
7. Mouris, D., & Tsoutsos, N. G. (2021). Zilch: A framework for deploying transparent zero-knowledge proofs. *IEEE Transactions on Information Forensics and Security*, *16*, 3269–3284. https://doi.org/10.1109/TIFS.2021.3074869.
8. Rasheed, A. A., Mahapatra, R. N., & Hamza-Lup, F. G. (2020). Adaptive group-based zero-knowledge proof-authentication protocol in vehicular ad hoc networks. *IEEE Transactions on Intelligent Transportation Systems*, *21*(2), 867–881. https://doi.org/10.1109/TITS.2019.2899321.
9. Zhang, X., Wang, J., Xu, J., & Gu, C. (2023). Detection of Android malware based on deep forest and feature enhancement. *IEEE Access*, *11*, 29344–29359. https://doi.org/10.1109/ACCESS.2023.3260977.
10. Yan, S., Ren, J., Wang, W., Sun, L., Zhang, W., & Yu, Q. (2023). A survey of adversarial attack and defense methods for malware classification in cyber security. *IEEE Communications Surveys & Tutorials*, *25*(1), 467–496. https://doi.org/10.1109/COMST.2022.3225137.
11. Dao, T. V., Sato, H., & Kubo, M. (2022). An attention mechanism for combination of CNN and VAE for image-based malware classification. *IEEE Access*, *10*, 85127–85136. https://doi.org/10.1109/ACCESS.2022.3198072.
12. Li, S., Li, Y., Wu, X., Otaibi, S. A., & Tian, Z. (2023). Imbalanced malware family classification using multimodal fusion and weight self-learning. *IEEE Transactions on Intelligent Transportation Systems*, *24*(7), 7642–7652. https://doi.org/10.1109/TITS.2022.3208891.
13. Wang, X., et al. (2023). Detection of thermal covert channel attacks based on classification of components of the thermal signal features. *IEEE Transactions on Computers*, *72*(4), 971–983. https://doi.org/10.1109/TC.2022.3189578.
14. Al-Andoli, M. N., Sim, K. S., Tan, S. C., Goh, P. Y., & Lim, C. P. (2023). An ensemble-based parallel deep learning classifier with PSO-BP optimization for malware detection. *IEEE Access*, *11*, 76330–76346. https://doi.org/10.1109/ACCESS.2023.3296789.
15. Khalil, A. H., Ghalwash, A. Z., Elsayed, H. A. G., Salama, G. I., & Ghalwash, H. A. (2023). Enhancing digital image forgery detection using transfer learning. *IEEE Access*, *11*, 91583–91594. https://doi.org/10.1109/ACCESS.2023.3307357.
16. Alshahrani, M., & Traore, I. (2019). Secure mutual authentication and automated access control for IoT smart home using cumulative keyed-hash chain. *Journal of Information Security and Applications*, *45*, 156–175. https://doi.org/10.1016/j.jisa.2019.02.003.

17. Liyanage, M., Pham, Q. V., Dev, K., Bhattacharya, S., Maddikunta, R. K. R., Gadekallu, T. R., & Yenduri, G. (2022). A survey on zero touch network and service management (ZSM) for 5G and beyond networks. *Journal of Network and Computer Applications, 203*, 103362. https://doi.org/10.1016/j.jnca.2022.103362.
18. Walshe, M., Epiphaniou, G., Al-Khateeb, H., Hammoudeh, M., Katos, V., & Dehghantanha, A. (2019). Non-interactive zero-knowledge proofs for the authentication of IoT devices in reduced connectivity environments. *Ad Hoc Networks, 95*, 101988. https://doi.org/10.1016/j.adhoc.2019.101988.
19. AlQadheeb, A., Bhattacharyya, S., & Perl, S. (2022). Enhancing cybersecurity by generating user-specific security policy through the formal modeling of user behavior. *Array, 14*, 100146. https://doi.org/10.1016/j.array.2022.100146.
20. Varma, I. M., & Kumar, N. (2023). A comprehensive survey on SDN and blockchain-based secure vehicular networks. *Vehicular Communications, 44*, 100663. https://doi.org/10.1016/j.vehcom.2023.100663.
21. Ferretti, L., Magnanini, F., Andreolini, M., & Colajanni, M. (2021). Survivable zero trust for cloud computing environments. *Computers & Security, 110*, 102419. https://doi.org/10.1016/j.cose.2021.102419.
22. Das, D., Dasgupta, K., & Biswas, U. (2023). A secure blockchain-enabled vehicle identity management framework for intelligent transportation systems. *Computers and Electrical Engineering, 105*, 108535. https://doi.org/10.1016/j.compeleceng.2022.108535.
23. Deva Priya Isravel, A. J., Sagayam, K. M., Bhushan, B., Sei, Y., Eunice, J., & Jennifer, E. (2023). Blockchain for healthcare systems: Architecture, security challenges, trends and future directions. *Journal of Network and Computer Applications, 215*, 103633. https://doi.org/10.1016/j.jnca.2023.103633.
24. Maurya, A. K., Das, A. K., Jamal, S. S., & Giri, D. (2021). Secure user authentication mechanism for IoT-enabled Wireless Sensor Networks based on multiple Bloom filters. *Journal of Systems Architecture, 120*, 102296. https://doi.org/10.1016/j.sysarc.2021.102296.
25. Isaja, M., Nguyen, P., Goknil, A., Sen, S., Husom, E. J., Tverdal, S., Anand, A., Jiang, Y., Pedersen, K. J., Myrseth, P., Stang, J., Niavis, H., Pfeifhofer, S., & Lamplmair, P. (2023). A blockchain-based framework for trusted quality data sharing towards zero-defect manufacturing. *Computers in Industry, 146*, 103853. https://doi.org/10.1016/j.compind.2023.103853.
26. Ferrag, M. A., Maglaras, L., Argyriou, A., Kosmanos, D., & Janicke, H. (2018). Security for 4G and 5G cellular networks: A survey of existing authentication and privacy-preserving schemes. *Journal of Network and Computer Applications, 101*, 55–82. https://doi.org/10.1016/j.jnca.2017.10.017.
27. Huang, L., et al. (2023). EAODroid: Android malware detection based on enhanced API order. *Chinese Journal of Electronics, 32*(5), 1169–1178. https://doi.org/10.23919/cje.2021.00.451.
28. Musikawan, P., Kongsorot, Y., You, I., & So-In, C. (2023). An enhanced deep learning neural network for the detection and identification of Android malware. *IEEE Internet of Things Journal, 10*(10), 8560–8577. https://doi.org/10.1109/JIOT.2022.3194881.
29. Vinayakumar, R., Alazab, M., Soman, K. P., Poornachandran, P., & Venkatraman, S. (2019). Robust intelligent malware detection using deep learning. *IEEE Access, 7*, 46717–46738. https://doi.org/10.1109/ACCESS.2019.2906934.

30. Won, D. O., Jang, Y. N., & Lee, S. W. (2023). PlausMal-GAN: Plausible malware training based on generative adversarial networks for analogous zero-day malware detection. *IEEE Transactions on Emerging Topics in Computing, 11*(1), 82–94. https://doi.org/10.1109/TETC.2022.3170544.
31. Sen, S., Aydogan, E., & Aysan, A. I. (2018). Coevolution of mobile malware and anti-malware. *IEEE Transactions on Information Forensics and Security, 13*(10), 2563–2574. https://doi.org/10.1109/TIFS.2018.2824250.
32. Peng, W., Li, F., Zou, X., & Wu, J. (2014). Behavioral malware detection in delay tolerant networks. *IEEE Transactions on Parallel and Distributed Systems, 25*(1), 53–63. https://doi.org/10.1109/TPDS.2013.27.
33. Kim, J., Paik, J. Y., & Cho, E. S. (2023). Attention-based cross-modal CNN using non-disassembled files for malware classification. *IEEE Access, 11*, 22889–22903. https://doi.org/10.1109/ACCESS.2023.3253770.
34. Jin, B., Choi, J., Hong, J. B., & Kim, H. (2023). On the effectiveness of perturbations in generating evasive malware variants. *IEEE Access, 11*, 31062–31074. https://doi.org/10.1109/ACCESS.2023.3262265.

CHAPTER 9

Enhancing Network Security with Zero-Trust Principles and Anonymous Identity Authentication

RAGHAVENDRA KULKARNI,[1*] CHETNA KAUSHAL,[2] ISMAIL KESHTA,[3] MUKESH SONI,[4] and RENATO R. MAALIW III[5]

[1]*Department of Computer Science, School of Science, Gandhi Institute of Technology & Management (GITAM) – Deemed to be University, Rudraram, Patancheru Mandal, Hyderabad, Telangana, India*

[2]*Chitkara University Institute of Engineering and Technology, Chitkara University, Punjab, India*

[3]*Computer Science and Information Systems Department, College of Applied Sciences, AlMaarefa University, Riyadh, Saudi Arabia*

[4]*Department of CSE, University Centre for Research & Development Chandigarh University, Mohali, Punjab, India*

[5]*College of Engineering, Southern Luzon State University, Lucban, Quezon, Philippines*

*Corresponding author

ABSTRACT

As a prime exemplar of the Internet of Things (IoT), the vehicle-to-vehicle network assumes a pivotal position in the realm of intelligent transportation. It provides various online services for vehicles and reduces the risk of accidents for drivers. However, during communication, the vehicle-to-vehicle network generates sensitive information, such as vehicle location and routes. Enhancing the anonymity of vehicle identities in secure services is a research interest in vehicle-to-vehicle network security, especially in Zero Trust network security.

Zero-Trust Learning: Applications in Modern Network Security. Chetna Kaushal, Amandeep Kaur, Mohit Angurala, Aryan Chaudhary (Eds.)

This article introduces an anonymous identity authentication scheme based on batch verification algorithms, leveraging the principles of Zero Trust security. It expands the scope of anonymous authentication methods for IEEE WAVE security services by incorporating techniques such as anonymous credentials and zero-knowledge proofs, in accordance with the tenets of the Zero Trust model. Furthermore, it offers a mechanism for identity recovery via a trusted third party, thereby establishing a holistic security framework. Experimental results indicate that when the number of signatures for batch verification exceeds 11, the computational cost of the proposed scheme is more efficient than some comparative schemes. Based on this, the article suggests the optimal cycle for batch verification in the DSRC's BSM and vehicle proximity payment applications while maintaining a zero-trust security posture.

9.1 INTRODUCTION

In the field of Intelligent Transportation Systems (ITSs), vehicular networking enables real-time information exchange between vehicles and infrastructure, playing a positive role in improving traffic safety, reducing environmental pollution, alleviating traffic congestion, and providing convenience [1]. Currently, standards involved in vehicular networking communication include ETSI 102, the United States Security Credential Management System (SCMS), the European Cooperative Intelligent Transport Systems Security Credential Management System (CCMS), the European C2C Communication Consortium, and the Vehicle Environment Wireless Access Standard (WAVE standard) proposed by IEEE. Because wireless communication technology is inherently broadcast-oriented, communication between vehicles is open; therefore, it is necessary to protect sensitive information such as vehicle location and trajectory generated during vehicular networking communication. Failure to do so can lead to privacy breaches and pose threats to the lives and property of drivers [2].

To address this issue, some scholars have proposed privacy protection through authorized access to personal data in WAVE G1609.2. However, absolute privacy protection is challenging to achieve, as some security applications rely on frequently broadcasted sensitive data, such as vehicle location. In vehicular networking, privacy can be categorized into three types: location privacy, trajectory privacy, and identity privacy [3–5]. Privacy breaches in vehicular networking stem from the disclosure of a vehicle's true identity. When attackers obtain a valid vehicle identity, they can engage in activities like impersonation, message replay, and modification attacks, disrupting the normal operation of vehicles in the vehicular network. Additionally, to prevent attackers from combining vehicle location and trajectory information with identity information, the identity information of vehicles must be isolated or obscured from their location and trajectory information. Privacy leakage issues also include “query-sampling” attacks and snapshot location attacks [6, 7]. In “query-sampling” attacks, attackers use a vehicle user’s location information to associate it with specific applications. In snapshot location attacks, attackers use a vehicle user’s past locations to infer potential paths.

WAVE G1609.2 employs anonymity to measure privacy; therefore, improving the anonymity of vehicle identity in WAVE security services is a hot topic in vehicular networking security. This paper, based on an anonymous authentication scheme proposed by our research team, aims to design an enhanced anonymous authentication scheme for WAVE security services with relatively good security and performance.

The primary aim of this study is to augment the levels of privacy and security in vehicular networking, particularly in relation to the Wireless Access in Vehicular Environments (WAVE) standard. Special emphasis is placed on safeguarding the identification of vehicles. The main objective is to develop an enhanced anonymous authentication system that guarantees the preservation of vehicle IDs’ anonymity, hence mitigating privacy breaches and security risks inside the vehicular network. The objective of this study is to make a valuable contribution to the progression of ITS through the enhancement of privacy and security measures in vehicle communication.

1. This study aims to investigate the crucial matter of privacy in vehicle networking, with a specific focus on three key aspects: location privacy, trajectory privacy, and identity privacy. The research makes a valuable contribution to the preservation of vehicle IDs’ privacy by establishing an improved anonymous authentication system. The implementation of this privacy upgrade is crucial in order to protect sensitive information that is created during communication inside vehicular networks, including data related to the location and trajectory of vehicles.
2. The suggested anonymous authentication approach serves to augment both privacy and security aspects of WAVE security services. The research addresses security concerns, such as impersonation, message replay, and modification attacks, by implementing measures to prevent attackers from acquiring legitimate vehicle identities. This aids in guaranteeing the regular and secure functioning of cars within the vehicular network.
3. The present study presents the notion of anonymity as a metric for quantifying privacy within the context of vehicular networking. The research contributes to the field’s knowledge by proposing an

enhanced anonymous authentication technique, so enhancing the understanding of how anonymity may be leveraged to efficiently safeguard vehicle identities. This statement enhances the existing theoretical framework pertaining to the preservation of privacy in the context of automotive networking.

4. Performance considerations are taken into account in addition to the advances made to security and privacy in the research. The objective of the proposed approach is to achieve a harmonious equilibrium between security and performance, with the primary goal of maintaining an efficient authentication procedure that does not impose substantial delays on vehicular communication. This enhances the solution's practicality for implementation in real-world ITS settings.

9.2 RELATED WORK

This chapter categorizes current mainstream identity authentication schemes into two types: those based on bilinear pairing schemes and those based on non-bilinear pairing schemes. Additionally, based on different underlying cryptographic methods, schemes are divided into four categories: public key-based, identity-based, group signature-based, and certificate less-based. Tables 9.1 and 9.2 provide an analysis and comparison of the security and performance of the proposed schemes. Table 9.2 lists the performance analysis of various schemes in the cases of single signature verification, n-times single signature verification, and batch verification of n signatures. Table 9.1 introduces the security analysis of the schemes as follows [8–10]:

1. **Conditional Privacy Preservation or Traceability:** The identity of malicious vehicles can be revealed and revoked by a third-party authority, which has the ability to prevent malicious vehicles from participating in VANET networks when necessary.
2. **Non-Repudiation:** After successful identity verification and traceability, the sender cannot deny having sent a disputed message.
3. **Unlinkability:** Attackers, malicious vehicles, RSUs, and third-party authorities cannot link any two signatures to the same signer. Additionally, attackers, and malicious vehicles cannot link any two or more messages to the same signer.
4. **Replay Attack:** Attackers may repeatedly inject previously received beacons and messages back into the network.

5. **Impersonation Attack:** Attackers attempt to impersonate legitimate vehicles to send false messages and attack other vehicles.

6. **Modification Attack:** Attackers send modified messages into the VANET network to achieve specific goals.

TABLE 9.1 Security Comparative Analysis of Relevant Schemes

Bilinear pairing technology	Underlying cryptography methods (pseudonym types)	Plan	Time	Communication mode	Pseudonym issuer	Security Analysis: Conditional privacy protection (traceability)	Non-repudiation	Unlink ability	Resist replay attacks	Resist impersonation attacks	Resist modification attacks
Based on bilinear pairing technology	Based on public key (public key)	EAAP [10]	2016	V2V	TA	√	-	-	-	√	√
		CPAV [11]	2018	V2V	TA	√	√	√	-	√	-
	Identity based (pseudonymous ID)	DAPPA [12]	2016	V2V	V2I TA	-	√	√	-	-	-
		Author [6]	2020	V2V	V2I TA	√	√	√	-	-	-
		Author [13]	2021	V2V	TA	-	-	-	√	√	-
	Based on group signature (group public key)	ABAH [14]	2016	V2V	V2R TA	√	√	-	-	-	-
		BAGKD [15]	2020	V2V	V2I TA	-	-	-	√	-	-
	Based on certificate less (pseudonymous ID)	[16]	2015	V2I	TRA	√	-	√	-	-	-
		CL-AS [17]	2019	V2V	V2I RSU	-	-	-	-	-	-
		Author [18]	2019	V2I	PKG	√	-	√	√	-	-
Based on non-bilinear pairing technology	Identity based (pseudonymous ID)	ID-MAP [19]	2018	V2V	V2R OBU	√	-	√	√	√	√
		EPA_CPPA [20]	2018	V2V	V2R TA	√	-	√	√	√	√
		Author [21]	2019	V2V	V2I TA	-	-	√	-	√	-
		SIPAR [22]	2020	V2V	V2I TPD	-	√	√	√	√	√
		EABAH [23]	2020	V2V	V2R TA	-	-	√	√	-	-
	Based on group signature (group public key)	PW-CPPA-GKA [24]	2018	V2V	TA	√	-	√	√	√	√
		HCPA-GKA [25]	2018	V2V	TA	√	-	√	√	√	√
		Author [26]	2021	V2V	V2R TPD	√	-	√	√	-	-
	Based on certificate less (pseudonymous ID)	Author [27]	2018	V2I	TRA	√	-	√	√	√	√
		Author [28]	2020	V2I	OBU	√	-	√	√	√	√
		Author [29]	2020	V2V	V2R TA	√	√	√	√	√	√

TABLE 9.2 Performance Comparative Analysis of Relevant Schemes

Plan	Signature Overhead	Signature verification overhead		
		Single Signature Verification	n Signature Verification	Batch Verification
EAAP [10]	$6U_{ep1}+U_p+4U_m$	$2U_p+4U_{epl}+U_{ep2}$	$2nU_p+4nU_{ep1}+nU_{ep2}$	$(1+n)\ U_p+4U_{epl}+nU_{ep2}$
CPAV [11]	$4U_{p1}+4U_{ep2}+U_h$	$4U_p+4U_{epl}+U_h$	$2nU_p+2nU_{ep1}+nU_h$	$(1+n)\ U_p+2nU_{epl}+nU_h$
Author [13]	–	–	–	nU_p+nU_m
ABAH [14]	–	–	–	$3U_p+(n+1)\ U_m$
Author [16]	$2U_{epl}$	$3U_p+U_{epl}+U_h$	$3nU_p+nU_{epl}+nU_h$	$3U_p+nU_{epl}+nU_h$
CLGAS [17]	$3U_{epl}$	$3U_p+U_{epl}$	$3nU_p+3nU_{epl}$	$4U_p+3nU_{epl}+2U_{mtp}$
Author [18]	$3U_{epl}$	$3U_p+U_{epl}+U_h$	$3nU_p+2nU_{epl}+nU_h$	$3U_p+2nU_{epl}+2U_h$
ID-MAP [19]	–	$5U_m$	$5nU_m$	$5\ (n/d)\ U_m$
EPA-CPPA [20]	U_m+2U_h	$4U_m+U_{e.a}+2U_h$	$4nU_m+nU_{e.a}+2nU_h$	$(2+2\ n)\ U_m+nU_{e.a}+2nU_h$
Author [21]	$2U_{mp}+2U_h$	U_h	nU_h	nU_h
SIPAR [22]	-	$4U_m$	$4nU_m$	$(2n+3)\ U_m+2nU_{e.sm}$
EABAH [23]	U_m+U_h	$3U_m+U_{e.a}+2U_h$	$3nU_m+2nU_{e.a}+2nU_h$	$2nU_{e.sm}+2nU_{e.a}+2nU_h(n+1)U_m$
PW-CPPA-GKA [24]	$6U_h$	$4U_h$	$4nU_h$	–
HCPA-GKA [25]	$2U_h+U_{emc}$	$4U_h+U_{dec}$	$4nU_h+nU_{dec}$	–
Tsao et al. [26]	$3U_m+2U_{e.a}$	$4U_m+2U_{e.a}$	$4nU_m+2nU_{e.a}$	–
Haddon [27]	$U_m+U_{e.a}+U_h$	$3U_m+2U_{e.a}+U_h$	$3nU_m+2nU_{e.a}+2nU_h$	$(n+2)\ U_m+2U_{e.sm}+(2+1)U_{e.a}+2nU_h$
Alshahrani & Traore [28]	$U_m+U_{e.a}$	$3U_m+2U_{e.a}$	$3nU_m+2nU_{e.a}$	$(n+2)\ U_m+(n+1)U_{e.a}$
Liyanage et al. [29]	U_m	$3U_m+2U_{e.a}$	$3nU_m+2nU_{e.a}$	$(2n+1)\ U_m+(3n-1)U_{e.a}$

9.2.1 IDENTITY AUTHENTICATION SCHEMES BASED ON BILINEAR PAIRING

Public key-based identity authentication schemes generally require vehicles to pre-load a substantial quantity of certificates and public-private key pairs, which incurs considerable storage expenses. In order to tackle this concern, the Sahraoui and Bilami [10] put forth an effective scheme for anonymous authentication that incorporates conditional privacy preservation (EAAP). When conflicts arise, a third-party authority divulges the true identity of the vehicle involved and records it in an Identity Revocation List (IRL). Without a third-party authority, vehicles, and RSUs can mutually authenticate anonymously, but as the number of vehicles within an RSU area increases, the issue of message loss may arise. Furthermore, this scheme was later shown to be vulnerable to internal attacks, false message attacks, frame attacks, and Sybil attacks. Maintenance of anonymous certificates and the IRL also resulted in time and space wastage.

In response to the shortcomings in EAAP's security, Rivas et al. [11] proposed an anonymous bidirectional authentication scheme that provides higher security levels for vehicles, enhancing the security and efficiency of vehicle communication. Along with anonymous batch message authentication and anonymous batch vehicle authentication, this scheme also reduced the cost of verification by authenticating a batch of messages sent by RSUs and verifying the authenticity of a group of vehicles within an RSU area in batch. However, this scheme did not consider communication overhead and message loss rates.

Traditional identity-based authentication schemes typically face key escrow issues. To address this problem, Hassan et al. [12] proposed an efficient distributed aggregate privacy-preserving authentication (DAPPA) scheme. Vehicles are required to generate one-time private keys and identity-based aggregate signatures during registration, leading to higher authentication latency and reduced communication efficiency.

Unlike the scheme in Ref. [12], the real-time anonymous batch authentication scheme proposed by Wu [13] in 2021 focuses on privacy protection and computation cost issues for entities such as vehicles and aims to alleviate RSU authentication burdens. This scheme is also intended for Vehicle-to-Vehicle (V2V) communication.

To address the issue of lengthy Certificate Revocation List (CRL) verification, Kouicem et al. [14] proposed an efficient anonymous batch authentication scheme in 2016 based on RSU partitioning and Hash-based Message

Authentication Code (HMAC). This scheme combines group signature and identity-based approaches to achieve anonymous message authentication using group signatures and ensure the authenticity of messages sent by RSUs using identity-based signatures. It improves verification efficiency through batch authentication. However, due to its complexity and high computational costs, the scheme may face message processing delays when an RSU needs to verify a large number of messages simultaneously. A similar problem exists in the literature [15], but it mitigates the risk of group key leakage by using a dynamic group key distribution mechanism.

Certificate-less public key encryption technology can address complex certificate issues, and many certificate-less aggregate signature schemes have been proposed based on this feature [16–18]. Among them, the scheme proposed by the Zhang et al. [18] for Vehicle-to-Infrastructure (V2I) communication achieves complete aggregation by using RSUs as aggregators to combine vehicle signatures, significantly reducing communication costs.

In 2020, our research team proposed a home-to-car interconnection service based on edge computing, using a novel anonymous authentication scheme based on efficient cryptographic primitives [6], which provides significant advantages in terms of communication overhead, computational cost, signaling overhead, service ratio, and authentication latency.

9.3 IDENTITY AUTHENTICATION SCHEMES BASED ON NON-BILINEAR PAIRING

To reduce RSU computational costs and improve authentication efficiency, some vehicle proxy-based identity authentication schemes have been proposed [19, 30]. In Liu's identity-based authentication scheme (Proxy-based Authentication Scheme, PBA) [30], proxy vehicles simultaneously verify identities with RSUs, reducing the verification time. However, Baudier et al. [19] argued that the scheme in [30] cannot guarantee message authenticity and is vulnerable to impersonation and modification attacks. To address these security issues, another new proxy vehicle-based authentication scheme (Identity-based Message Authentication Scheme using Proxy Vehicles, ID-MAP) was proposed in Azad [20], and this scheme has been proven to be secure against adaptive chosen message attacks.

To address security issues related to system private keys in the Conditional Privacy Preserving Authentication (CPPA) scheme [31, 32], Azad [20] introduced an efficiently provably secure CPPA scheme, which enhances message authentication performance through batch verification. In 2019, Usman and Gutierrez [21] proposed an anonymous mutual authentication scheme. In this scheme, vehicles need to mutually authenticate with a third-party authority before communication to prevent impersonation attacks. Additionally, vehicles prevent malicious vehicles from obtaining useful information during communication through side-channel attacks by regularly updating information stored in a Tamper-Proof Device (TPD). Due to its low communication and computational costs, this scheme is more suitable for the multi-vehicle communication environment of vehicular networks. The schemes proposed by Rizi and Seno [22] and Ziegler et al. [23] avoid directly storing the master key in the TPD, providing a more direct defense against side-channel attacks. In the Extensible and Effective Anonymous Batch Authentication Scheme (EABAH) [23], a method for fast CRL distribution is employed to effectively prevent attackers from using unexpired certificates to access valid messages. Additionally, this scheme uses Elliptic Curve Cryptography (ECC) to reduce the computational overhead of batch authentication and employs hash seed revocation of vehicles issued by a third-party authority.

Chen et al. [24] proposed a conditional privacy-preserving authentication scheme using HMAC. Unlike other group signature-based schemes, this scheme designed and implemented a secure group key generation and management protocol. Typically, group signature-based identity authentication schemes require the determination of group keys through key agreement schemes. Tsao et al. [26] proposed an anonymous authentication scheme based on non-singular elliptic curves and introduced an efficient group key agreement scheme. This plan utilizes ECC to improve both computational and communication efficiency. In a prior study [33], conditional privacy preservation was introduced, employing hash functions (known as HCPA-GKA, Hash Function-based Conditional Privacy-Preserving Authentication and Group Key Agreement Schemes). Additionally, it introduced a group key agreement approach utilizing the Chinese Remainder Theorem, eliminating the need for intricate encryption and decryption procedures to grant group members access to group keys, thus lowering computational burdens. Das et al. [34] proposed a scheme based on elliptic curve discrete logarithm problems using aggregate signature technology, reducing computational and communication overhead. Andrew et al. [35] introduced a certificate-less aggregate signature scheme based on non-singular elliptic curve cryptosystems. However, this scheme requires the maintenance of vehicle registration tables and pseudo-identity revocation lists (IRLs), incurring significant communication and storage costs. Maurya et al. [36] proposed an efficient and secure authentication scheme with lower computational and communication overhead.

Based on the analysis from Tables 9.1 and 9.2, along with the previous discussion, it is evident that schemes based on bilinear pairings have relatively simple verification and support batch verification of multiple messages. This makes them suitable for vehicular networks where a large number of messages need to be verified simultaneously. Therefore, this chapter proposes a bilinear pairing-based anonymous batch verification scheme to address authentication latency issues in vehicle communications.

9.4 ANONYMOUS AUTHENTICATION SCHEME

In this section, we will first provide an overview of the network model and the scheme to be improved, and introduce the enhanced batch verification algorithm. Then, we will discuss the process of applying the improved scheme to the WAVE security protocol and provide a comparison of the Secured Protocol Data Unit (SPDU) structure before and after the improvement.

9.4.1 NETWORK MODEL

Vehicular networks are a special type of mobile ad-hoc network with vehicles as nodes. Their uniqueness is primarily manifested in several aspects: (i) nodes have high mobility; (ii) node mobility follows regular patterns; (iii) node information is private; (iv) frequent information exchange; (v) variations in traffic scenarios; (vi) relatively abundant node resources; and (vii) unstable wireless communication channels [33–36]. Vehicular networks typically consist of roadside units (RSUs), on-board units (OBUs), and third-party authorities, enabling Vehicle-to-Vehicle (V2V) and Vehicle-to-Infrastructure (V2I) communication, as shown in Figure 9.1.

Here is an introduction to the entities in vehicular networks:

1. **Roadside Unit (RSU):** In vehicular networks, RSUs are base stations installed on roadsides. They use Dedicated Short-Range Communication (DSRC) technology to communicate with vehicles. As an intermediate entity between OBUs and third-party authorities, RSUs possess centralized storage and computational capabilities.

2. **On-Board Unit (OBU):** OBUs are installed on vehicles and periodically broadcast traffic-related status information such as position

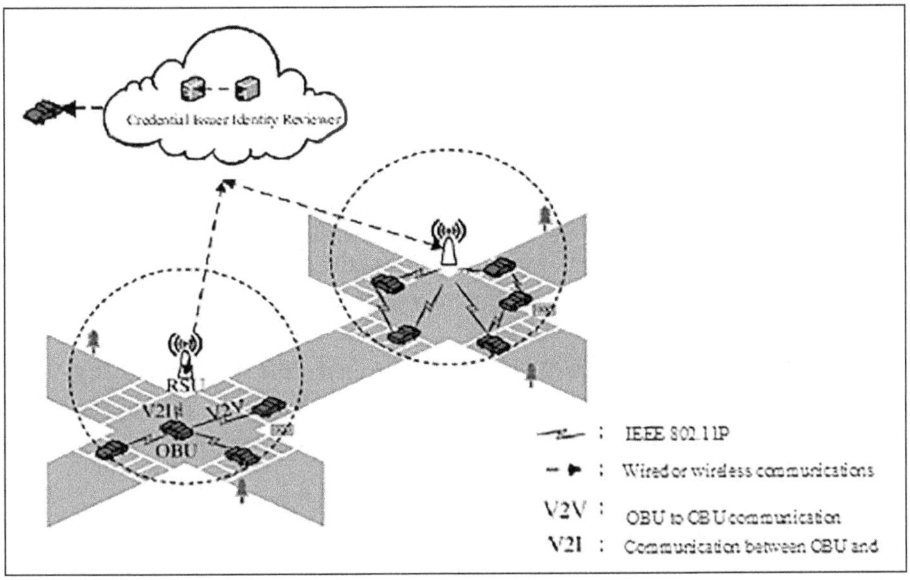

FIGURE 9.1 Vehicle-to-infrastructure (V2I) and vehicle-to-vehicle (V2V) communication model in WAVE.

and speed. OBUs also use DSRC technology to communicate with RSUs and other vehicles.

3. **Credential Issuer (referred to as "Issuer"):** The Issuer is a trusted third party with significant computational and storage capabilities. It is responsible for vehicle registration, issuing anonymous credentials, and other tasks to ensure privacy protection in V2V and V2I communication.

4. **Identity Verifier (referred to as "Verifier"):** The Verifier is another trusted third party with substantial computational and storage capabilities. Its primary role is to perform identity verification for vehicles.

9.5 SCHEME FRAMEWORK

The team on which this chapter relies proposed an anonymous authentication scheme for vehicular networks [6]. This scheme involves four roles: Issuer, Prover, Verifier, and Examiner, each operating on different entities within the vehicular network. It achieves anonymous communication between vehicles by creating an anonymous region within the vehicular network. The original scheme can be summarized in four main steps: system initialization and key generation, anonymous credential issuance, token generation, and token verification.

In the system initialization and key generation phase, the Issuer first generates system parameters {Q, R, H_1, H_2, HU, e, p} and defines a hash function $J_\perp : G_U \rightarrow H_1, J: \rightarrow H_1 \times \{0,1\}^* \rightarrow A_r^*$. Random values $s_{is} \in A_r^*$ are selected, and the Issuer computes their private key $y_{is} = s_{is}Q \in H_1$ and public key $Y_{is} = e(Q, R)^{Sis} \in G_U$. The Prover selects a random value $s_{kq} \in A_r^*$ and computes their public key $Yq = s_{kq}Q \in H_1$. The Examiner selects a random value $\alpha \in RA_r^*$ as their private key and computes public keys $J_1 = \alpha Q \in H_1$ and $J_2 = -\alpha R \in H_2$.

In the anonymous credential issuance phase, the Issuer utilizes their private key y_{is}, the Prover's public key J_1, and the Verifier's public key Y_q to compute an anonymous credential for the Prover:

$$D_1 = y_{is} + Y_q + sJ_1, D_2 = -sQ,$$

where r is a random value. Upon receiving the anonymous credential, the Verifier uses properties of bilinear mapping to determine if $e(D_1, R) = Y_{is} * e(Y_q, R) * e(D_2, J_2)$ holds true, thereby verifying the validity of the anonymous credential.

In the token generation phase, the Prover first generates a random value s and calculates a commitment $N_q = Y_q + sJ_1$ on group H_1, as well as a commitment $N_\perp Q = e(N_Q, Q) \in G_U$. The Prover also calculates $\sigma \leftarrow J_\perp (N_{\perp Q})^{(-\alpha)} \in H_1$ using the $J_\perp$ hash function. Subsequently, the Prover selects another random value s' and computes $D_1' = D_1 + (s + s')J_1$ and $D_2' = D_2 - sQ$. Alongside these calculations, the Prover determines parameters for a zero-knowledge proof based on their private key skq, the Verifier's public key J_1, and the given system parameters $\{Q, R, H_1, H_2, HU, e, q\}$, which include:

$$E = yQ + tJ_1 \in H_1, e' = J(E \,||\, \text{nonce}), u = y + e' * s_{kq}, \text{ and } w = t + e' * s,$$

where y and t are random values. Finally, the Prover combines the anonymous credential (D_1,' D_2') and generates an indistinguishable token $Q = (u, v, D, e'), D' = (D_1, 'D_2')$.

In the token verification phase, the Verifier calculates $e' = e_1$, and verify

$$e(D_1' + \sigma, R) = Y_{is} \cdot e_\perp^{e}(D_2' + J_\perp^{(e\perp)}, J_2)E + e'N_q = uQ + wJ_1.$$

The Verifier checks if this equation holds true to verify the token's validity. If the token is valid, an anonymous session between vehicles is successfully established.

When it becomes necessary to reveal a vehicle's true identity, the Examiner calculates $\omega = \alpha N_{\perp}$ and sends it to the Issuer. The Issuer then computes $YQ = N + \omega$ to obtain the vehicle's public key. The identity recovery process is depicted in Figure 9.2.

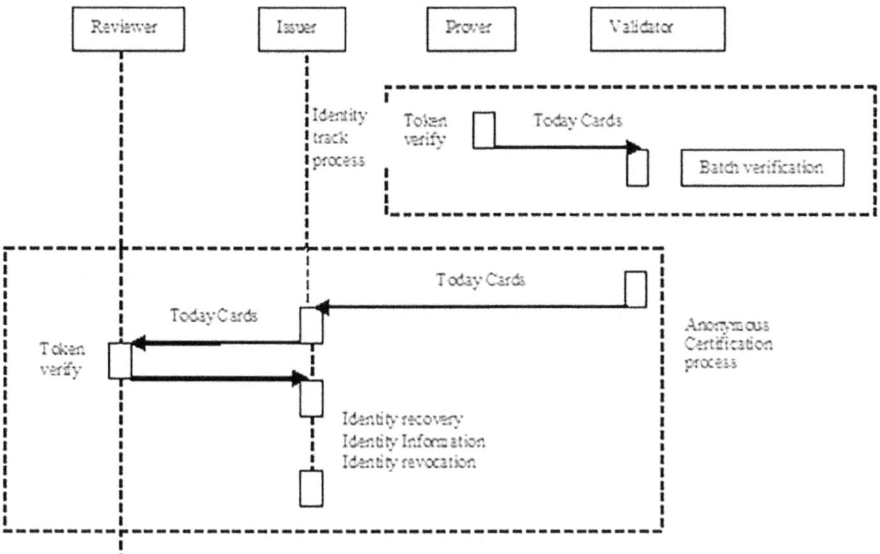

FIGURE 9.2 Flowchart of the anonymous authentication and identity recovery process.

9.6 BATCH VERIFICATION ALGORITHM

To address the issue of signature delay, batch verification algorithms were proposed in references [37, 38]. Batch verification algorithms enable vehicles to simultaneously verify multiple received messages without individually verifying each one, thereby reducing the signature verification delay. Moreover, algorithms constructed using bilinear pairings can better facilitate the development of batch verification algorithms, thus reducing computational overhead.

To reduce the verification cost and improve verification efficiency [39], this chapter enhances identity verification using batch verification to reduce authentication delay in V2V communication. Compared to other batch verification algorithms, the algorithm proposed in this chapter is constructed using zero-knowledge proofs, which enhances algorithm performance while maintaining security. The batch verification algorithm is presented in Algorithm 1, where n represents the number of messages received within a given verification cycle. The verification process is illustrated in Figure 9.2.

Algorithm 1: Batch_Verify()

Input: n u, w, E_i, e_i', (D'_{1i}, D'_{2i}), R, J_2, Y_{is}, nonce, N_i, σ_i

Output: accept/reject

For each of the *n* received messages do

1. e_i= J(E_i||$nonce_i$)
2. $\hat{e}_i$ = e(N_i, R) (given hash function $\hat{J}: H_U \rightarrow H_1$)

End for

3. $\text{If}(e_i' = e_i) \text{ and } e(\sum_{i=1}^{n}(D_{1i}' + \sigma_i), R) = \prod Y_{is} \cdot e(\sum_{i=1}^{n} N_i, R) \cdot e(\sum_{i=1}^{n}(D_{2i}' + \hat{H}(\hat{e_i})), H_2)$ $\text{and} \sum_{i=1}^{n}(E_i + e_i' N_i) = (\sum_{i=1}^{n} u_i) \cdot Q + (\sum_{i=1}^{n} w_i) \cdot J_1 \textit{then}$
4. Accept
5. Else
6. Reject
7. End if

The correctness proof of the above algorithm is as follows:

$$(e_i' = J(E_i \,\|\, \text{nonce}_i)$$

$$= e_i \prod_{i=1}^{n} e(D_{1i}' + \sigma_i, R)$$

$$= e(\sum_{i=1}^{n} (D_{1i}^{n} + \sigma_{i,} R)$$

$$= \prod_{i=1}^{n} Y_{is} \prod_{i=1}^{n} e(N_{i,} R) \prod_{i=1}^{n} e(D_{2i}' + \hat{J}(\hat{e_i}), J_2)$$

$$= \prod_{i=1}^{n} J_{is} \cdot e(\sum_{i=1}^{n} N_{i,} R) \cdot e(\sum_{i=1}^{n} (D_{2i}' + \hat{J}(\hat{e_i})), J_2)$$

$$= (\sum_{i=1}^{n} (E_i + e_i' N_i)$$

$$= \sum_{i=1}^{n} (y_i Q + J_1 t_i + e_{i'} sk_q Q + e_{i'} r J_1)$$

$$= \sum_{i=1}^{n} (y_i + e_{i'} sk_q) \cdot Q + \sum_{i=1}^{n} (s_i + e_{i'} r) \cdot J_1$$

$$= \sum_{i=1}^{n} u_i \cdot Q + \sum_{i=1}^{n} w_i \cdot J_1$$

9.7 EXTENDED APPLICATIONS IN WAVE SECURE SERVICES

WAVE 1609.2 employs the Elliptic Curve Digital Signature Algorithm (ECDSA) specified in FIPS 186–4 for implementing Secure Data Services (SDS). The Signed Secured Protocol Data Unit (SignedSPDU) is a data structure introduced in WAVE 1609.2 to achieve secure information transmission. Before information transmission, the Secure Data Exchange Entity (SDEE) uses SDS by invoking it multiple times to add security envelopes to SDEE, transforming the insecure protocol data units exchanged between entities into SPDU.

Roman et al. [6] deploy an anonymous authentication scheme in the application layer of vehicular networks. To support more applications, this chapter integrates the enhanced scheme with the WAVE security protocol. The token generation and verification algorithms of this enhanced scheme are placed in the MAC layer of the vehicular network, replacing the standard ECDSA signature algorithm in WAVE 1609.2. Figure 9.3 provides a breakdown of the process for generating a signed SPDU in WAVE 1609.2. In SDS, the SDEE first signs the To Be Signed Data in the vehicular network using the ECDSA algorithm and then assembles it into the signed SPDU to achieve secure transmission of the To Be Signed Data.

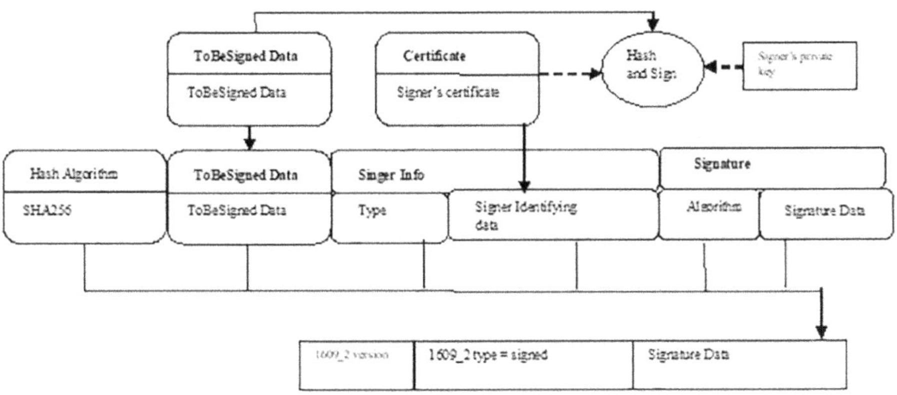

FIGURE 9.3 Decomposition of signed SPDU in WAVE.

Figure 9.4 illustrates the process of replacing the ECDSA algorithm with the token generation algorithm to generate a token for the To Be Signed Data and assemble it into a signed SPDU. When verifying the signed SPDU in WAVE 1609.2, the signed SPDU is divided into four parts:

message, signature, signer certificate, and other certificates. Relevant data is extracted from the signer certificate, and the ECDSA signature algorithm is used to verify the signature. Similarly, upon receiving an SPDU that uses the improved signature algorithm, it is divided into four parts: transmitted message, token, signer certificate, and other data. Information is extracted from the signer certificate and other data, and then the Batch_verify() function is used to efficiently verify the received token.

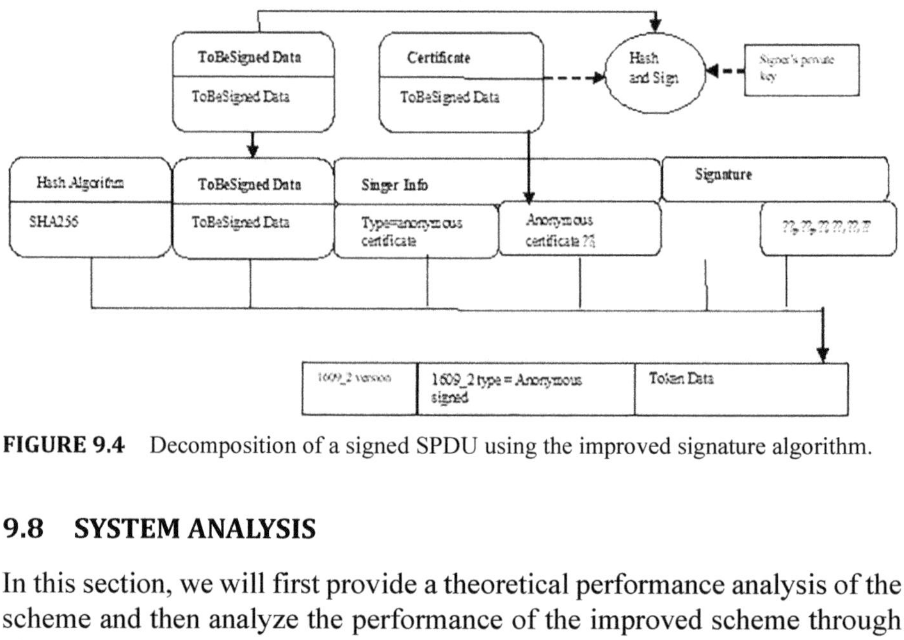

FIGURE 9.4 Decomposition of a signed SPDU using the improved signature algorithm.

9.8 SYSTEM ANALYSIS

In this section, we will first provide a theoretical performance analysis of the scheme and then analyze the performance of the improved scheme through simulation experiments.

9.8.1 THEORETICAL PERFORMANCE ANALYSIS

The performance of the scheme based on the basic cryptographic operations employed. Our analysis includes the following basic cryptographic operations: point multiplication in groups H_1 and H_2, hash operation $ê:H_U \rightarrow H_1$, and bilinear pairing e: $H_1 \times H_2 \rightarrow H_U$. To achieve a security level equivalent to the 1024-bit RSA algorithm, we define bilinear pairing on a super singular elliptic curve $F/G_q:x^2=y^3+y$ with embedding degree 2. The parameters and their lengths involved are as follows:

Length of elements in groups H_1, H_2, and H_U denoted as l_{H1}, l_{H2}, l_{HU}, with a length of 1024 bits.

- Length of elements in A_q denoted as l_{Aq}, with a length of 160 bits.
- Length of the hash function denoted as l_J, with a length of 256 bits.

Table 9.3 provides an explanation of the symbols representing the computation time involved in the system and their theoretical runtime. U_{mm} represents the computation time for multiplication in A^*_r and serves as the basis for our calculations, based on reference.

TABLE 9.3 Explanation of Computation Time Symbols and Theoretical Runtime

Symbol Description Theoretical Running Time	Description	Theoretical Running Time
U_m	H_1, H_2 dot multiplication operation time	29 U_{mm}
U_{mtp} hash function $H_U \rightarrow H_1$ time $29U_{mm}$	hash function GT →G1 time	29 U_{mm}
U_Q	The operation time of bilinear pairing e	87 U_{mm}

9.8.1.1 COMMUNICATION OVERHEAD ANALYSIS

The communication overhead during verification is generally influenced by the message size that is transmitted over the vehicular network. Regarding the scheme for anonymous authentication described in this article, the sender will send a 5696G bits token ($N \in H_1$, $\sigma \in H_1$, $d \in H_1$, $e' \in \{0, 1\}256$, $u \in Ar$, $v \in Ar$, $D_1' \in H_1$, $H_1' \in H_1$) to the verifier in a non-interactive manner for token verification.

9.8.1.2 COMPUTATIONAL OVERHEAD AND AUTHENTICATION DELAY ANALYSIS

The computational overhead in this chapter is represented by the verifier's cost of verifying token validity using the Batch_Verify() algorithm. The computational overhead in Roman et al. [6] is denoted as U_{ver}, and when the number of messages to be verified is n, the computational overhead in the original scheme is nU_{ver}. In the present scheme, the computational overhead for batch verification of n messages is denoted as U_{ber}. The calculations for nU_{ver} and U_{ber} are as follows:

$$nU_{ver} = 3nU_m + 3nU_q$$

$$U_{ber} = 3U_m + (n + 3)U_q = (87n + 174)U_{mm}$$

Based on Rizi et al. [22] and Liyanage et al. [29], this chapter ignores operations with minor effects on computational overhead, such as scalar multiplication based on small exponents and point addition on elliptic curves. According to Marinescu [16] and Ziegler et al. [23], actual simulation results, the impact of U_{mtp} on computational overhead is minimal, and therefore, this operation is omitted when calculating overhead.

The present scheme requires n + 3 bilinear pairings, which is more than the schemes based on bilinear pairings in Table 9.2. However, the required number of point multiplications is 3, which is fewer than the schemes in Table 9.2. In actual experiments, the number of point multiplication operations required for computational overhead is consistently less than in the comparative schemes. Hence, the actual computational overhead depends on the time required for bilinear pairing operations and point multiplication operations.

9.9 SIMULATION EXPERIMENTS AND PERFORMANCE ANALYSIS

In this research, we carried out simulation experiments employing the open-source vehicular network simulation framework known as Veins. This framework operates within an event-based network simulator called OMNeT++ and interfaces with the road traffic simulator SUMO. Various elements within Veins are tasked with configuring, executing, and overseeing the simulation. The Veins simulation platform is built upon the IEEE 802.11p and IEEE 1609.4 DSRC/WAVE network layers, providing a comprehensive model.

In the simulation experiments, we selected an area of 1000m × 1000m around the Northeastern University Hunnan Campus in Shenyang City using OpenStreetMap as the map. Road files were generated using the netconvert script provided by SUMO, vehicle behavior files were generated using the randomTrips script, and terrain files were generated using the polyconvert script. Since the simulation configuration in SUMO GUI can only read files in the sumocfg format, we configured the road file, vehicle behavior file, and terrain file in the sumocfg file. The vehicle behavior file includes the generation points and paths of vehicles, and vehicle generation and movement are based on the randomTrips script.

In the Veins framework, a basic Wave Short Message (WSM) message class is defined, and we placed different data generated by the scheme into corresponding fields, such as using the wsmData field in WSM to store the generated tokens. For signature generation, the simulation framework does not provide specific functions, so we wrote a token generation function based on the token generation algorithm and used the constructor DemoSafetyMessage() of the WSM message class to generate WSM. We placed the generated token into the wsmData field and then used the sendDown() interface function provided by the simulation framework to broadcast WSM to other vehicles. For signature verification, the simulation framework defines a verification function onWSM() for WSM by specifying the IEEE 802.11p DemoBaseApGpLayer class. We inherited the DemoBaseAppLayer class in the simulation implementation and redefined the verification function onWSM() based on the token verification algorithm, enabling OBUs in the network to verify tokens. Using the above methods, we implemented the scheme's application in the WAVE protocol in the simulation. The simulation platform and its parameters are listed in Table 9.4.

TABLE 9.4 Simulation Platform and Parameter Values

Parameter	Value
Hardware platform	CPU: 3.9GHzIntel(R)Core(TM)i5G11600K,8GBRAM
Operating system	Debian9.4
Traffic simulation generator	SUMO1.8.0
Network simulator	OMNET++ 5.6.2
Simulation platform	Veins5.1Gi2
Simulation area size	1000m × 1000m
Data transfer rate/Mbps	6
Transmission power/mW	20
Noise level/dBm	−89
Simulation time/s	500
Number of vehicles	20~80

Figure 9.5 illustrates the comparison of batch verification overhead between this scheme and schemes based on bilinear pairings in Table 9.2 for different numbers of verifications. From Figure 9.5, it can be observed that when the number of signatures to be batch-verified is less than or equal to 2, this scheme's batch verification overhead is superior to those in references [14, 16, 18]. When the number of verifications exceeds 8, this scheme's batch verification overhead outperforms those [11, 13]. When the number of verifications exceeds 11, this scheme's batch verification overhead is better than that in reference [10]. While [14, 16, 18] perform better than other schemes, analysis from Table 91 shows that [14] lacks support for unlink

ability and has lower security compared to other schemes. By calculation, this scheme's computational overhead is 104% higher than that of [14], but this scheme satisfies unlink ability. Therefore, compared to [14], this chapter incurs higher computational overhead in exchange for improved security.

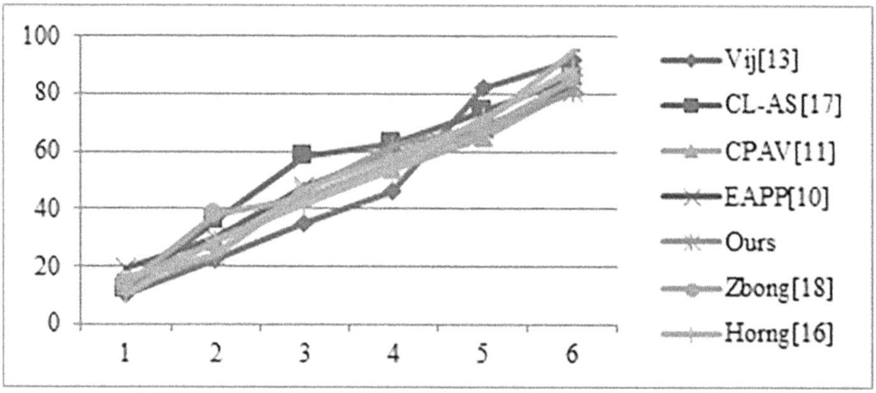

FIGURE 9.5 Comparison of computational overhead for different numbers of verifications.

9.10 APPLICATION SCENARIO ANALYSIS

As described in reference, assuming $U_{auth\text{-}end}$ is the end time for batch verification of n messages, and Ti send is the sending time of message m_i, the average authentication delay is calculated as AgAuthDelay $= 1/n\sum_{i=1}^{n}(U_{auth\text{-}end} - U^i_{send})$. Following reference, the verification period in this experiment ranges from 20 to 90 ms, and the number of vehicles per square kilometer varies from 20 to 80. Figure 9.6 presents the average authentication delay under different verification periods and different numbers of vehicles, along with the minimum authentication delay requirements in two different application scenarios.

9.10.1 DSRC APPLICATION SCENARIO BASED ON BASIC SAFETY MESSAGE (BSM)

A BSM is a secure data exchange used for sharing vehicle status information based on the DSRC standard. Vehicles typically broadcast BSMs every 300 ms [37]. When applying the proposed scheme to the BSM scenario in simulations, it is observed that, with fewer than 40 vehicles per square kilometer, authentication delays meet BSM requirements for verification periods

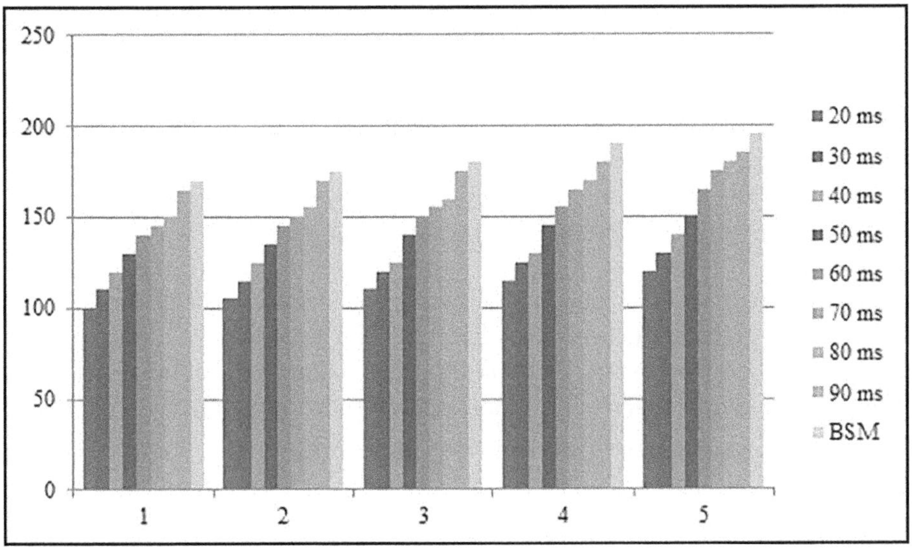

FIGURE 9.6 Average authentication delay under different verification periods.

ranging from 20 to 90 ms. when there are 40 vehicles, and the verification period is less than 70 ms, authentication delays can satisfy BSM requirements. When there are 50 vehicles, and the verification period is less than 50 ms, authentication delays can meet BSM requirements.

9.10.2 NEAR-FIELD PAYMENT APPLICATION FOR VEHICLES

Near-field payment for vehicles refers to using vehicles as payment terminals, indirectly sending character commands to banking and financial institutions through information exchange between vehicles and RSUs, resulting in currency payments and fund transfers, thus enabling in-vehicle payment functionality. This application requires system delays to be below 500 ms. When applying the proposed scheme to the scenario of near-field payment for vehicles in simulations, it is found that, with no more than 80 vehicles per square kilometer and verification periods not exceeding 90 ms, authentication delays meet the requirements of near-field vehicle payments.

9.11 CONCLUSION

This chapter presents a batch verification algorithm that enhances the security of WAVE systems by replacing the ECDSA-based scheme in the WAVE standard. Computational overhead and authentication delay are compared in simulations under different verification periods. The suggested scheme is compared to schemes that are based on bilinear pairings in Table 9.2 [10, 11, 14, 16, 18] and schemes based on elliptic curves [22, 29] for verification overhead. When the number of verifications exceeds 11, the proposed scheme's batch verification overhead outperforms some of the schemes. Additionally, this chapter provides verification periods for the proposed scheme that demonstrate the system's robustness for BSM and near-field vehicle payment scenarios. Although the proposed scheme meets the requirements of the given application scenarios, it still has room for improvement in terms of authentication delay due to its heavy reliance on bilinear pairings. Therefore, future research directions include reducing authentication delay while ensuring security, drawing from other schemes.

KEYWORDS

- **anonymous identity authentication**
- **intelligent transportation**
- **internet of things (IoT)**
- **network security**
- **vehicle-to-vehicle network**
- **zero trust security**

REFERENCES

1. Quamara, S., & Singh, A. K. (2022). A systematic survey on security concerns in cryptocurrencies: State-of-the-art and perspectives. *Computers & Security, 113*, 102548. https://doi.org/10.1016/j.cose.2021.102548.
2. Nasr, T., Torabi, S., Bou-Harb, E., Fachkha, C., & Assi, C. (2022). Power jacking your station: In-depth security analysis of electric vehicle charging station management systems. *Computers & Security, 112*, 102511. https://doi.org/10.1016/j.cose.2021.102511.
3. Mukta, R., Paik, H. Y., Lu, Q., & Kanhere, S. S. (2022). A survey of data minimisation techniques in blockchain-based healthcare. *Computer Networks, 205*, 108766. https://doi.org/10.1016/j.comnet.2022.108766.

4. Werner, J., Westphall, C. M., & Westphall, C. B. (2017). Cloud identity management: A survey on privacy strategies. *Computer Networks, 122*, 29–42. https://doi.org/10.1016/j.comnet.2017.04.030.
5. Kumaresan, M., Gopal, R., Mathivanan, M., & Poongodi, T. (2022). Amalgamation of blockchain, IoT, and 5G to improve security and privacy of smart healthcare systems. In S. Tanwar (Ed.), *Blockchain Applications for Healthcare Informatics* (pp. 283–312). Academic Press. https://doi.org/10.1016/B978-0-323-90615-9.00015-3.
6. Roman, R., Zhou, J., & Lopez, J. (2013). On the features and challenges of security and privacy in distributed Internet of Things. *Computer Networks, 57*(10), 2266–2279. https://doi.org/10.1016/j.comnet.2012.12.018.
7. Roman, R., Lopez, J., & Mambo, M. (2018). Mobile edge computing, Fog et al.: A survey and analysis of security threats and challenges. *Future Generation Computer Systems, 78*(Part 2), 680–698. https://doi.org/10.1016/j.future.2016.11.009.
8. Kalla, A., de Alwis, C., Porambage, P., Gür, G., & Liyanage, M. (2022). A survey on the use of blockchain for future 6G: Technical aspects, use cases, challenges, and research directions. *Journal of Industrial Information Integration, 30*, 100404. https://doi.org/10.1016/j.jii.2022.100404.
9. Can, O. (2022). The security and privacy aspects in semantic web-enabled IoT-based healthcare information systems. In S. Tiwari, F. O. Rodriguez, & M. A. Jabbar (Eds.), *Intelligent Data-Centric Systems: Semantic Models in IoT and eHealth Applications* (pp. 89–116). Academic Press. https://doi.org/10.1016/B978-0-32-391773-5.00011-X.
10. Sahraoui, S., & Bilami, A. (2015). Efficient HIP-based approach to ensure lightweight end-to-end security in the Internet of Things. *Computer Networks, 91*, 26–45. https://doi.org/10.1016/j.comnet.2015.08.002.
11. Rivas, D. A., Barceló-Ordinas, J. M., Guerrero Zapata, M., & Morillo-Pozo, J. D. (2011). Security on VANETs: Privacy, misbehaving nodes, false information, and secure data aggregation. *Journal of Network and Computer Applications, 34*(6), 1942–1955. https://doi.org/10.1016/j.jnca.2011.07.006.
12. Hassan, M. U., Rehmani, M. H., & Chen, J. (2019). Privacy preservation in blockchain-based IoT systems: Integration issues, prospects, challenges, and future research directions. *Future Generation Computer Systems, 97*, 512–529. https://doi.org/10.1016/j.future.2019.02.060.
13. Wu, A. (2023). A comprehensive approach for the evaluation of the impact of blockchain on the photovoltaic supply chain using a hybrid data analytic method. *Optik, 291*, 171361. https://doi.org/10.1016/j.ijleo.2023.171361.
14. Kouicem, D. E., Bouabdallah, A., & Lakhlef, H. (2018). Internet of things security: A top-down survey. *Computer Networks, 141*, 199–221. https://doi.org/10.1016/j.comnet.2018.03.012.
15. Ouaddah, A. (2019). A blockchain-based access control framework for the security and privacy of IoT with strong anonymity, unlinkability, and intractability guarantees. In S. Kim, G. C. Deka, & P. Zhang (Eds.), *Advances in Computers* (Vol. 115, pp. 211–258). Elsevier. https://doi.org/10.1016/bs.adcom.2018.11.001.
16. Marinescu, D. C. (2018). Cloud security. In D. C. Marinescu (Ed.), *Cloud Computing (Second Edition)* (pp. 405–437). Morgan Kaufmann. https://doi.org/10.1016/B978-0-12-812810-7.00015-7.

17. Mathur, S., Kalla, A., Gür, G., Bohra, M. K., & Liyanage, M. (2023). A survey on the role of blockchain for IoT: Applications and technical aspects. *Computer Networks, 227*, 109726. https://doi.org/10.1016/j.comnet.2023.109726.
18. Zhang, P., Schmidt, D. C., White, J., & Dubey, A. (2019). Consensus mechanisms and information security technologies. In S. Kim, G. C. Deka, & P. Zhang (Eds.), *Advances in Computers* (Vol. 115, pp. 181–209). Elsevier. https://doi.org/10.1016/bs.adcom.2019.05.001.
19. Baudier, P., Kondrateva, G., & Ammi, C. (2023). Can blockchain enhance motivation to donate: The moderating impact of religion on donors' behavior in the USA's charity organizations. *Technological Forecasting and Social Change, 191*, 122524. https://doi.org/10.1016/j.techfore.2023.122524.
20. Azad, T. B. (2008). Security guidance for operating systems and terminal services. In T. B. Azad (Ed.), *Securing Citrix Presentation Server in the Enterprise* (pp. 69–156). Syngress. https://doi.org/10.1016/B978-1-59749-281-2.00002-0.
21. Usman, A. B., & Gutierrez, J. (2018). Toward trust-based protocols in a pervasive and mobile computing environment: A survey. *Ad Hoc Networks, 81*, 143–159. https://doi.org/10.1016/j.adhoc.2018.07.009.
22. Rizi, M. H. P., & Seno, S. A. H. (2022). A systematic review of technologies and solutions to improve security and privacy protection of citizens in the smart city. *Internet of Things, 20*, 100584. https://doi.org/10.1016/j.iot.2022.100584.
23. Ziegler, G., Farkas, C., & Lőrincz, A. (2006). A framework for anonymous but accountable self-organizing communities. *Information and Software Technology, 48*(8), 726–744. https://doi.org/10.1016/j.infsof.2005.08.007.
24. Chen, H., Luo, X., Shi, L., Cao, Y., & Zhang, Y. (2023). Security challenges and defense approaches for blockchain-based services from a full-stack architecture perspective. *Blockchain: Research and Applications, 4*(3), 100135. https://doi.org/10.1016/j.bcra.2023.100135.
25. Frimpong, S. A., Han, M., Boahen, E. K., Sosu, R. N. A., Hanson, I., Larbi-Siaw, O., & Senkyire, I. B. (2023). RecGuard: An efficient privacy preservation blockchain-based system for online social network users. *Blockchain: Research and Applications, 4*(1), 100111. https://doi.org/10.1016/j.bcra.2022.100111.
26. Tsao, K. Y., Girdler, T., & Vassilakis, V. G. (2022). A survey of cyber security threats and solutions for UAV communications and flying ad-hoc networks. *Ad Hoc Networks, 133*, 102894. https://doi.org/10.1016/j.adhoc.2022.102894.
27. Haddon, D. A. E. (2021). Zero Trust networks, the concepts, the strategies, and the reality. In H. Jahankhani, L. M. O'Dell, G. Bowen, D. Hagan, & A. Jamal (Eds.), *Strategy, Leadership, and AI in the Cyber Ecosystem* (pp. 195–216). Academic Press. https://doi.org/10.1016/B978-0-12-821442-8.00001-X.
28. Alshahrani, M., & Traore, I. (2019). Secure mutual authentication and automated access control for IoT smart home using cumulative Keyed-hash chain. *Journal of Information Security and Applications, 45*, 156–175. https://doi.org/10.1016/j.jisa.2019.02.003.
29. Liyanage, M., Pham, Q-V., Dev, K., Bhattacharya, S., Maddikunta, R. K. R., Gadekallu, T. R., & Yenduri, G. (2022). A survey on Zero-touch network and Service Management (ZSM) for 5G and beyond networks. *Journal of Network and Computer Applications, 203*, 103362. https://doi.org/10.1016/j.jnca.2022.103362.
30. Walshe, M., Epiphaniou, G., Al-Khateeb, H., Hammoudeh, M., Katos, V., & Dehghantanha, A. (2019). Non-interactive zero knowledge proofs for the authentication

of IoT devices in reduced connectivity environments. *Ad Hoc Networks, 95*, 101988. https://doi.org/10.1016/j.adhoc.2019.101988.

31. AlQadheeb, A., Bhattacharyya, S., & Perl, S. (2022). Enhancing cybersecurity by generating user-specific security policy through the formal modeling of user behavior. *Array, 14*, 100146. https://doi.org/10.1016/j.array.2022.100146.
32. Varma, I. M., & Kumar, N. (2023). A comprehensive survey on SDN and blockchain-based secure vehicular networks. *Vehicular Communications, 44*, 100663. https://doi.org/10.1016/j.vehcom.2023.100663.
33. Ferretti, L., Magnanini, F., Andreolini, M., & Colajanni, M. (2021). Survivable zero trust for cloud computing environments. *Computers & Security, 110*, 102419. https://doi.org/10.1016/j.cose.2021.102419.
34. Das, D., Dasgupta, K., & Biswas, U. (2023). A secure blockchain-enabled vehicle identity management framework for intelligent transportation systems. *Computers and Electrical Engineering, 105*, 108535. https://doi.org/10.1016/j.compeleceng.2022.108535.
35. Andrew, J., Isravel, D. P., Sagayam, K. M., Bhushan, B., Sei, Y., & Eunice, J. (2023). Blockchain for healthcare systems: Architecture, security challenges, trends and future directions. *Journal of Network and Computer Applications, 215*, 103633. https://doi.org/10.1016/j.jnca.2023.103633.
36. Maurya, A. K., Das, A. K., Jamal, S. S., & Giri, D. (2021). A secure user authentication mechanism for IoT-enabled Wireless Sensor Networks based on multiple Bloom filters. *Journal of Systems Architecture, 120*, 102296. https://doi.org/10.1016/j.sysarc.2021.102296.
37. Isaja, M., Nguyen, P., Goknil, A., Sen, S., Husom, E. J., Tverdal, S., Anand, A., Jiang, Y., Pedersen, K. J., Myrseth, P., Stang, J., Niavis, H., Pfeifhofer, S., & Lamplmair, P. (2023). A blockchain-based framework for trusted quality data sharing towards zero-defect manufacturing. *Computers in Industry, 146*, 103853. https://doi.org/10.1016/j.compind.2023.103853.
38. Ferrag, M. A., Maglaras, L., Argyriou, A., Kosmanos, D., & Janicke, H. (2018). Security for 4G and 5G cellular networks: A survey of existing authentication and privacy-preserving schemes. *Journal of Network and Computer Applications, 101*, 55–82. https://doi.org/10.1016/j.jnca.2017.10.017.
39. Yu, Y., Peng, W., & Lu, J. (2022). Wireless network security game based on conditional privacy policy. *Computer Communications, 184*, 96–106. https://doi.org/10.1016/j.comcom.2021.12.011.

CHAPTER 10

Evolutionary Approach for Cybersecurity Situational Assessment in Zero-Trust Networks

ARIJEET CHANDRA SEN*

Joint Director, Government of India, MTech Cyber Security, BITS Pilani, Delhi, India

*Corresponding author

ABSTRACT

Currently, in the cyber safety world, traditional machine learning models suffer from the drawbacks of lengthy training times and high sensitivity to redundant features, making them ill-suited to address the growing complexity of the cyber security landscape, particularly in the context of the Zero Trust network paradigm. To tackle the vast and high-dimensional nature of cyber security elements and enhance the precision and efficiency of cyber security situational assessment within a Zero Trust network framework, a novel approach has been introduced. This approach leverages PRF-RFECV feature selection and GA-Light-BM to develop a cyber security situational assessment model. It begins by determining feature importance through parallel random forests (PRFs) and then selects the optimal feature set using looping aspect elimination with cross-authorization. Subsequently, a genetic algorithm is employed to perform a global search for the best parameters of the Light Gradient Boosting Machine (Light GBM) exemplary for arrangement within the context of Zero Trust network architecture. Experimental simulations illustrate that this model surpasses traditional cyber security situational assessment algorithms in both accuracy and F_1 score, all while maintaining higher efficiency within the Zero Trust network framework.

Zero-Trust Learning: Applications in Modern Network Security. Chetna Kaushal, Amandeep Kaur, Mohit Angurala, Aryan Chaudhary (Eds.)

10.1 INTRODUCTION

As the operational capabilities of network-centric warfare continue to enhance and in-depth research on technologies such as cloud computing, big data, and the Internet of Things (IoT) progresses, today's society places higher demands on the definition and understanding of cybersecurity in cyberspace. Cybersecurity issues have also garnered wider attention [1]. Timely and accurate comprehension of the overall security information of network systems and strengthening vulnerable and susceptible attack points are critical indicators influencing the future development of the Internet. Network security situational assessment, also known as network security situational understanding, holds a significant position throughout the process of perceiving the cybersecurity situation. Its essence lies in gathering security elements within network systems, establishing suitable models based on a constructed security indicator system, and moving away from researching isolated security events. Instead, it comprehensively evaluates the security threats that the entire network system has faced and will face in the future. It dynamically responds to the actual operational status of the network, obtains a macroscopic view of network security situations, assesses the security level of network systems, and assists managers in timely reinforcement and prevention to achieve complex decision-making objectives [2].

Extensive academic research has been devoted to the subject of situational evaluation in network security in recent years. Literature [3] aims to examine the shortcomings of current network security situational evaluation methods, specifically in terms of low efficiency and poor reliability. To solve these difficulties, a novel approach is provided, which involves the utilization of an enhanced BP neural network for quantitative assessment purposes. This method combines the Cuckoo search algorithm aimed at enhancing the rate of convergence and assessment accuracy but has the problem of low assessment stability. To enhance network security situational assessment performance, Literature [4] proposed a network security situational assessment model that combines KNN and Support Vector Machine (SVM) algorithms to solve the problem of SVM misclassification near the hyperplane, reducing SVM's misjudgment rate and improving the accuracy of network security situational assessment. However, it consumes too much time for the calculation of massive multi-dimensional data, sacrificing real-time assessment. Literature [5] shows that in order to overcome deficiencies in accurately assessing network security situations within complicated network information scenarios, a novel model for network security situational awareness was put forth. This model incorporates the element of time and utilizes a compound Convolutional Neural Network (CNN) structure. By utilizing the advantages of these two structures and forming a serial compound unit structure, they significantly reduced network parameters and shortened network runtime [6–9].

Finally, machine learning has emerged as the predominant approach in the domain of network security situation evaluation. However, with the continuous increase in network users and the growing complexity of network topology, network security elements are growing explosively, leading to a significant decrease in model training speed and evaluation accuracy. Therefore, traditional machine learning models can no longer match the high real-time and large accuracy requirements of current network security. To resolve these problems, this article proposes a network security situation assessment framework based on Parallel Random Forest (PRF) and Recursive Feature Elimination with Cross-Validation (RFECV) for feature selection, followed by parameter optimization using Genetic Algorithm (GA), and evaluation using the lightweight Gradient Boosting Machine (LightGBM) model. This framework, denoted as PRF-RFECV-GA-LightGBM, not only effectively mitigates the issue of overfitting but also demonstrates notable advantages in terms of prediction efficacy, training time, and generalization ability. It can be broadly applied to network security condition testing problems.

10.2 NETWORK SECURITY SITUATION ASSESSMENT

10.2.1 DATA FUSION

Data fusion is a multi-level, comprehensive data processing method that runs through the entire network security situation system. Its main purpose is to integrate security elements from multiple data sources, making the results obtained through data analysis more accurate and reliable. According to the level of information abstraction, data fusion can be categorized into three primary tiers: feature-level fusion, data-level fusion, and decision-level fusion. In terms of real-time performance and defect tolerance, decision-level data fusion is considered the most optimal. The KDDCup99 dataset is collected from nine weeks of network connection data from a simulated US Air Force LAN. It contains a total of 41 feature attributes and one label attribute [10–12]. Nine feature attributes are discrete, while the rest are continuous. The label attribute includes one normal type and 22 common network attack types, and the feature attributes consist of both discrete and continuous numerical types. Due to the different measurement methods for continuous feature attributes, to avoid influencing the evaluation results and achieve decision-level fusion, it is necessary to improve training efficiency to increase real-time performance. This chapter not only standardizes and normalizes continuous feature attributes but also extracts four secondary situations from the 22 attack types, as listed in Table 10.1.

TABLE 10.1 Decision-Level Data Fusion

Level 2 Situation	Level 1 Situation
DOS	back, land, neptune, pod, smurf, teardrop
R2L	ftp_write,guess_passwd,imap,multihop,phf,spy,warezclient,warezmaster
U2R	buffer_overflow, loadmodule, perl, rootkit
Probing	ipsweep, nmap, portsweep, satan

Among them, DOS stands for Denial-of-Service attacks, R2L represents unauthorized entry from distant machines, U2R indicates illegal attempts by regular users to access local superuser privileges, and Probing refers to port monitoring and other probing activities.

10.2.2 METRIC CONSTRUCTION

The acquisition of the entire network security situation requires analyzing network threats and attack methods according to network security standards from massive network data and extracting network attack types from network traffic. This chapter combines the Common Vulnerability Scoring System (CVSS) to score attack types in the KDDCup99 dataset. It quantifies the impact of the four major attack types on the entire network system, which are Probing (finding target processes), R2L (intrusions from remote to local), U2R (local privilege escalation), and DOS (resource consumption for Denial-of-Service attacks). These attack types are ranked in order of their impact on network security, from low to high, as listed in Table 10.2.

TABLE 10.2 Network Attack Hazard Classification

Attack Type	Attack Impact
Probing	0.33
R2L	0.44
U2R	0.55
DOS	0.77

As per the information provided in Table 10.2, it is necessary to compute the network security position values for each host inside the network system, taking into account the various types of attacks they have encountered and the corresponding quantities of such attacks. Suppose there are N hosts in the network, with attack types denoted as $i(1\leqslant i\leqslant 4)$ and corresponding impact factors as s_i. If the j-th host receives m_i instances of the ith attack type, then the situation value for the j-th host is calculated as follows (Equations 10.1 and 10.2):

$$M_j = \frac{1}{m_{ij}} \sum_{i=1}^{4} s_i m_{ij} \tag{10.1}$$

Calculate the situation values for each of the N hosts individually, and then obtain the overall network system safety condition value by summing up the weighted values.

$$M = \sum_{j=1}^{N} q_j M_j \tag{10.2}$$

where, q represents the importance level of the jth host within the entire network system, serving as the weight.

The overall network system security situation value calculated using the above formula can be normalized to a specified range, as outlined in the topic of interest that pertains to the development and implementation of a comprehensive National Emergency Response Plan specifically designed to address sudden public events. The process of normalization is employed to categorize the security status of the network system into five distinct categories, as outlined in Table 10.3.

TABLE 10.3 Classification of Network Safety Condition Levels

Network Security Posture Level	Network Safety Condition Value Range
Safety	[0.00,0.31]
Small risk	[0.31,0.61]
Intermediate risk	[0.61,0.91]
Large risk	[0.91,1.21]
Huge risk	[1.21,1.51]

10.2.3 MODEL FRAMEWORK

The GA-G Light-GBM-based network security situational assessment model with PRF-GRFECV feature selection proposed in this chapter comprises three primary components, as shown in Figure 10.1: the Situation Extraction

Module (SEXM), the Situation Assessment Module (SASM), and the Situation Visualization Module (SVIM).

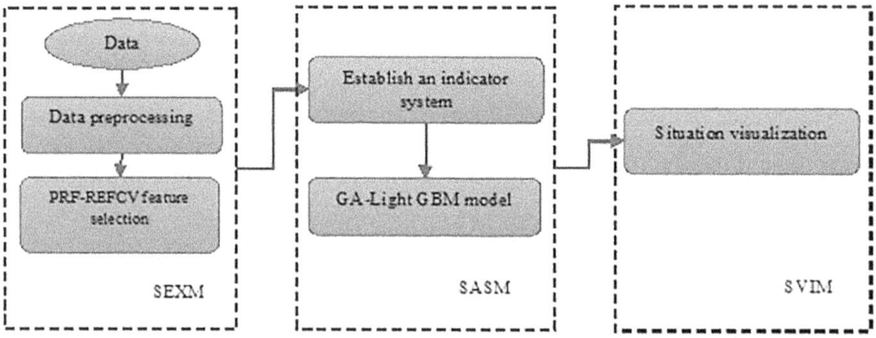

FIGURE 10.1 PRF-RFECV-GA-LightGBM model framework.

10.2.3.1 DATA PROCESSING MODULE

After preprocessing the dataset, including normalization and data conversion, the model performs deep-level data fusion and decision-level fusion. It fuses first-level situations into representative second-level situations such as DOS, R2L, U2L, and Probing, reducing data dimensionality and enhancing the real-time capabilities of the classification model. Subsequently, it constructs multiple decision trees using PRFs to mine potential relationships between features, assess their importance, and finally, with cross-validation, automatically selects the optimal feature subset. This process eliminates less impactful and irrelevant features to avoid computational explosions and low classification accuracy caused by feature redundancy.

10.2.3.2 SITUATION ASSESSMENT MODULE

The module selects an authoritative indicator system, removes factors with minimal impact on the entire network system, and standardizes quantitative indicators to mitigate the effects of multi-indicator comprehensive processing. Given the complexity and diverse meanings of Light-GBM parameters, to prevent relying on subjective experience for parameter tuning and falling into a locally optimal tuning range, a genetic algorithm based on evolutionary theory and population genetics is used to optimize the critical parameters of Light-GBM. This approach leverages global spatial information adaptively to approach the optimal state.

10.2.3.3 *SITUATION VISUALIZATION MODULE*

By synthesizing the information from Tables 10.2 and 10.3, it reflects the macro network security situational value. When combined with management personnel's input, it enables timely security risk warnings and reduces asset losses.

10.3 LIGHTWEIGHT GRADIENT BOOSTING MACHINE

10.3.1 LIGHTGBM ALGORITHM

The LightGBM algorithm, proposed by the author in 2017, is a distributed gradient boosting decision tree (GBDT) algorithm based on the Histogram algorithm [13]. The fundamental concept underlying the Histogram algorithm involves the process of discretizing continuous floating-point feature values into a set of k integers. This discretization is carried out simultaneously with the construction of a histogram, which has a width of k. During the process of traversal, the discretized values are employed as indices to accumulate statistical quantities within the histogram. The determination of the ideal split point is then predicated on the discrete values. The LightGBM algorithm incorporates the Histogram algorithm's concept, resulting in significant reductions in storage space, memory consumption, and computation time. Since GBDT requires multiple passes through the entire dataset during each iteration, conventional GBDT algorithms like XGBoost cannot meet the needs of users when dealing with massive datasets.

LightGBM departs from the level-wise tree growth strategy commonly used in most GBDT algorithms and instead employs the leaf-wise algorithm with depth limitations. Compared to level-wise, leaf-wise provides better accuracy under the same number of splits. Simultaneously, it ensures efficiency and anti-overfitting when constrained by maximum depth.

Assuming a given training dataset $D=\{(x_1,y_1),\cdots,(x_N,y_N)\}$, the boosting tree model can be represented as an additive model using decision trees as base learners [9] (Equations 10.3–10.6).

$$F(x)=\sum_{t=1}^{T}\alpha_t f_t(x) \tag{10.3}$$

where $f_t(x)$ represents a regression tree, T is the number of trees that need to be built in the gradient boosting decision tree, and α is the corresponding weight for the tth tree.

The objective of the LightGBM algorithm is to minimize an objective function.

$$L^{(t)} = \arg min \sum_{i=1}^{N} L(y_i, \hat{y}^{(t)}) + \Omega(f_t)$$
$$= \arg min \sum_{i=1}^{N} L(y_i, \hat{y}^{(t-1)} + f_t(x_i) + \Omega(f_t) \quad (10.4)$$

where L represents the loss incurred when building the t-th tree based on the true values and predicted values, and Ω is the corresponding regularization term. According to LightGBM, the regularization term is chosen as follows:

$$\Omega(f_t) = \gamma^T + \frac{1}{2} \| w \|^2 \quad (10.5)$$

where w is a vector composed of the predicted values at the leaf nodes, and γ and λ (both greater than or equal to 0) are hyperparameters used to control their weights. During the model learning process, the following regularization strategy is employed to mitigate overfitting:

$$F_t(x) = F_{t-1}(x) + v f_t(x) \quad (10.6)$$

where $0 < v \leq 1$ is the learning rate used to update the model. A smaller learning rate requires generating more regression trees, which results in higher accuracy but also increases training time. Therefore, you can control parameters like γ, λ, v, and so forth, to build a fast and accurate LightGBM model.

It uses a leaf-wise growth algorithm with depth constraints to avoid unnecessary computation costs and prevent overfitting. Through optimized feature parallelism and data parallelism, it accelerates computation, improving evaluation efficiency and accuracy. This makes the model well-suited for network security situational assessment, which often involves assessing the efficiency of network security elements due to the massive amount of data.

10.3.2 FEATURE IMPORTANCE ASSESSMENT

Due to the high dimensionality of features in today's network traffic, some features have little impact on network security situational assessment results and may even introduce noise, potentially disrupting experimental accuracy. Moreover, high-dimensional features are more likely to lead to overfitting, and a large amount of data can increase model complexity, ultimately reducing evaluation model accuracy. Therefore, feature selection from the original feature set in the dataset is necessary.

Random forest algorithms can calculate feature importance based on the Gini index, discover mutual relationships between features, and optimize downstream algorithm performance [14, 15].

Let N be the total number of samples, and there are B features $X_1, X_2, X_3, ..., X_B$. Feature importance is assessed based on the Gini index, and the formula for calculating the Gini index is as follows (Equations 10.7–10.11):

$$GI_m(p) = \sum_{k=1}^{K} p_{mk}(1 - p_{mk}) = 1 - \sum_{k=1}^{K} p_{mk}^2 \tag{10.7}$$

In the given context, the variable k denotes the overall count of categories, while p_mk indicates the sample weight of class k in node m.

The significance of feature X_j at node m, denoted by the difference in the Gini index before and after the split at node m, can be computed using the following formula:

$$VIM_{jm}^{gini} = GI_m - GI_l - GI_r \tag{10.8}$$

where Gl_l and Gl_r represent the Gini indices of the two new nodes created after branching to the left and right, respectively.

If feature X_j appears in nodes within the set M in decision tree i, then the importance of feature X_j in the i-th decision tree is calculated as follows:

$$VIM_{ij}^{gini} = \sum_{(m \in M)} VIM_{jm}^{gini} \tag{10.9}$$

Assuming there are a total of n decision trees in the random forest, the feature importance of feature X_j is calculated as follows:

$$VIM_j^{gini} = \sum_{i=1}^{n} VIM_j^{gini} \tag{10.10}$$

In conclusion, each of the acquired importance scores is normalized.

$$\mathrm{VIM}_j = \frac{\mathrm{VIM}_j}{\sum_{i=1}^{B} VIM_i} \tag{10.11}$$

Feature importance is the sum of importance scores for a feature across all decision trees. In theory, a random forest can achieve strong resistance to overfitting and high accuracy by constructing a sufficient number of decision trees. However, building many trees can lead to inefficiency. PRF [16–19] leverages characteristics such as the ability to calculate each decision tree in a random forest independently and constructs a PRF. This allows multiple decision trees to be computed in parallel, avoiding the problem of low training efficiency caused by constructing many decision trees. The principle involves the "master" distributing the complete dataset into several data chunks along with corresponding executable programs to each process. After parallel computation, the master collects each result and organizes them into the desired format, as illustrated in Figure 10.2.

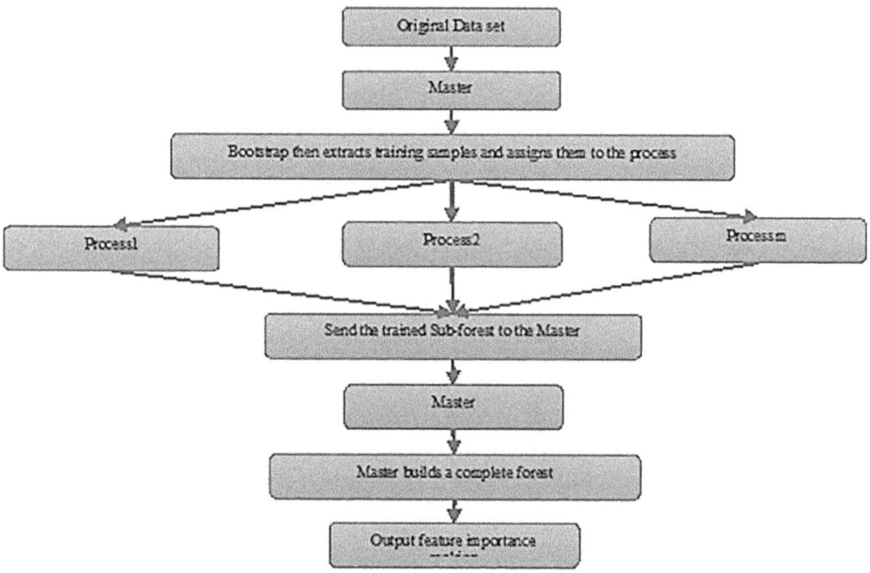

FIGURE 10.2 Example of parallel random forest workflow.

10.3.3 FEATURE SUBSET SELECTION

The recursive feature elimination (RFE) method can be described as iteratively training a base model, recording the loss value in each iteration, and eliminating the least important features. Then, the next round of training is conducted based on the new feature set until only one feature remains, which represents the elimination order. RFECV builds upon RFE by performing cross-validation on different feature combinations in parallel and calculating the validation error for all subsets using the base model. It selects the subset with the least error rate as the optimal feature subset [20–23]. The algorithm is as follows: first, it applies a PRF algorithm to the preprocessed dataset to construct multiple decision trees and calculates the importance of each feature based on the Gini index, sorting them in descending order. Then, it sequentially removes one or more features with the lowest importance until the number of features in the subset is one, recording the feature subset after each iteration and using it as the new feature subset. Finally, it performs a 10-fold cross-validation on all feature subsets based on the base model. The optimal feature set is determined by selecting the feature subset with the highest score.

Algorithm 1: PRF-RFECV

Input: *D, C*, step

Output: *F*

1. if $|C| > 1$ do:
2. Feature_list = PRF(*D*)
3. sub = Feature_list – Feature[–step:]
4. C_list, appent(sub)
5. Endif
6. For each sub in C_list:
7. rfc = RandomForesClassifier()
8. scores = cross_val_score(rfc, *D*, sub, CV = 10)
9. End for
10. F = argmax(scores).

10.3.4 GENETIC OPTIMIZATION

Genetic algorithms are algorithms that draw inspiration from natural selection and genetic mechanisms in the biological world. They possess characteristics such as high parallelism, randomness, and adaptive search. In the context of the situational assessment module, to enhance the precision and generalization ability of the model while avoiding overfitting, genetic algorithms are employed to optimize parameters related to LightGBM.

First, the potential solutions to the problem are encoded as floating-point numbers based on Equations 10.5 and 10.6, and the mapping relationship is as follows $(v_1, v_2, \ldots, v_m)^{\mathrm{T}}$ (Equations 10.12–10.16):

$$\boldsymbol{X} = (x_1, x_2, \ldots, x_n)^{\mathrm{T}} \Rightarrow V = (v_1, v_2, \ldots, v_m)^{\mathrm{T}} \tag{10.12}$$

where X represents the phenotype representing the decision vector, V represents the genotype representing the chromosome, vi represents one gene within the chromosome, m equals the number of decision variables n, and the maximum evolutionary generation is set, initializing the first generation population as P(0).

In order to assess the quality of prediction outcomes, the fitness function chosen is Mean Absolute Percentage Error (MAPE). The formula for calculating MAPE is as follows (Equations 10.13 and 10.14):

$$M(y) = \frac{1}{n}\sum_{i=1}^{n} \frac{|y_i - \hat{y}_i|}{y_i} \tag{10.13}$$

Where y_i is the true value, $\hat{Y}_i$ is the predicted value, and to satisfy requirements such as non-negativity, continuity, and maximization, the fitness function can be transformed into:

$$F(y) = \frac{1}{1 + M(y)} \tag{10.14}$$

To avoid falling into local optima, a widely applicable and efficient tournament selection algorithm [13] is chosen. The best individuals are repeatedly selected into the offspring population based on their fitness, forming the new generation.

The Simulated Binary Crossover is selected as the crossover operator, and boundary mutation serves as the mutation operator. In order to mitigate the risk of premature convergence during the genetic process resulting from excessively low probabilities, the crossover and mutation probabilities are represented as Pc and Pm, respectively. These probabilities are constrained within the intervals [Pcmin, Pcmax], and [Pmmin, Pmmax], where Pcmin = 0, Pcmax = 0.9, Pmmin = 0.01, and Pmmax = 0.1.

Assuming the average fitness of all individuals in the population is f_avg, the fitness after crossover and mutation are denoted as f' and f, respectively, yielding [14] (Equations 10.15 and 10.16):

$$P_c = \begin{cases} P_{cmax} - \dfrac{(P_{cmax} - P_{cmin})(f_{max} - f')}{f_{max} - f_{avg}}, & f' \geq f_{avg} \\ P_{cmax}, & f' < f_{avg} \end{cases} \tag{10.15}$$

$$P_m = \begin{cases} P_{mmax} - \dfrac{(P_{mmax} - P_{m\,min})(f_{max} - f')}{f_{max} - f_{avg}}, & f' \geq f_{avg} \\ P_{mmax}, & f' < f_{avg} \end{cases} \tag{10.16}$$

10.4 EXPERIMENT AND SIMULATION

10.4.1 EVALUATION METRICS

In cybersecurity situational awareness, there is often an issue of uneven data distribution among security elements. Therefore, to accurately reflect the true performance of the constructed model, appropriate evaluation metrics should be selected. The confusion matrix is commonly used to assess relevant metrics of network security models [24, 25], as listed in Table 10.4.

TABLE 10.4 Confusion Matrix

True Category	Prediction Category	
	Positive	**Negative**
Positive	True Positive	False Negative
Negative	False Positive	True Negative

A true positive (TP) is defined as the accurate classification of a positive sample as positive. A true negative (TN) is the accurate classification of a negative sample as negative. A false positive (FP) transpires when an incorrectly classified negative sample is identified as positive. A false negative (FN) transpires when an accurately positive sample is incorrectly categorized as negative. To accurately and effectively evaluate the model's performance in a real environment, the following metrics are introduced to assess its effectiveness. Precision represents the proportion of correctly detected positive samples to true positive samples and is calculated using the following formula (Equations 10.17–10.20):

$$Precision = \frac{\sum_{i=1}^{N} TP_i}{\sum_{i=1}^{N} (TP_i + FP)} \tag{10.17}$$

Recall denotes the ratio of negative samples that were accurately identified to the total number of negative samples. The formula for calculation is:

$$\text{Recall} = \frac{\sum_{i=1}^{N} TP_i}{\sum_{i=1}^{N} (TP_i + FN_i)} \tag{10.18}$$

Accuracy is quantified as the ratio of negative and positive samples that were identified with precision out of the overall sample count. The formula for calculation is:

$$\text{Accuracy} = \frac{\sum_{i=1}^{N} TP_i + TN_i}{\sum_{i=1}^{N} (TP_i + TN_i + FP_i + FN_i)} \tag{10.19}$$

The F1 score is a metric often employed in the evaluation of classification models to provide a thorough assessment of their performance. It is calculated as the harmonic mean of precision and recall. The formula used for calculation is as follows:

$$F1 - \text{score} = \frac{2 \times \text{Precision} \times \text{Recall}}{\text{Precision} + \text{Recall}} \tag{10.20}$$

where N is the number of categories. Due to imbalanced sample distribution, as seen in real network traffic data where normal samples are abundant and abnormal samples are scarce, to avoid the influence of common categories, and to treat each category equally, the above metrics are evaluated using macro-averaging.

10.4.2 EXPERIMENTAL ENVIRONMENT AND DATA

To accurately test the precision and efficiency of the PRF, GRFECV, GA, and Light-GBM algorithms, the experimental environment is set up with the Ubuntu operating system, Python 3.8 programming language, and Anaconda 3 toolkit. Detailed hardware specifications include an Intel(R) Core(TM) i7-10750H processor, an RTX 2060 graphics card, 4 cores of CPU, 64GB of main memory, and 16GB of internal memory.

The dataset employed in this research is the KDD Cup 99 dataset [26], comprising network connection data gathered over a nine-week duration and obtained from a simulated U.S. Air Force LAN. A subset of unlabeled test data and a collection of labeled training data comprise the dataset. The test dataset contains attack types absent from the training dataset, resulting in a divergence of probability distributions between the two. Consequently, the test dataset is more suitable for experimental purposes in network threat detection models. The investigation utilizes a dataset comprising 494,010 entries from the KDD Cup 99. Among these, 396,734 are attack flow samples and 97,276 are normal flow samples. The ratio of training to test samples is 3:1. The distribution of data is detailed in Table 10.5.

TABLE 10.5 Data Distribution of Training Set and Test Set

Attack Type	Training Samples	Test Sample	Training Set Proportion (%)	Test Set Proportion (%)
Normal	97,279	60,594	20.69	20.48
DOS	391,459	229,854	80.24	74.9
U2R	53	229	0.02	0.08
R2L	1127	16,190	0.24	5.21
Probing	4108	4167	0.84	1.35

10.4.3 FEATURE SELECTION

To eliminate irrelevant features, avoid overfitting, and enhance evaluation accuracy, the dataset is normalized and transformed. PRF and RFE with cross-validation (RFECV) are used to assess feature importance and construct optimal feature subsets. LightGBM is employed to evaluate accuracy and efficiency before and after feature selection. The specific training results of the LightGBM model based on PRF and RFECV feature selection on the KDD Cup 99 training dataset are listed in Table 10.6.

TABLE 10.6 Experimental Comparison of Different Optimal Feature Numbers

Number	Accuracy	Precision	Recall	F1	Time/s
28	0.99999	0.97001	0.97266	0.97133	13
36	0.98688	0.93539	0.96478	0.94986	15
42	0.99714	0.96438	0.96441	0.96439	17

From Table 10.6, it is evident that when only normalizing the data without feature selection, that is, with a total of 41 features, the accuracy evaluated by the LightGBM model is only 0.99694, and the training and evaluation time reaches 16 seconds. When the number of selected features exceeds 27 and reaches 35 based on feature importance, all evaluation metrics decline. According to the experiment, training on the optimal feature subset (27 features) obtained through the PRF and RFECV algorithms for automatic feature selection not only improves the accuracy from 0.99694 to 0.99979 but also outperforms all evaluation metrics compared to the full feature evaluation. This indicates that the algorithm is effective and feasible, and the use of parallel computing significantly leverages the available computing resources, greatly improving the model's training speed.

10.4.4 RESULTS COMPARISON

10.4.4.1 EVALUATION PERFORMANCE COMPARISON OF ALGORITHM OPTIMIZATION

To analyze the impact of PRF, RFECV, and GA on the evaluation performance of the LightGBM model, seven sets of training samples from Monday to Sunday are used for training LightGBM and PRF, RFECV, GA, and LightGBM models for network security situation assessment. Based on the evaluation results from seven sets of test datasets and actual assessment results, stable training models and their related parameters are provided. The accuracy and F1 score values for each of the seven test datasets are presented, along with a two-level situation trend chart within a week, aiding network administrators in timely reinforcement. Specific training results are listed in Table 10.7.

TABLE 10.7 Comparison of Simulation Experiments Between LightGBM and GA-LightGBM

Date	Light GBM		GA-Light GBM	
	Efficiency	F1-Tally	Efficiency	F1-Tally
Monday	0.9967	0.7624	0.9994	0.9691
Tuesday	0.9964	0.8065	0.9995	0.9785
Wednesday	0.9944	0.864	0.9995	0.9865
Thursday	0.9991	0.786	0.9999	0.9823
Friday	0.9978	0.7829	0.9998	0.9534
Saturday	0.9901	0.7933	0.9999	0.9668
Sunday	0.9995	0.7891	0.9998	0.9568

From Table 10.6, it can be observed that among the seven sets of experimental data, the GA-GA-LightGBM algorithm achieves higher network security situation assessment accuracy and F1 scores compared to the LightGBM algorithm alone. The GA-GA-LightGBM algorithm achieves an assessment accuracy of above 0.999 and F1 scores above 0.95 for each group. The F1 values for LightGBM in Table 6 indicate that due to the scarcity of abnormal samples in the test set, the trained LightGBM model is not sensitive to abnormal samples, resulting in F1 scores below 0.9 for each group. In contrast, the GA-GA-LightGBM model achieves F1 scores above 0.95 for all datasets, demonstrating higher precision and recall, and overall better model quality.

Next, a situation trend chart is constructed based on the condition value indicators to compare the network safety condition values computed by GA-GA-LightGBM with the actual values within a week. The visual results are shown in Figure 10.3.

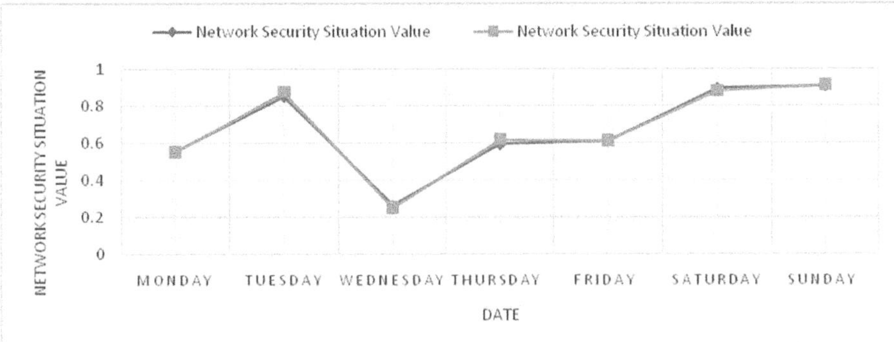

FIGURE 10.3 Comparison of network security situation assessment values.

According to the findings presented in Figure 10.3, the network security situation assessment values obtained through the employment of GA-GA-Light GBM exhibit a strong alignment with the corresponding actual values throughout the span of a week, with only a minor deviation on Saturday. When the network security situation value exceeds 0.6, it indicates the need to initiate network security risk warnings, and when it exceeds 0.9, network security control measures should be activated. Visualizing network security situation values helps network security managers grasp the overall network security risk situation, enabling them to promptly issue warning and control information and reinforce security systems.

10.4.4.2 COMPARISON OF EVALUATION PERFORMANCE OF DIFFERENT ALGORITHMS

In order to enhance the credibility of the PRF-GA-RFECV-LightGBM model for assessing network security situations, comparative experiments are carried out on the KDD Cup 99 dataset utilizing the Bayesian Network (BN) [27], KNN [28], SVM [29], and BP neural network [30]. The results presented in Table 10.8 indicate that the PRF-GA-RFECV-LightGBM algorithm exhibits superior performance in terms of assessment accuracy across all categories when compared to alternative algorithms. With an assessment accuracy of 91.56%, even the "U2R" category, comprising a mere 0.07% of the total samples, surpasses the performance of the KNN, SVM, and BP algorithms by a significant margin.

TABLE 10.8 Comparison of Evaluation Accuracy of 5 Algorithms in Each Category

Category	Proportion	Accuracy				
		KNN	SVM	BN	BP	GA-Light GBM
Normal	19.49	85.7	95.46	96.35	99.52	99.96
DOS	73.91	86.73	96.89	94.68	99.86	99.98
U2R	0.08	80.87	9.66	93.22	60.39	91.57
R2L	5.21	79.84	6.17	92.18	80.5	99.76
Probing	1.35	83.24	78.55	94.63	90.64	99.02

Additionally, due to the LightGBM algorithm's use of optimization algorithms such as Histogram, Goss, and EFB, it significantly reduces computational overhead. Training LightGBM on the KDD Cup 99 dataset with 494,010 data points in a known experimental environment takes only 16 seconds, whereas training with the PRF-GA-RFECV-LightGBM model requires up to 300 seconds. Moreover, LightGBM can be used with parallel and distributed computing, while SVM and KNN algorithms become increasingly challenging to train with data volumes in the hundreds of thousands. Therefore, the final experimental results demonstrate that the GA-GA-LightGBM model based on RF-GA-RFECV feature selection outperforms other evaluation models and is more suitable for network security situation assessment scenarios with massive and high-dimensional data.

10.5 CONCLUSION

This chapter employed the PRF-GA-RFECV algorithm to select the optimal feature set and used the GA algorithm to optimize the crucial parameters of LightGBM. Through experimentation, it was demonstrated that feature selection with the PRF-GA-RFECV algorithm, which removes redundant features, can improve assessment accuracy and efficiency. Additionally, optimizing LightGBM's important parameters with the GA algorithm enhances the accuracy and stability of model evaluation. In contrast to prevalent algorithms utilized for assessing network situations, the PRF-GA-RFECV-LightGBM model, which is introduced in this manuscript, demonstrates exceptional efficacy and precision in its evaluations. However, the introduction of the GA algorithm resulted in a significant decrease in model evaluation performance. Therefore, the next step will involve optimizing the GA algorithm or using better parameter optimization techniques to enhance the overall evaluation performance of network security situation awareness, enabling real-time dynamic monitoring of the actual state of network security.

KEYWORDS

- **cyber security**
- **evolutionary algorithm**
- **feature selection**
- **genetic algorithm**
- **LightGBM algorithm**
- **recursive feature elimination**
- **situation assessment module**
- **zero trust network**

REFERENCES

1. Cao, Y., Li, S., Lv, C., Wang, D., Sun, H., Jiang, J., Meng, F., Xu, L., & Cheng, X. (2023). Towards cyber security for low-carbon transportation: Overview, challenges, and future directions. *Renewable and Sustainable Energy Reviews, 183*, 113401. https://doi.org/10.1016/j.rser.2023.113401.
2. Ayodeji, A., Mohamed, M., Li, L., Di Buono, A., Pierce, I., & Ahmed, H. (2023). Cyber security in the nuclear industry: A closer look at digital control systems, networks and human factors. *Progress in Nuclear Energy, 161*, 104738. https://doi.org/10.1016/j.pnucene.2023.104738.
3. Syed, N. F., Shah, S. W., Shaghaghi, A., Anwar, A., Baig, Z., & Doss, R. (2022). Zero Trust Architecture (ZTA): A comprehensive survey. *IEEE Access, 10*, 57143–57179. https://doi.org/10.1109/ACCESS.2022.3174679.
4. Chen, B., et al. (2021). A security awareness and protection system for 5G smart healthcare based on zero-trust architecture. *IEEE Internet of Things Journal, 8*(13), 10248–10263. https://doi.org/10.1109/JIOT.2020.3041042.
5. Lin, C., Luo, M., Huang, X., Choo, K. K. R., & He, D. (2022). An efficient privacy-preserving credit score system based on noninteractive zero-knowledge proof. *IEEE Systems Journal, 16*(1), 1592–1601. https://doi.org/10.1109/JSYST.2020.3045076.

6. Zhu, X., He, D., Bao, Z., Luo, M., & Peng, C. (2023). An efficient decentralized identity management system based on range proof for social networks. *IEEE Open Journal of the Computer Society, 4*, 84–96. https://doi.org/10.1109/OJCS.2023.3258188.
7. Wang, Y., Li, J., Zhao, S., & Yu, F. (2020). Hybridchain: A novel architecture for confidentiality-preserving and performant permissioned blockchain using trusted execution environment. *IEEE Access, 8*, 190652–190662. https://doi.org/10.1109/ACCESS.2020.3031889.
8. Feng, W., Yan, Z., Yang, L. T., & Zheng, Q. (2022). Anonymous authentication on trust in blockchain-based mobile crowdsourcing. *IEEE Internet of Things Journal, 9*(16), 14185–14202. https://doi.org/10.1109/JIOT.2020.3018878.
9. Szymanski, T. H. (2022). The "cyber security via determinism" paradigm for a quantum safe zero trust deterministic Internet of Things (IoT). *IEEE Access, 10*, 45893–45930. https://doi.org/10.1109/ACCESS.2022.3169137.
10. Song, W., et al. (2020). Android data-clone attack via operating system customization. *IEEE Access, 8*, 199733–199746. https://doi.org/10.1109/ACCESS.2020.3035089.
11. Sun, X., Yu, F. R., & Zhang, P. (2022). A survey on cyber-security of connected and autonomous vehicles (CAVs). *IEEE Transactions on Intelligent Transportation Systems, 23*(7), 6240–6259. https://doi.org/10.1109/TITS.2021.3085297.
12. Bhutta, M. N. M., et al. (2021). A survey on blockchain technology: Evolution, architecture and security. *IEEE Access, 9*, 61048–61073. https://doi.org/10.1109/ACCESS.2021.3072849.
13. Ray, P. P. (2022). Web3: A comprehensive review on background, technologies, applications, zero-trust architectures, challenges and future directions. *Internet of Things and Cyber-Physical Systems, 3*, 213–248. https://doi.org/10.1016/j.iotcps.2023.05.003.
14. Kowalski, M., & Mazurczyk, W. (2023). Toward the mutual routing security in wide area networks: A scoping review of current threats and countermeasures. *Computer Networks, 230*, 109778. https://doi.org/10.1016/j.comnet.2023.109778.
15. Liyanage, M., et al. (2022). A survey on Zero touch network and service management (ZSM) for 5G and beyond networks. *Journal of Network and Computer Applications, 203*, 103362. https://doi.org/10.1016/j.jnca.2022.103362.
16. Walshe, M., et al. (2019). Non-interactive zero knowledge proofs for the authentication of IoT devices in reduced connectivity environments. *Ad Hoc Networks, 95*, 101988. https://doi.org/10.1016/j.adhoc.2019.101988.
17. AlQadheeb, A., et al. (2022). Enhancing cybersecurity by generating user-specific security policy through the formal modeling of user behavior. *Array, 14*, 100146. https://doi.org/10.1016/j.array.2022.100146.
18. Varma, I. M., & Kumar, N. (2023). A comprehensive survey on SDN and blockchain-based secure vehicular networks. *Vehicular Communications, 44*, 100663. https://doi.org/10.1016/j.vehcom.2023.100663.
19. Heiding, F., Katsikeas, S., & Lagerström, R. (2023). Research communities in cyber security vulnerability assessments: A comprehensive literature review. *Computer Science Review, 48*, 100551. https://doi.org/10.1016/j.cosrev.2023.100551.
20. Govindarajan, U. H., Singh, D. K., & Gohel, H. A. (2023). Forecasting cyber security threats landscape and associated technical trends in telehealth using bidirectional encoder representations from transformers (BERT). *Computers & Security, 133*, 103404. https://doi.org/10.1016/j.cose.2023.103404.

21. Hasan, M. K., et al. (2022). Review on cyber-physical and cyber-security system in smart grid: Standards, protocols, constraints, and recommendations. *Journal of Network and Computer Applications, 209*, 103540. https://doi.org/10.1016/j.jnca.2022.103540.
22. Safaei Pour, M., et al. (2023). A comprehensive survey of recent internet measurement techniques for cyber security. *Computers & Security, 128*, 103123. https://doi.org/10.1016/j.cose.2023.103123.
23. Kampourakis, V., Gkioulos, V., & Katsikas, S. (2023). A systematic literature review on wireless security testbeds in the cyber-physical realm. *Computers & Security, 133*, 103383. https://doi.org/10.1016/j.cose.2023.103383.
24. Ferretti, L., Magnanini, F., Andreolini, M., & Colajanni, M. (2021). Survivable zero trust for cloud computing environments. *Computers & Security, 110*, 102419. https://doi.org/10.1016/j.cose.2021.102419.
25. Das, D., Dasgupta, K., & Biswas, U. (2023). A secure blockchain-enabled vehicle identity management framework for intelligent transportation systems. *Computers and Electrical Engineering, 105*, 108535. https://doi.org/10.1016/j.compeleceng.2022.108535.
26. J, A., Isravel, D. P., Sagayam, K. M., Bhushan, B., Sei, Y., & Eunice, J. (2023). Blockchain for healthcare systems: Architecture, security challenges, trends and future directions. *Journal of Network and Computer Applications, 215*, 103633. https://doi.org/10.1016/j.jnca.2023.103633.
27. Maurya, A. K., Das, A. K., Jamal, S. S., & Giri, D. (2021). A secure user authentication mechanism for IoT-enabled wireless sensor networks based on multiple Bloom filters. *Journal of Systems Architecture, 120*, 102296. https://doi.org/10.1016/j.sysarc.2021.102296.
28. Isaja, M., Nguyen, P., Goknil, A., Sen, S., Husom, E. J., Tverdal, S., Anand, A., Jiang, Y., Pedersen, K. J., Myrseth, P., Stang, J., Niavis, H., Pfeifhofer, S., & Lamplmair, P. (2023). A blockchain-based framework for trusted quality data sharing towards zero-defect manufacturing. *Computers in Industry, 146*, 103853. https://doi.org/10.1016/j.compind.2023.103853.
29. Ferrag, M. A., Maglaras, L., Argyriou, A., Kosmanos, D., & Janicke, H. (2018). Security for 4G and 5G cellular networks: A survey of existing authentication and privacy-preserving schemes. *Journal of Network and Computer Applications, 101*, 55–82. https://doi.org/10.1016/j.jnca.2017.10.017.
30. Yu, Y., Peng, W., & Lu, J. (2022). Wireless network security game based on conditional privacy policy. *Computer Communications, 184*, 96–106. https://doi.org/10.1016/j.comcom.2021.12.011.

CHAPTER 11

Leveraging AI and Zero-Trust Principles for Improved Security

GUNSEERAT KAUR*

School of Computer Science and Engineering, Lovely Professional University, Punjab, India

*Corresponding author

ABSTRACT

Security in computing systems holds ample significance; new methods are devised to provide better protection against pertinent risks and threats. The zero-trust principle is a noteworthy strategy for providing security. Tactically, zero-trust architecture accommodates protective measures for securing hardware components and software. Importantly, the security boundary should include devices, data, identities, network environments, and application workloads. The benefit of this technique is that the level of granular approach towards security allows identities to inherit access instead of trusting users. Since artificial intelligence (AI)-based technologies and algorithms are providing solutions in various fields, they are diligently experimented with for the provisioning of security and privacy in computing systems. Zero-trust security as a practice requires strong scrutiny while implementing robust objectives. It is a carefully highlighted set of practices that allows for a comprehensive approach to impenetrable security perimeters. The execution of such a framework with advanced techniques from AI can create better monitoring, authenticating, and verifying mechanisms. AI techniques are particularly well-established in controlled outputs and improvements from past memory. Hence, they provide a strong basis for the possibility of improving zero-trust principles. This chapter mainly focuses on highlighting practices adopted from AI to enhance zero-trust principles.

Zero-Trust Learning: Applications in Modern Network Security. Chetna Kaushal, Amandeep Kaur, Mohit Angurala, Aryan Chaudhary (Eds.)

11.1 INTRODUCTION

Security holds different meanings for different individuals, apparently, practices of ensuring security also differ from individual to individual. Defining a secure system is theoretically easy, but hard to maintain practically. Growing threat maps and increasing risks have raised requirements for better mitigation practices. Provisioning security at different stages in digital devices is a complex task [1]. There are layered approaches to provide practices against risks and threats. The term security covers different aspects in different digital scenarios, Figure 11.1 describes the set of areas where security can be provided.

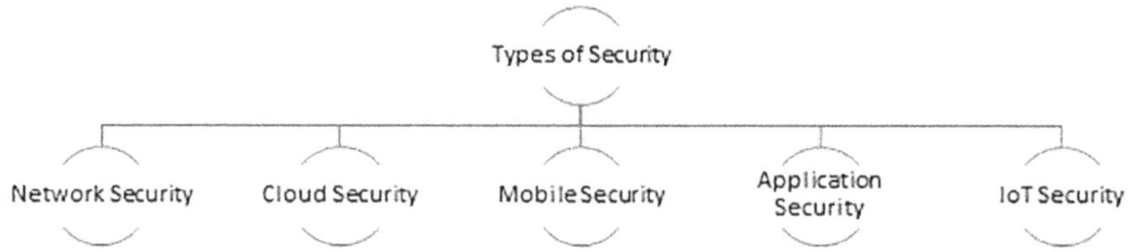

FIGURE 11.1 Areas of security provisioning.

Source: Adapted from Ref. [3, 5]

Network security is a very critical aspect of security, as most attacks occur through networks. Solutions and mitigation methods are designed with extra attention paid to network protocols, data, and access controls, enforcing stricter policies. Cloud security is a strategy to improve security solutions for cloud-based implementations. As a vast number of applications are hosted on the cloud or use cloud-based resources, it is essential to protect against different attacks and threats [2]. Hand-held devices often store sensitive information about their users, requiring security against threats to confidentiality, integrity, and availability. These devices are more susceptible to app-based attacks, phishing, and social engineering. An increasing number of smart devices has also created multiple loopholes for attacking users, exposing vulnerabilities in IoT devices [4]. The data in these devices can be classified as sensitive in various contexts; therefore, tighter security protocols are required in these edge solutions. As security is not easy to implement, previous studies and research have suggested security models. These models provide a framework or blueprint to establish security methods for devices, users, and mediums of communication between them [3]. The strategies include norms against unauthorized access, disclosure of sensitive information, disruption of services, modification of messages, or destruction of information. The aim is to preserve CIA goals. Figure 11.2 shows the different policies on which these models are based.

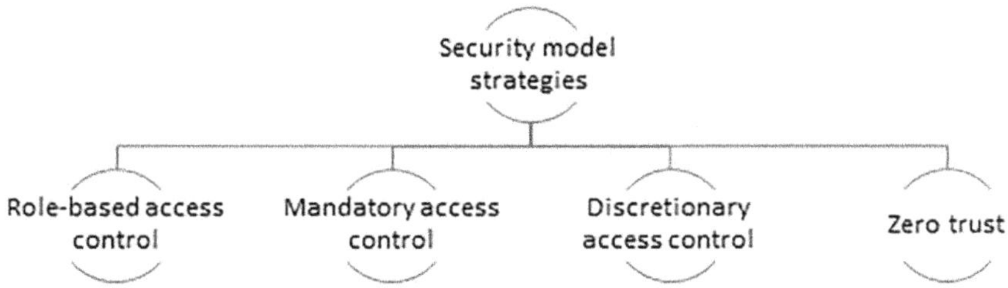

FIGURE 11.2 Strategies in a security model.

Source: Adapted from Ref. [5, 6]

Role-based access control (RBAC) provides fine-grained modular control over roles and permissions to users. As a practice in this strategy, it is not recommended to provide permissions for each user separately, such a process will be tedious and require complete updates on whether the person has been promoted to a more permission role or has left the position altogether [6]. Therefore, role-based access needs to give permissions to roles and not users. This provides a better method to classify, grant, and revoke permissions in time of need. Role assignments are less error-prone and manageable, Figure 11.3 shows an arrangement in role-based access.

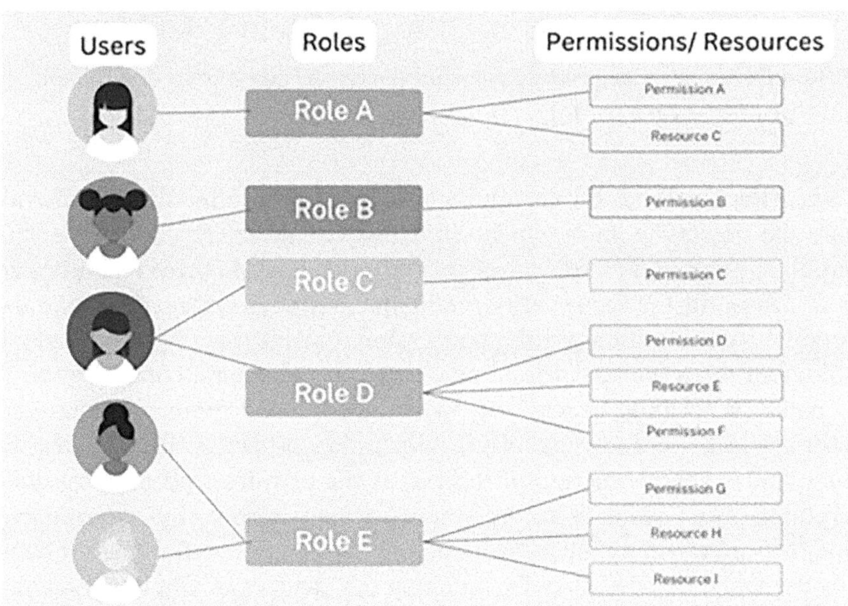

FIGURE 11.3 Arrangement of roles, permissions, and responsibilities.

Source: Adapted from Ref. [7, 8]

Mandatory access control is performed based on the classification of data in relation to the organization. The classification involves data labeled as public, internal, sensitive, and highly sensitive [6, 9]. Post-labeling, the extent of data sharing is a further step to determine which roles or individuals can access what data. This is an approach to information clearance available to different individuals; this strategy is used for organizations that believe in data classification with respect to objects or targets [5, 8]. Discretionary access control is rather different, as it allows access to certain types of users for a certain resource. This method does not create any roles and gives direct permissions to users on a discretionary basis. Figure 11.4 shows how mandatory access control and discretionary access control differ in implementation.

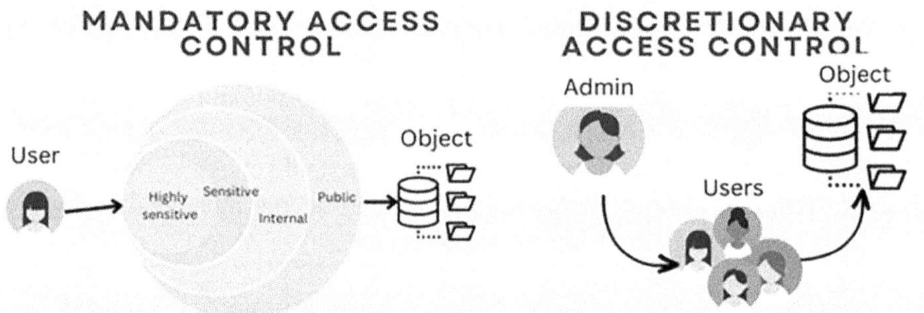

FIGURE 11.4 Security models based on discretionary and mandatory access control.
Source: Adapted from Ref. [2, 10]

The above methods provide granular control over different roles, permissions, and objects in an organization. However, there are very few discussions about the types of devices or people connecting to these resources and objects. Zero trust is the security model that assumes that no device or user should be trusted with any role, permission, or object [1, 8]. It extends the ideology of the principle of least privilege rule. The main consideration for this policy is to grant access only when needed and remove it afterward. Zero trust principles are useful in maintaining proper security goals. The main quality is the reduction of threats, as the zero-trust principle assumes everything to be hostile; therefore, it reduces many risks by considering them malicious and avoiding them. Secondly, it consistently performs verification of its users and devices, thereby reducing the chances of any impersonation [1]. To implement such strong initiatives, it is equally important to align security architecture with the consideration of the zero-trust principle. Manually performing actions leads to chances of error, thereby defeating the purpose of revoking access to objects [8]. In the automation of processing zero trust, AI is deployed. This chapter discusses how AI can help in securing the landscape with zero-trust principles while using AI-based tools. This chapter is divided into five major sections: the introduction to the zero-trust architecture and principle takes precedence in section one, followed by how researchers have devised AI-based techniques in security through a review of literature. In the third section, ZTA is highlighted with best practices in different aspects. Section four will highlight comparisons of results and analysis of practices deployed. Finally, section five consists of a conclusion.

11.2 LITERATURE REVIEW

The zero-trust principle directly and implicitly disallows any associations with networks, users, or objects. The term was coined in 1994 by Stephen Paul, who has worked with artificial agents to identify trusted devices and their traffic to understand how trusting a device could lead to issues in the long run [11]. Zero trust principles are further used for different applications, implemented through norms of micro-segmentation, identity access management, and periodic monitoring. NIST has provided norms for implementing this approach for data and service protection; however, this can help with other aspects. Zero trust policies can be applied to a setup of microservices, as service workloads are exchanged over data centers and cross-service dependencies arise. Zaheer et al. [10] have demonstrated the implementation of the zero-trust principle to control the perimeter by assigning control to workloads instead of services and reducing the threat map. The implementation of eZtrust also showed two to five times less latency when used over microservice architecture. Cloud-based services are often at risk due to high data exchange within the network, and relevant layers are involved. Mehraj and Banday [12] have discussed customization and aggregation of trust information while evaluating the system. They have listed seven layers of the trust engine to identify and map data and determine access control and zero-trust perimeter. Application security is provided for cloud-based applications, and regular checks for zero trust are conducted through automation and orchestration. Campbell [2] has shown similar regulations and the usage of AI to identify and verify devices momentarily. An extended version of zero trust has been discussed which highlights policies applicable for supply chain and third-party partners or any other regulatory platforms. Teerakanok et al. [27] have shown the usage of zero trust principles along with blockchain to secure the medical records of patients. The method involves dual protection while sharing medical records on the sender and receiver's side. At the sender's side, double verifications are performed before sending data, which include logging in with a password, health check for the device, and encryption over records. On the side of the receiver, log in with a password and two-factor authentication of the receiver, followed by decryption of records shared by the sender. Along with the verification, blockchain will be used to share proof of work (PoW) for every node. Another case study is discussed by Collier and Sarkis [14], who elaborated usage of zero-trust principles while ensuring security in the supply chain. The four main issues identified in supply chain management are protecting resources, authenticating, and authorizing assets and users, securing communications, and setting up a perimeter. Dhar and Bose [15] have highlighted methodology to implement zero trust for IoT-based devices where sensors and actuators are implicitly trusted with data transfer, they have used blockchain to secure nodes, and IoT devices to connect to an extended trust network in managing IoT better. Ahmed et al. [4] have established that sensitive information should be protected in a zero-trust environment, their fundamental idea is to control access with a proxy network to verify the sources continuously. They have analyzed data according to application, service, and user type to apply zero-trust policies. In this book [16], Zero-trust security has been categorized and explained for multiple case theories, mainly for network security policies they have listed architectures to protect firewalls, security operations, and cloud-based infrastructure using the zero-trust model. The fundamental part of establishing zero trust is to verify with more enhanced methods involving AI-based systems. Chen et al. [17] has provided a protection system for 5G-based smart healthcare; their four-dimensional model takes a security awareness approach. The four dimensions include subject, object, environment, and behavior. Regulation of identity for access control, the least privilege rule for resource allocation, and continuous evaluation of trust are essential. The system also uses AI-based behavior anomaly detection to verify the subjects and objects. Shah et al. [18] has discussed a lightweight continuous device-to-device authentication system that can be implemented to always verify the nodes. Their system sends a secret key and identifiers initially; edge devices are then allowed to perform either device-to-device or device-to-gateway communication with initial mutual authentication and continuous authentication while the protocol is used. Neale et al. [8] has discussed the implications of the zero-trust model for conducting digital forensics; their findings include the formation of zero trust to ensure tamper-free proofs. Verification methods can include techniques like artifact verification and tool verification. Olson and Keller [3] have discussed the implementation of proper network orchestration while exchanging data with third-party applications, their proposed model uses operation, administration, and maintenance-based protocols along with network security protocols to establish zero trust security in orchestration. Wang et al. [6] discussed the approach of ZTA in the power grid with a secure access method created with fingerprint biometrics and identification verification through the public-key algorithm, continuous verification is implemented to curb flood attacks and code attacks. Anderson et al. [5] have discussed the trust model for connected and autonomous vehicles that uses a distinguished approach that enables a zero-trust controller that will enforce policy to an incoming agent through sensors. The context manager will verify and allow or disallow requests. They have experimented with their model using a controller area network for internal vehicles within the trust boundary. Researchers have applied zero-trust principles in establishing trust boundaries across various requirements. They suggest that these principles implicitly verify every node within communication. However, whether it is zero trust or zero trust extended, stricter norms are needed to verify the devices to ensure proper security.

11.3 METHODOLOGY

Security architecture in the zero-trust model holds a layered approach. The architecture's layers allow implementation in a precise manner, and the focus remains on the verification techniques. Specifically, zero-trust can be established during multi-factor authentication (MFA), where the device or application asks the user to verify with one or two more methods of authentication of an individual's identity. Apart from MFA, it is also used in conditional access to an application or user. The conditions can include time-based or role-based access. Micro-segmentation of the network can also allow isolation of components, and modular forms of communication can be managed well. Finally, it is important to verify with continuous assessments. Zero-trust is a complex form of application that ensures powerful security. As the applications of AI are growing, there is also a rise in ensuring zero-trust architecture using AI-based tools and techniques. This section will discuss how AI-based tools can collaborate to ensure zero-trust architectures. To explore zero-trust architecture, Figure 11.5 shows a defined set of active components in any system. These components have responsibilities depending upon the task assigned; the following section explains the responsibility of all these components.

11.3.1 COMPONENTS OF ZERO-TRUST ARCHITECTURE

11.3.1.1 DATA

Data for an organization or an individual is an important entity, therefore, the classification of data is a must. While classifying data it is utterly important to keep sensitive information separate from public information. It is equally important to label and encrypt the classified data. The focus while handling data should be on restriction and resource authorization. Liu et al. [7] have shown methods of using neural networks to automatically classify sensitive information from public information to preserve rouge access to sensitive internal data from external users. Like this, Wahab et al. [19] has shown the usage of federated learning in the multi-level classification of data to process information and grant access accordingly.

11.3.1.2 USERS

Users are an integral part of any organization and can connect or communicate in multiple ways. The users might use an authenticated application on their smart devices or connect from an internet browser. The security of these methods is provided through user authentication, authorization after correct authentication, and verification when sensitive data is accessed. Sharma et al. [20] has discussed the usage of machine learning in access control over 5G-driven IoT devices. Khilar et al. [21] has used the K-nearest neighbor and decision tree algorithms to classify access control for objects based on features like user behaviors, bogus requests, and unauthorized requests. Their study shows the implementation of access control with machine learning (Figure 11.5).

11.3.1.3 DEVICES

Devices are an integral part of any architecture; usage of personal devices for official work has also soared, due to more remote work allocation. Trust factor is tough to establish in such scenarios, Syed et al. [26] have shown a five-step method to establish trust factor. Upon access request, all basic information regarding the requester such as system details, system patches, and applications are verified. User identification, attributes, and privileges are verified in the next stage. Location-based access control is also

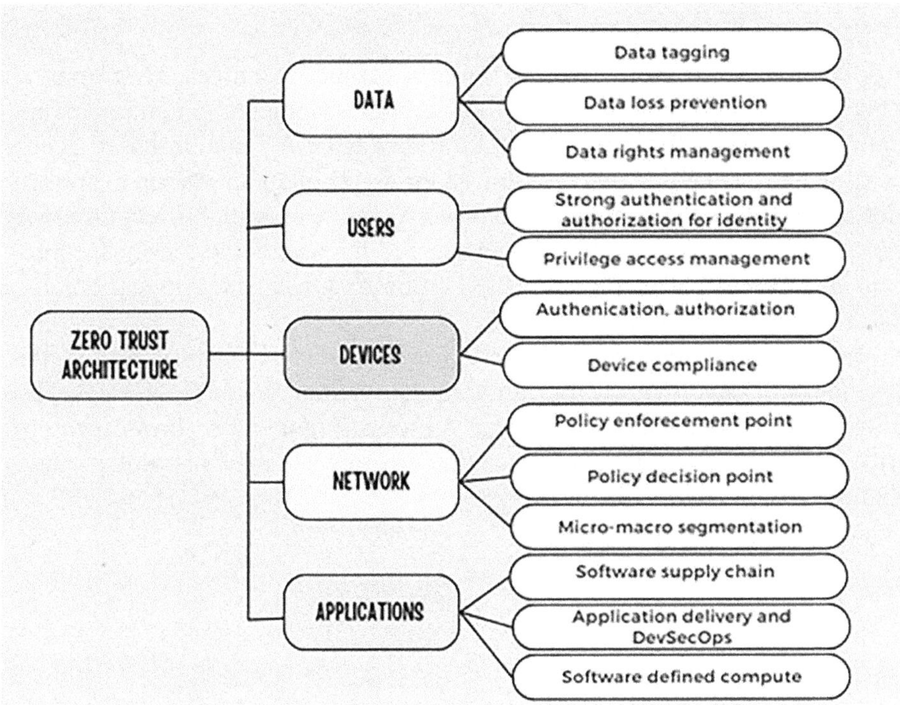

FIGURE 11.5 Components of a zero-trust architecture.

Source: Adapted from Ref. [22]

established, and MFA is used to re-verify the identity in separate intervals [11]. Threat intelligence is established in devices to perform context-based malware detection.

11.3.1.4 NETWORK

Network-based attacks account for major security issues [23] and have shown automation for networks to preserve the principles of zero-trust. They have implemented authentication on appliances at the first packet exchange and transport access control. They have used RESTful APIs to generate a token that will establish identity and will be controlled during access management. Another layer of implementation includes a dynamic firewall to control the trusted network [22]. It is important to have trusted parties verified to maintain the trust principle. A demilitarized zone can be created, isolating internal and external network subnets.

11.3.1.5 APPLICATIONS

Applications need more scrutiny while accessing resources. Hosney et al. [24] has explored the approach of using micro-segmentation and software-defined networks for applications to access data. A function-based access control approach is discussed to restrict the rights of authorization to applications; this allows the application to re-verify while making different requests for resources. Yang et al. [25] has discussed the usage of Security Information and Event Management (SIEM) orchestration to run applications in a controlled and secure manner.

In reviewing the literature and previous studies, it is expected that providing a zero-trust model requires automation in multiple ways. The automation can be established using AI tools. Figure 11.6 shows the zero-trust model in conjunction with machine learning and deep learning methods that can help in easing the task.

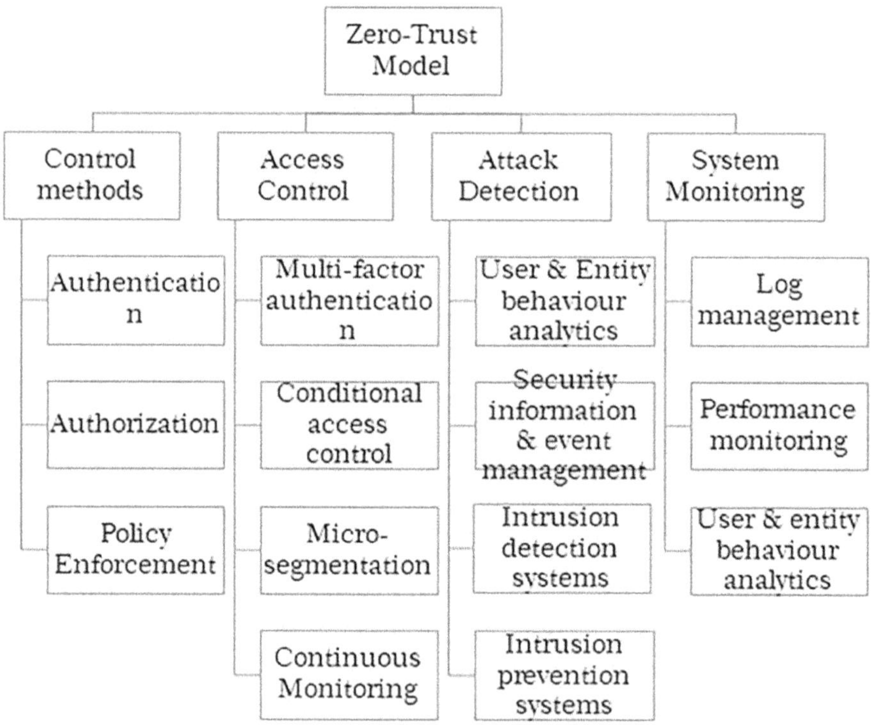

FIGURE 11.6 Zero-trust model.

Source: Adapted from Ref. [5, 26, 27]

While considering the zero-trust model, there are multiple methods of using AI-based tools to help in applying the policies. Upon evaluation, four main aspects of the trust model come to light, namely the control plane, which is the decision-making aspect of the zero-trust model [28]. The main function is to assess the request and decide, based on given security policies and norms, whether to deny or allow access. The method of access control critically allows the implementation of methods to control access for users, applications, data, and devices. It generally monitors access maps for on-premises or off-premises resources to improve security posture [21]. Increasing attacks through networks have compelled better attack detection techniques. The attacks need to be detected and mitigated; if mitigation is not critically handled, it makes it difficult to understand the extent of the attack and the affected users, data, or components of a system.

AI-based techniques for attack detection have been gaining confidence with increasing accuracy, requiring less human intervention. According to the zero-trust policy, no device can be trusted [7]. Continuous monitoring and evaluation of risks require tools based on user behavior, information management, intrusion detection, and prevention systems. Information on users, such as their location, device, and application versions, allows for the detection of any suspicious behavior. Intrusion detection systems (IDSs) mainly use a mixture of anomaly-based and signature-based methods to detect any unusual activity in the network or among users [29]. This also accounts for a similar approach to system monitoring, where logs are maintained, checked, and scrutinized for issues relevant to any congestion zones or different patterns. System monitoring is advantageous as it provides improved visibility of the overall system. The risks are reduced, allowing for improved compliance with proper regulations and checks maintained for the standards [16]. These steps allow for the creation of a more robust space for implementing the zero-trust model. Table 11.1 discusses, in brief, the techniques in AI that are used in zero-trust architecture. AI-based techniques are categorized into two broad categories: statistical and symbolic. Symbolic AI is based on logic and rules, an earlier technique that has been superseded by statistical analysis. Statistical analysis is based on knowledge gained from data from large datasets. The methodology of AI includes techniques like machine learning, deep learning, natural language processing, computer vision, and robotics [30] (Table 11.1).

TABLE 11.1 AI-Techniques Used in Zero-Trust Architecture

Zero-Trust Technology	AI-Based Mechanism Used	Scope	Objective	Authors
Trust evaluation	Reinforcement learning, federated learning, Deep learning	Contractual trust building, policy granting	Classification	[6, 12, 13]
Trust algorithm	Federated learning	Trust and policy decision point	Classification	[12]
Access control	Support Vector Machine, Recurrent Neural Network,	User authentication, Device authentication	Classification	[4, 32]
Authorization	Deep learning, Support Vector Machine, Convolutional neural network	User authorization, Device authorization	Authorization	[27, 33]
Biometric detection	Convolutional neural network, Computer Vision	Facial recognition, Finger print recognition	Classification	[9]
User identification	Convolutional neural network, Deep learning, Decision Trees, Computer Vision	User behavior analytics, Finger gesture, Human activity	Authentication	[3, 12, 22]
Threat intelligence	Reinforcement learning, Recurrent neural network, Convolutional neural network	Device login and authentication	Classification	[1, 5, 6]
Log analysis	K-nearest neighbors, Deep learning, Decision trees	Feature engineering, traffic grouping	Classification	[34]

11.4 CONCLUSION

AI-based techniques help create ease of scrutinizing users, devices, networks, and data in terms of security. When establishing security, it is important to keep track of devices connecting to networks, data exchanges, and users entering or exiting the network. To secure these exchanges, it is important to implement strong rules. The zero-trust principle is an ideology to implement this very approach, and the zero-trust model is adopted over this approach. The zero-trust model takes single-handed action based on "never trust, always verify." To establish this verification, continuous monitoring is needed, which requires sophisticated tools and critical analysis. AI-based tools learn from large datasets to understand patterns, recognize logic, and follow rules and algorithms with accuracy and precision. Leveraging AI for applying zero-trust principles for provisioning security is a better approach.

KEYWORDS

- **AI-based tools**
- **artificial intelligence**
- **cloud security**
- **machine learning**
- **security**
- **zero trust principle**
- **zero-trust security**

REFERENCES

1. Rose, S., Borchert, O., Mitchell, S., & Connelly, S. (2020). Zero trust architecture. https://doi.org/10.6028/NIST.SP.800-207.
2. Campbell, M. (2020). Beyond zero trust: Trust is a vulnerability. *Computer, 53*(10), 110–113. https://doi.org/10.1109/MC.2020.3011081.
3. Olson, K., & Keller, E. (2021). Federating trust: Network orchestration for cross-boundary zero trust. *Proceedings of the 2021 SIGCOMM 2021 Poster and Demo Sessions, Part of SIGCOMM 2021*, 48–49. https://doi.org/10.1145/3472716.3472865.
4. Ahmed, I., Nahar, T., Urmi, S. S., & Taher, K. A. (2020). Protection of sensitive data in zero trust model. *ACM International Conference Proceeding Series*. https://doi.org/10.1145/3377049.3377114.
5. Anderson, J., Huang, Q., Cheng, L., & Hu, H. (2023). A Zero Trust Architecture for Connected and Autonomous Vehicles. *IEEE Internet Computing*. https://doi.org/10.1109/MIC.2023.3304893.
6. Wang, Z. hua, Jin, M. hui, Jiang, L., Feng, C. jia, Cao, J. yi, & Yun, Z. (2023). Secure access method of power internet of things based on zero trust architecture. *Lecture Notes in Computer Science (Including Subseries Lecture Notes in Artificial Intelligence and Lecture Notes in Bioinformatics), 13969 LNCS*, 386–399. https://doi.org/10.1007/978-3-031-36625-3_31/COVER.
7. Liu, B., Ding, M., Shaham, S., Rahayu, W., Farokhi, F., & Lin, Z. (2021). When machine learning meets privacy. *ACM Computing Surveys (CSUR), 54*(2). https://doi.org/10.1145/3436755.
8. Neale, C., Kennedy, I., Price, B., Yu, Y., & Nuseibeh, B. (2022). The case for zero trust digital forensics. *Forensic Science International: Digital Investigation, 40*, 301352. https://doi.org/10.1016/J.FSIDI.2022.301352.
9. Martin, E., & Tao, X. (2006). Inferring access-control policy properties via machine learning. *Proceedings–Seventh IEEE International Workshop on Policies for Distributed Systems and Networks, Policy 2006*, 2006, 235–238. https://doi.org/10.1109/POLICY.2006.19.

10. Zaheer, Z., Chang, H., Mukherjee, S., & Van Der Merwe, J. (2019). EZTrust: Network-independent zero-trust perimeterization for microservices. *SOSR 2019–Proceedings of the 2019 ACM Symposium on SDN Research,* 49–61. https://doi.org/10.1145/3314148.3314349.
11. Marsh, S. P. (1994). Formalising trust as a computational concept. http://dspace.stir.ac.uk/handle/1893/2010 (accessed on 11 November 2024).
12. Mehraj, S., & Banday, M. T. (2020a). Establishing a zero trust strategy in cloud computing environment. *2020 International Conference on Computer Communication and Informatics, ICCCI 2020*. https://doi.org/10.1109/ICCCI48352.2020.9104214.
13. Sultana, M., Hossain, A., Laila, F., Taher, K. A., & Islam, M. N. (2020). Towards developing a secure medical image sharing system based on zero trust principles and blockchain technology. *BMC Medical Informatics and Decision Making, 20*(1), 1–10. https://doi.org/10.1186/S12911-020-01275-Y/TABLES/2.
14. Collier, Z. A., & Sarkis, J. (2021). The zero trust supply chain: Managing supply chain risk in the absence of trust. *International Journal of Production Research, 59*(11), 3430–3445. https://doi.org/10.1080/00207543.2021.1884311.
15. Dhar, S., & Bose, I. (2021). Securing IoT devices using zero trust and blockchain. *Journal of Organizational Computing and Electronic Commerce, 31*(1), 18–34. https://doi.org/10.1080/10919392.2020.1831870.
16. Garbis, J., & Chapman, J. W. (2021). *Zero Trust Security: An Enterprise Guide*. https://doi.org/10.1007/978-1-4842-6702-8/COVER.
17. Chen, B., Qiao, S., Zhao, J., Liu, D., Shi, X., Lyu, M., Chen, H., Lu, H., & Zhai, Y. (2021). A security awareness and protection system for 5G smart healthcare based on zero-trust architecture. *IEEE Internet of Things Journal, 8*(13), 10248–10263. https://doi.org/10.1109/JIOT.2020.3041042.
18. Shah, S. W., Syed, N. F., Shaghaghi, A., Anwar, A., Baig, Z., & Doss, R. (2021). LCDA: Lightweight continuous device-to-device authentication for a zero trust architecture (ZTA). *Computers & Security, 108*, 102351. https://doi.org/10.1016/J.COSE.2021.102351.
19. Wahab, O. A., Mourad, A., Otrok, H., & Taleb, T. (2021). Federated machine learning: Survey, multi-level classification, desirable criteria and future directions in communication and networking systems. *IEEE Communications Surveys and Tutorials, 23*(2), 1342–1397. https://doi.org/10.1109/COMST.2021.3058573.
20. Sharma, P., Jain, S., Gupta, S., & Chamola, V. (2021). Role of machine learning and deep learning in securing 5G-driven industrial IoT applications. *Ad Hoc Networks, 123*, 102685. https://doi.org/10.1016/J.ADHOC.2021.102685.
21. Khilar, P. M., Chaudhari, V., & Swain, R. R. (2019*). Trust-Based Access Control in Cloud Computing Using Machine Learning,* 55–79. https://doi.org/10.1007/978-3-030-03359-0_3.
22. Yan, X., & Wang, H. (2020). Survey on zero-trust network security. *Communications in Computer and Information Science, 1252 CCIS*, 50–60. https://doi.org/10.1007/978-981-15-8083-3_5/COVER.
23. Eidle, D., Ni, S. Y., Decusatis, C., & Sager, A. (2017). Autonomic security for zero trust networks. *2017 IEEE 8th Annual Ubiquitous Computing, Electronics and Mobile Communication Conference, UEMCON 2017*, 2018-January, 288–293. https://doi.org/10.1109/UEMCON.2017.8249053.

24. Hosney, E. S., Halim, I. T. A., & Yousef, A. H. (2022). An artificial intelligence approach for deploying zero trust architecture (ZTA). *5th International Conference on Computing and Informatics, ICCI 2022*, 343–350. https://doi.org/10.1109/ICCI54321.2022.9756117.
25. Yang, J., Zhang, J., & Wang, H. (2021). Urban traffic control in software defined internet of things via a multi-agent deep reinforcement learning approach. *IEEE Transactions on Intelligent Transportation Systems, 22*(6), 3742–3754. https://doi.org/10.1109/TITS.2020.3023788.
26. Syed, N. F., Shah, S. W., Shaghaghi, A., Anwar, A., Baig, Z., & Doss, R. (2022). Zero trust architecture (ZTA): A comprehensive survey. *IEEE Access, 10*, 57143–57179. https://doi.org/10.1109/ACCESS.2022.3174679.
27. Teerakanok, S., Uehara, T., & Inomata, A. (2021). Migrating to zero trust architecture: Reviews and challenges. *Security and Communication Networks, 2021*. https://doi.org/10.1155/2021/9947347.
28. Cao, Y., Pokhrel, S. R., Zhu, Y., Ram Mohan Doss, R., & Li, G. (2022). Automation and orchestration of zero trust architecture: Potential solutions and challenges. https://doi.org/10.36227/TECHRXIV.21385929.V1.
29. Li, S., Iqbal, M., & Saxena, N. (2022). Future industry Internet of Things with zero-trust security. *Information Systems Frontiers*, 1–14. https://doi.org/10.1007/S10796-021-10199-5/FIGURES/6.
30. Liu, H., & Lang, B. (2019). Machine learning and deep learning methods for intrusion detection systems: A survey. *Applied Sciences, 9*(20), 4396. https://doi.org/10.3390/APP9204396.
31. Bandara, E., Liang, X., Shetty, S., Mukkamala, R., Rahman, A., & Keong, N. W. (2022). Skunk – A blockchain and zero trust security enabled federated learning platform for 5G/6G network slicing. *Annual IEEE Communications Society Conference on Sensor, Mesh and Ad Hoc Communications and Networks Workshops, 2022*, 109–117. https://doi.org/10.1109/SECON55815.2022.9918536.
32. Song, H. (2022). Research on security technology requirement for zero trust security product. *Smart Innovation, Systems and Technologies, 258*, 161–165. https://doi.org/10.1007/978-981-16-5168-7_21/COVER.
33. Dimitrakos, T., Dilshener, T., Kravtsov, A., La Marra, A., Martinelli, F., Rizos, A., Rosett, A., & Saracino, A. (2020). Trust aware continuous authorization for zero trust in consumer Internet of Things. *Proceedings–2020 IEEE 19th International Conference on Trust, Security and Privacy in Computing and Communications, TrustCom 2020*, 1801–1812. https://doi.org/10.1109/TRUSTCOM50675.2020.00247.
34. Andrews, D. K., Agrawal, R. K., Matthews, S. J., & Mentis, A. S. (2019). Comparing machine learning techniques for Zeek log analysis. *2019 IEEE MIT Undergraduate Research Technology Conference, URTC 2019*. https://doi.org/10.1109/URTC49097.2019.9660501.

CHAPTER 12

Machine Learning-Based Zero-Trust Framework for Advanced Network Security

VIKRAM SINGH[1,*] and SANJAY TYAGI[2,*]

[1]*Computer Science and Engineering, Chaudhary Devi Lal University, Sirsa, Haryana, India*

[2]*Department of CSA, Kurukshetra University, Kurukshetra, Haryana, India*

*Corresponding author

ABSTRACT

In an era marked by the evolving nature of cyber threats and the increasing reliance on digital technologies, protecting network security is more crucial than ever. The complex and changing threat landscape might be too much for traditional security measures to handle. Machine learning techniques are now widely used for network traffic classification and filtering. On another networking front, zero-trust modeling has been presented as a high-level security and protection technique that does not place any inherent trust in the devices or users requesting access to a network resource. It is important to look into the possibilities for using machine learning and zero trust to address network security issues and concerns. The security architecture that develops will be capable of anticipatorily identifying and mitigating risks while guaranteeing that access controls are continuously enhanced in light of shifting risk considerations. The chapter's objective is to thoroughly explain to readers how machine learning enhances security in a dynamic threat environment. This chapter will cover a wide range of aspects of machine learning integration, including continuous authentication, threat detection, and risk

Zero-Trust Learning: Applications in Modern Network Security. Chetna Kaushal, Amandeep Kaur, Mohit Angurala, Aryan Chaudhary (Eds.)

assessment, to mention a few. We'll go over the benefits and challenges of this integration, as well as actual use cases and implementation advice.

12.1 INTRODUCTION

With cyber threats increasing in frequency, sophistication, and impact in today's digitally connected world, network security is a crucial concern [1]. To protect against emerging threats, traditional security methods, which mainly rely on perimeter defenses and assume confidence within the network, are rapidly proving insufficient [2]. Due to this insufficiency, security techniques have undergone a paradigm shift that emphasizes the "never trust, always verify" tenet, sometimes known as the zero-trust framework [3].

By promoting constant verification of every user and device attempting to access the network, regardless of their location or level of assumed confidence, the zero-trust framework opposes the conventional security model [4]. Recognizing that dangers may originate from within the network and calling for a re-evaluation of trust presumptions, this proactive approach to warding against prospective risks has gained popularity [2].

To strengthen network security against a variety of cyberattacks, this chapter investigates the integration of machine learning capabilities within the zero-trust architecture [5]. The field of artificial intelligence's (AIs) machine learning provides strong tools and methodologies to identify patterns, abnormalities, and potential dangers [6]. By integrating learning algorithms and approaches within the zero-trust framework, a more dynamic and efficient security paradigm can be achieved [2].

The potential for machine learning and the zero-trust architecture to work together is enormous. Machine learning algorithms are able to identify trends in massive datasets and anticipate security threats, allowing for proactive defense against cyberattacks [7]. Anomaly detection powered by machine learning also improves the zero-trust model's capacity to quickly adjust to changing threat environments, thereby lowering risks [4].

The integration of machine learning capabilities into the zero-trust security architecture will be thoroughly discussed, as well as how it will alter security [8]. It examines the guiding concepts of zero trust and clarifies how machine learning may support them. A proactive, adaptable security approach that surpasses the rigidity of traditional security measures is ensured by the incorporation of machine learning-powered threat detection [1].

The zero-trust framework's fundamental principles will be covered in detail in the parts that follow, with an emphasis on how machine learning can improve each principle [9]. Various machine learning algorithms, methods, and applications that support the zero-trust security model will also be covered [1]. Some of the difficulties encountered in integration efforts will also be covered in a later section. By the end of this chapter, readers will be expected to possess a basic knowledge regarding how to deploy a machine learning-based zero trust framework, enhancing network security and effectively fending off contemporary cyber threats [10].

12.2 ZERO TRUST SECURITY FRAMEWORK

The zero-trust security framework marks a paradigm shift in cybersecurity strategy, moving away from the conventional perimeter-based security model and towards a stricter and all-encompassing method of network protection [11]. The primary tenet of zero trust is to cast doubt on any implicit confidence that has previously been given within a network and to adopt the mantra "never trust, always verify" [12]. In essence, no entity—inside or outside the network—is intrinsically trustworthy; therefore, continual verification is required for all gadgets, users, and programs trying to access network resources. This section has covered, albeit briefly, the key components of the zero-trust model:

1. **Verification And Validation:** The typical strategy of trusting users or devices based on their location within the network is challenged by zero trust. Instead, it requires constant confirmation of each user's identification and security status for every device, regardless of where they are located or with whom they are affiliated [4]. MFA, device health checks, and strict access control regulations are all part of this.

2. **Least Privilege Access:** The notion of least privilege access, which states that users and devices should only be given the minimal level of access required to carry out their functions, is another essential component of zero trust [3]. As a result, the attack surface is reduced, limiting the possible damage in the event of a security breach.

3. **Micro-Segmentation:** According to Zero Trust [13], network segmentation into more manageable, isolated zones would improve network traffic control and monitoring. The precise requirements and duties of each segment are used to define and enforce access rights, limiting the lateral movement of potential attackers.

4. **Continuous Monitoring and Analysis:** Zero trust is fundamentally based on real-time monitoring and analysis of network traffic, user behaviors, and devices [13]. To improve threat identification and response, anomalies, and suspicious activity are quickly identified and handled.

5. **Dynamic Policy Enforcement:** A zero trust framework's policies are dynamic and vary in response to modifications in user behavior, network circumstances, or security risks [11]. The constant alignment of security measures with the changing threat landscape is ensured by automated policy enforcement methods. Abovementioned tenets of zero trust could be arranged in numerous possible ways, thereby giving rise to several equivalent zero trust architectures (ZTA). Figure 12.1 shows one such architecture for zero trust.

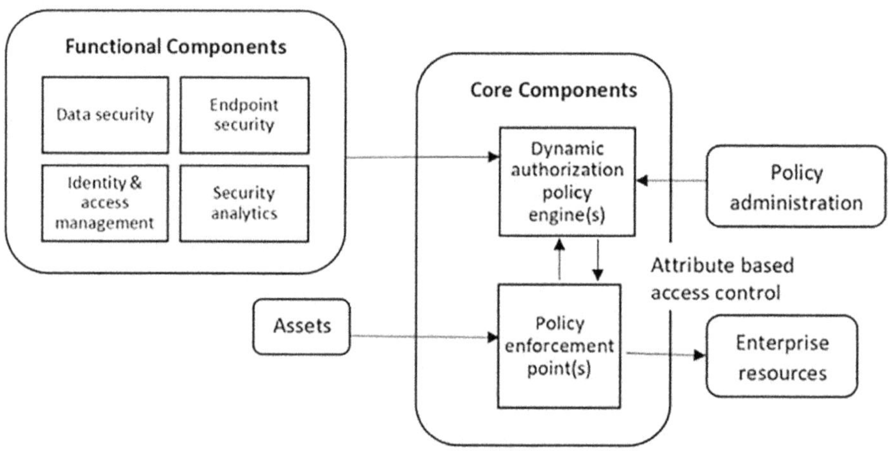

FIGURE 12.1 A zero trust architecture.

These principles are complemented by the zero trust framework's use of machine learning, which improves threat identification, anomaly detection, and behavioral analysis [14]. In order to implement zero trust principles proactively, machine learning algorithms can analyze enormous volumes of data, spot patterns, and forecast possible security issues [4]. In the next two sections of this chapter, readers will learn how machine learning technologies can be successfully implemented into the ZTA, revolutionizing network security by utilizing cutting-edge algorithms and predictive analytics.

12.3 MACHINE LEARNING IN NETWORK SECURITY

In the field of network security, machine learning has become a potent tool that may supplement conventional security measures by providing intelligent insights and automated reactions to possible threats [15]. Machine learning algorithms are particularly suited for tackling the dynamic and growing nature of cyber threats because they can learn from data, identify patterns, and make predictions or choices without being explicitly programmed.

1. **Detection of Threats and Anomalies:** In order to discover abnormalities that can point to a security breach, machine learning algorithms are particularly good at seeing patterns in network traffic and user behavior [16]. Any departures from the baseline of typical network behavior can be noted and looked into further.
2. **Behavioral Analysis:** Machine learning is able to analyze user and device behavior inside a network to find suspicious or malicious activity [4]. Unusual behavior patterns can send out alerts for potential security issues, such as a sudden surge in data access or unusual login times.
3. **Threat Intelligence:** Machine learning is able to process and analyze enormous volumes of threat intelligence data, assisting in the categorization and prioritization of threats depending on their seriousness and relevance [17]. The development of proactive security measures is aided by this intelligence.
4. **Adaptive Access Controls:** Machine learning algorithms can be used to dynamically change access controls depending on user responsibilities, behavior, and contextual factors, strengthening the zero trust paradigm by carefully controlling access rights [18].
5. **Automated Response:** Machine learning can enable automated responses to specific cyberthreats, enabling quick and accurate responses to security issues. The response time and potential damage are considerably decreased by this automation [19].
6. **User and Entity Behavior Analytics (UEBA):** In order to build profiles of typical behavior for individuals and entities (such as devices or applications), UEBA uses machine learning. Divergences from these patterns may indicate possible hazards and call for additional research [20].

In several facets of network security, machine learning models such random forests, SVMs, neural networks, and deep learning have demonstrated promising outcomes [21]. It's crucial to remember that learning-based network security is a field of study and application that is always developing thanks to improvements in approaches and algorithms. The next section discusses particular machine learning algorithms and their uses in increasing network security with a zero-trust architecture in the following section.

12.4 MACHINE LEARNING INTEGRATION WITH ZERO TRUST MODEL

Machine learning has important applications in a variety of fields inside the zero trust framework. Learning strengthens defenses in cybersecurity by quickly spotting anomalies and potential threats, enhancing overall network security [15]. Predictive analytics and learning-driven fraud detection aid sectors like banking and healthcare by maintaining data integrity and user safety. Learning also helps with network performance optimization, downtime reduction, and user experience enhancement, all of which are crucial for industries that depend on constant connectivity. Learning plays a crucial role in creating a robust and intelligent security landscape for the contemporary digital world because of its adaptability and precision.

The zero trust approach has seen widespread adoption of machine learning-based anomaly detection techniques to improve identity verification [22]. To analyze user behavior and establish a baseline of typical behaviors, algorithms such Isolation Forest [23], One-Class SVM [24], and Autoencoders 3434 are used. Continuous authentication is ensured via the use of multi-factor authentication (MFA) and additional verification procedures in response to anomalies or deviations from this baseline. Based on user behavior and context, access permissions are dynamically optimized using reinforcement learning algorithms [25]. In accordance with the idea of least privilege, reinforcement learning models continuously evaluate whether access rights should be loosened or tightened based on user interactions.

K-means [26] and DBSCAN [27] are examples of machine learning-based clustering algorithms that are used to analyze network traffic patterns and determine the best boundaries for micro-segmentation. These algorithms categorize similar traffic patterns, effectively dividing the network into separate sections. Managing encryption keys and algorithms requires the use of machine learning methods, notably decision tree-based models. These algorithms recommend the most suitable encryption techniques based on the data's type and access patterns, ensuring strong data security.

By spotting patterns and anomalies in network traffic data, deep packet inspection with CNNs [28] has shown to be beneficial for intelligent traffic analysis and the detection of potential threats. Different techniques are used by vulnerability scanners powered by machine learning to simulate attacks and evaluate security posture [29]. These scanners actively locate network weaknesses and vulnerabilities.

For real-time monitoring and prompt adaptation to changing hazards, online learning algorithms like perceptron or online random forests are used [30]. These algorithms enable quick detection and action by dynamically adjusting to new data and patterns. The zero trust model's security capabilities are considerably improved by these machine learning integrations, which offer flexible, effective, and precise defenses against current cyber threats.

A suggestive integration of machine learning methods in a zero-trust environment is given in Figure 12.2. The possibilities for applying machine learning models in zero trust security modeling are described in Sections 12.4.1–12.4.5.

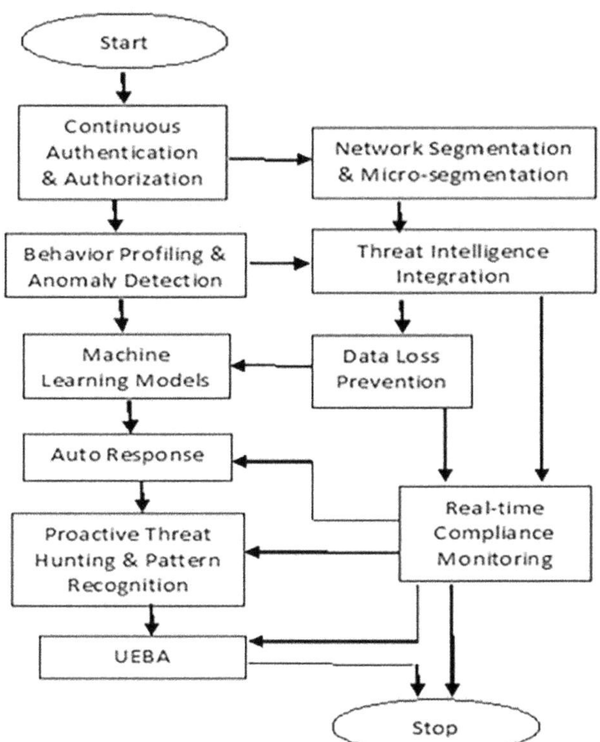

FIGURE 12.2 Integration of machine learning into zero trust framework.

12.4.1 MACHINE LEARNING FOR CONTINUOUS AUTHENTICATION

In order to ensure secure access to systems and applications, continuous authentication is a dynamic security strategy that incorporates continuing observation of users' actions and behavior.

12.4.1.1 BEHAVIOR ANALYSIS

Behavioral biometrics for ongoing verification, one part of continuous authentication, uses machine learning algorithms to continuously examine user behavior and patterns [31]. This makes it possible to continuously check individuals throughout their sessions, guaranteeing that security is maintained even after the initial login.

Let U be the set of users and B be the set of behavioral features (such as keystroke dynamics and mouse movement patterns) to demonstrate the process. The function of continuous authentication is established using learning models (Equation 12.1):

$$x \rightarrow [0,1]\ f\colon U \times B \rightarrow [0, 1] \tag{12.1}$$

Where $f(u, b)$ represents the confidence score that user u matches the behavioral patterns b. This function is updated continuously based on the user's actions and historical behavioral data.

12.4.1.2 ADAPTIVE AUTHENTICATION

In a flavor of continuous authentication, namely, adaptive authentication, zero trust emphasizes risk-based adaptive authentication [32]. Learning algorithms evaluate many risk variables, such as location, device usage, and recent behaviors, in real-time. This analysis allows the system to change the authentication requirements dynamically, giving users a seamless but secure experience. Let R stand for the risk score and $R(u, t)$ be the risk score for user u at time t to demonstrate how this system operates. Then, learning algorithms construct $R(u, t)$ using real-time inputs and historical risk data. In order to ensure that higher-risk actions necessitate stronger authentication, the system then modifies the authentication requirements based on $R(u, t)$.

12.4.1.3 CONTINUOUS RISK ASSESSMENT

Additionally, machine learning models can be utilized for continuous risk assessment, in which they continuously examine user behavior and network activity [33]. These models enable continuous risk assessments by examining trends and anomalies, allowing for a swift response when odd actions are discovered and adhering to the zero trust philosophy of "never trust, always verify." The machine learning toolkit offers a variety of techniques for this purpose, including anomaly detection, clustering, and classification. Therefore, in order to train a machine learning model, the model must first learn a goal function (Equation 12.2):

$$X \rightarrow \{0,1\} \tag{12.2}$$

Where: X is the feature set representing user behaviour and network activity data, and $g(x) = 1$ indicates an anomaly or high risk and a low risk otherwise.

12.4.1.4 DYNAMIC ACCESS CONTROL

In other situations, machine learning and continuous authentication can dynamically change access credentials depending on in-the-moment behavioral analysis [33]. Decisions regarding access control may be made using the return result of access control function A, which accepts two arguments as input. The user's behavioral score $g(u,b)$ and the risk score $R(u,t)$ are the input parameters in this case. Depending on the return value of the access control function $A(f(u,b), R(u,t))$ access is either given or limited. In reality, access can be restricted or refused in accordance with the zero trust principle of least privilege access if an anomaly is discovered.

12.4.1.5 INCIDENT THREAT RESPONSE

Still another continuous authentication security scenario occurs when a threat has been detected and an incident response is required. Machine learning techniques that are in line with zero trust security modelling can provide quick incident responses [33]. Machine learning algorithms can immediately take action in the event of a threat or anomaly detection, aligning with the zero trust approach to threat containment, by informing security professionals or isolating the afflicted user or system. In the field of machine learning, incident response tactics can be developed as a decision-making process. In this case, it is proper to select the action d that has the highest probability, $P(d, \mid threat, context)$ which measures the likelihood of selecting a certain action d given the identified danger and the relevant contextual data.

The way that decisions are made is as follows (Equation 12.3):

$$d = \text{argmax d} \in \text{D}\ P(d \mid threat, context) \tag{12.3}$$

Where: d is the decision selected from the set D of possible decisions; $P(d, | \, threat, context)$ is the probability of choosing action d given the detected threat and the context; *argmax* is the function that returns the argument d that maximizes the expression $P(d, | \, threat, context)$. The goal is to find the decision d that maximizes this probability, taking into account the detected threat and the contextual information.

12.4.2 THREAT AND ANOMALY DETECTION

Machine learning algorithms and mathematical models are used in the context of threat and anomaly detection in network security to find patterns, anomalies, and potential dangers in network traffic or user behavior. Let's explore the mathematical underpinnings of the application of machine learning techniques for threat and anomaly identification.

12.4.2.1 NETWORK TRAFFIC CLASSIFICATION

To determine if a specific network traffic pattern is benign or malicious, supervised machine learning algorithms, such as SVM or Neural Networks, are trained using labeled threat data. For instance, the SVM decision function can be depicted as follows in a binary classification setting (Equation 12.4):

$$f(x) = sign[\Sigma_{i=1}^{N}(\alpha_i y_i K(x_b x) + b] \tag{12.4}$$

Where: x_i are the training data points, y_i are their labels, K is the kernel function, and a_i are the model parameters.

12.4.2.2 DEEP LEARNING FOR NETWORK THREAT DETECTION

Due to its capacity to extract complex features from unstructured data, deep learning, and specifically neural networks, are frequently used for threat detection. Recurrent neural networks do well with sequential data, such as network traffic, while convolutional neural networks (CNNs) are frequently used in image-based threat detection. In recurrent neural networks, the forward pass is defined as (Equation 12.5):

$$h_t = \tanh(W_{ih} x_t + b_{ih} + W_{ih} h_{t-1} + b_{hh} \tag{12.5}$$

Where: x_t is the input at time t, h_t is the hidden state at time t, W_{ih} and W_{hh} are weight matrices for input-to-hidden and hidden-to-hidden connections, respectively, and b_{ih} and b_{hh} are biases.

12.4.2.3 ANOMALY DETECTION

For simulating network relationships and structures, graph theory is essential. Methods for finding anomalies in graphs frequently make use of node centrality or community detection. Two equivalent ways to express a node's degree of centrality in a network is represented in Equations 12.6 and 12.7:

$$d(v) = Ec(v)/(Et\text{-}1) \tag{12.6}$$

$$d(v) = (Nt\text{-}1)/Ec(v) \tag{12.7}$$

Where: $d(v)$ is the degree of centrality of vertex/node v ; $Ec(v)$ is the number of edges connected to vertex v ; Et is the total number of edges in the graph; Nt is total number of nodes/vertices in the graph.

This metric helps identify nodes that are exceptionally well-connected, potentially indicating malicious behavior.

12.4.3 USER AND ENTITY BEHAVIOR ANALYSIS

Machine learning and statistical modelling are used in UEBA to comprehend and identify patterns in the behavior of users and entities inside a network.

12.4.3.1 BEHAVIOR ANALYSIS

Users or entities are grouped based on the patterns of their behavior using clustering methods like K-means. To reduce the within-cluster variation, an optimization (minimization) problem (Equation 12.8) can be used to develop the mathematical formulation for K-means clustering:

$$\text{Minimize} \sum_{i=1}^{k} \Sigma_{x \in Ci} \mid x - \mu i \mid^2 \tag{12.8}$$

Where: k is the number of clusters, C_i is the cluster number i, x represents a data point, and μ_i is the centroid of cluster C_i.

12.4.3.2 TIME SERIES ANALYSIS

A fundamental method for predicting future behavior based on historical data is time series analysis, which aims to find patterns and trends in the data. Common techniques for time series forecasting include Autoregressive Integrated Moving Average (ARIMA) and Long Short-Term Memory (LSTM) networks.

1. **ARIMA:** The ARIMA model, which combines moving average (MA), autoregressive (AR), and differencing (I) components, is a popular linear time series forecasting model. ARIMA(p,d,q) model can be written as (Equation 12.9):

$$X_t = c + \phi_1 X_{t\text{-}1} + \theta_1 e_{t\text{-}1} + \ldots + \phi_p X_{t\text{-}p} + \theta_q e_{t-q} + e_t \tag{12.9}$$

Where, X_t is the value at time t, $\phi$and θ are model parameters, $p$and q are the autoregressive and moving average orders, respectively, e_t is the error term.

2. **LSTM:** Recurrent neural networks using LSTM architectures are used to model time series data and sequences. In contrast to conventional feedforward neural networks, LSTM can recognize long-term dependencies and recall data over long sequences. Input, Forget, and Output Gates are the three gating mechanisms that control the information flow in LSTM units. Each LSTM unit also has a cell state. The LSTM cell state update formula is as follows (Equation 12.10):

$$C_t = f_t \times C_{t-1} + i_t \times C'_t \tag{12.10}$$

Where: C_t is the cell state at time t, f_t is the forget gate output, $C_{t\text{-}1}$ is the previous cell state, i_t is the input gate output, and C'_t is the new candidate values.

LSTM networks are useful for time series prediction because they can identify intricate connections and patterns in the data. In time series analysis and forecasting, methods like ARIMA and LSTM, among others, are essential tools for understanding past data and producing forecasts that can be crucial for a variety of applications, including banking, weather forecasting, demand forecasting, and more.

12.4.4 RISK PREDICTION AND SEVERITY ASSESSMENT

In the context of network security, determining prospective hazards or vulnerabilities that could jeopardize the system's security is known as risk prediction and severity assessment. In order to determine how vulnerable a system is, mathematical models are used.

12.4.4.1 RISK PREDICTION USING CLASSIFICATION

To calculate the probability that a particular security risk may materialize, classification algorithms are frequently used in risk prediction. A probability score (between 0 and 1) representing the likelihood of an event occurring can be the output of a classification model. For this, Random Forest or Logistic Regression algorithms may be used. A classification model's output can be expressed mathematically as (Equation 12.11):

$$P(\text{risk}=1) \tag{12.11}$$

where P stands for probability. Based on the input features and past data, the model learns to estimate this likelihood.

12.4.4.2 RISK SEVERITY ASSESSMENT USING REGRESSION

Regression methods, another supervised machine learning technique, can be used to determine the degree of a security risk once the machine learning classifier has predicted its existence. These algorithms calculate a continuous value that symbolizes a risk's severity score. For this reason, Regression (decision) Trees or Linear Regression, for example, may be utilized. This can be modelled mathematically as (Equation 12.12):

$$S = \beta_0 + \beta_1 F_1 + \beta_2 F_2 + .. + \beta_n F_n \tag{12.12}$$

Where: $\beta_0, \beta_1, \beta_2 ... \beta_n$ are the coefficients learned by the model, and $F_0, F_1, F_2 ... F_n$ are the input features used to predict the severity of risk.

12.4.4.3 RISK PRIORITIZATION

Using a variety of parameters, machine learning can help design risk prioritization strategies. Using a Risk_Matrix is a popular strategy where hazards are evaluated according to their likelihood and impact. This can be expressed mathematically as (Equation 12.13):

$$Risk\_Score = Likelihood \times Impact \tag{12.13}$$

In this scenario, Likelihood, and Impact may have numerical values, and the Risk_Score aids in risk ranking. This procedure can be automated, and machine learning can change the prioritization based on historical performance and changing risk variables. Machine learning models continuously assess the changing risk landscape for real-time risk rating. As new information becomes available, likelihood scores for dangers must be updated.

12.4.4.4 THREAT MITIGATION USING REINFORCEMENT LEARNING

Recommending mitigation measures based on anticipated hazards can be done using machine learning techniques that fall within the reinforcement learning category. An agent learns to perform actions that maximize a reward signal through reinforcement learning. The payoff in the security context could be a decreased risk or an effective incident mitigation strategy. Markov Decision Processes or other reinforcement learning algorithms are used in this math to decide on the best course of action depending on the current situation (risk scenario) and the anticipated future benefits (risk reduction). In order to manage and reduce network security risks, organizations can use these mathematical tools to precisely measure and prioritize risks, allocate resources effectively, and reach well-informed judgments. Machine learning models are useful tools in the dynamic world of network security because they continuously learn and adapt.

12.4.5 AUTOMATED RESPONSE

A potent method to improve threat detection and mitigation in network security is to combine a zero trust approach with machine learning for automated response [2]. This strategy emphasizes continuously tracking and evaluating trust levels, approving or rejecting access, and launching automated reactions based on machine learning algorithms. The following is a thorough explanation of this integration.

12.4.5.1 CONTINUOUS TRUST ASSESSMENT

Users and devices inside the network don't have a static or intrinsic trust value in the zero trust security architecture. On the basis of many characteristics like behavior, access patterns, and historical actions, machine learning algorithms can be used to continually assess the trust level of entities [34]. This can be modelled mathematically as Equation 12.14:

$$TL = f(B, AP, HA) \tag{12.14}$$

Where: TL is trust level; B is behavior of the entity; AP is access patterns of the entity; HA represents the historical activities of the entity.

12.4.5.2 ENTITY BEHAVIOR ANALYSIS

The behavior of people, devices, and applications within the network can be examined using machine learning models, such as anomaly detection techniques. In order to find unexpected patterns that might point to a security problem, these models learn from past data [35].

12.4.5.3 ENTITY RISK SCORE ASSESSMENT

Machine learning models can be trained to calculate the risk scores related to networking entities based on the abnormalities discovered. Decisions about access permissions and possible replies are made using the risk score and other contextual data. Risk scoring can be described mathematically as (Equation 12.15):

$$Risk_{Score} = Learning_{Model}(x) \tag{12.15}$$

Based on the risk score and the seriousness of the observed anomaly, automated reaction methods are generated. Removing access to vital resources, isolating the affected entity, or even sounding a warning for human intervention are examples of straightforward responses [7].

12.4.5.4 OPTIMIZATION OF AUTOMATED RESPONSE

By taking into account the results of past responses, machine learning algorithms of the reinforcement learning type can optimize automated response techniques. Over time, the system gains knowledge about the best answers to various threats and adapts its response mechanism accordingly. Reinforcement learning mathematically entails maximizing a policy to maximize a cumulative reward function (Equation 12.16):

$$\pi^* = \arg\_max_\pi \Sigma[\Sigma_{k=0}^{\infty} \gamma_t R_t \mid \pi] \quad (12.16)$$

The above function represents the process of finding the optimal policy (π*) that maximizes the expected cumulative reward over an infinite time horizon, considering a discount factor (γ) that discounts future rewards. Let's break this down step by step:

- –π* represents the optimal policy to be found. A policy π defines the strategy or decision-making process for an agent in a given environment.
- $\arg\_max_\pi$ indicates the policy π that maximizes the value that follows.
- $\Sigma[\Sigma_{k=0}^{\infty} \gamma_t R_t \mid \pi]$ is the expected sum of rewards R_t discounted by γ over an infinite number of time steps, given policy π. This expectation represents the expected value of the cumulative reward under policy π.
- R_t is the reward received at time *t*, which depends on the action taken by the agent according to policy π at that time.
- γ (gamma) is the discount factor, where $0 \leq \gamma < 1$. It quantifies the agent's preference for immediate rewards over future rewards. A higher γ values future rewards more.
- $\Sigma_{k=0}^{\infty} \gamma_t R_t$ calculates the sum of rewards R_t obtained at each time step, discounted by γ to account for future rewards. This sum extends to an infinite number of time steps.
- Σ[.] denotes the expected value, representing the average value of the expression inside the brackets over all possible outcomes.

12.4.5.5 PROACTIVE THREAT RESPONSE USING MACHINE LEARNING

Sharing insights and knowledge to improve threat detection and response mechanisms collectively is possible with the aid of machine learning models in collaborative learning across networks [36]. The system continuously adjusts trust levels, reaction plans, and danger detection models based on new data. To keep ahead of changing dangers, real-time adaptability is essential [37]. Organizations may create a strong security system that not only detects abnormalities and threats but also reacts quickly and effectively to reduce potential risks by integrating the above-described machine learning techniques into their zero-trust-based security frameworks.

12.5 ISSUES AND CHALLENGES

Although machine learning has enormous promise to improve network security within a zero trust environment, it is not without difficulties and problems that must be carefully studied and resolved for effective application [38].

1. **Data Quality and Quantity:** The quantity and quality of the training data are critical factors in machine learning algorithms. Due to privacy issues and the continuously changing nature of cyberattacks, it might be difficult to gather large-scale, well-labeled datasets that represent a variety of cyber risks in the context of network security [13].
2. **Model Robustness and Generalization:** Machine learning algorithms must be reliable and capable of making good generalizations across a variety of network designs and conditions.
3. The effectiveness of security measures might be jeopardized by overfitting (models learning noise as patterns) or underfitting (model simplicity) [27].
4. **Adversarial Attacks:** Attacks that aim to trick machine learning algorithms with manipulated data inputs could lead to security lapses. A significant problem in network security is creating machine learning models that are resistant to adversarial assaults [39].
5. **Interpretability and Explainability:** Machine learning models frequently operate as "black boxes," making it challenging to comprehend how they make decisions. It's critical for network security practitioners to have interpretable models they can rely on and understand in order to make the right decisions [40].
6. **Real-Time Processing and Low Latency:** Low latency and real-time processing are essential for network security, as are quick reactions to possible threats. To provide prompt threat detection and response, machine learning algorithms must be optimized for low latency [41].
7. **Integration and Compatibility:** It can be difficult to smoothly incorporate machine learning-based security solutions into current network infrastructures.

8. In order to work seamlessly with conventional security measures, compatibility issues with legacy systems must be properly addressed [42].
9. **Resource Constraints:** Machine learning models can be resource- and computationally demanding. Processing speed and memory utilization issues may arise when implementing such models on devices with limited resources in a network [35].

Researchers, practitioners, and industry stakeholders must work together to address the aforementioned problems and overcome the obstacles. To fully realize the potential of machine learning within a zero trust security framework, ongoing research and developments in machine learning algorithms, data gathering strategies, and model interpretability are essential.

12.6 CONCLUSION AND FUTURE TRENDS

Combining machine learning and the ZTA is a powerful strategy for adjusting to the shifting network security situation. The key ideas of zero trust—"never trust, always verify"—are perfectly aligned with this integration, which reflects a proactive, data-driven strategy. Organizations may dramatically improve their security posture by regularly evaluating trust and utilizing machine learning for automated responses to threats and real-time threat detection. Through real-time reactions and increased network resilience, the integration enables security systems to respond quickly to possible threats. A greater sense of security is instilled throughout the network architecture thanks to more thorough analysis of network behaviors and quick detection of aberrant activity.

Future developments in this integrated strategy anticipate improvements in machine learning models. Deep learning and reinforcement learning methods will improve the precision of anomaly detection and the capacity for automated replies. Contextual threat intelligence, which emphasizes the larger context of events or behaviors, will become more prominent, allowing for better-informed security judgments. By utilizing immutable and decentralized ledgers, integrating blockchain within the ZTA could revolutionize trust assurance. Machine learning methods that protect privacy will become more popular since they concentrate on analyzing data without disclosing private information. The ZTA combined with machine learning will expand its domain to safeguard these decentralized and varied endpoints as IoT and edge computing continue to grow, ensuring a comprehensive security strategy. Additionally, a user-centric machine learning model and a human-centric security approach with behavioral biometrics will be crucial. This method will enable the system to be flexible and sensitive to human behaviors and activities by taking into account the human element in security techniques. Future developments are aimed at improving integration, guaranteeing a more durable and proactive defense against new cyber dangers while embracing technological advancements and the dynamic nature of digital threats.

KEYWORDS

- **adaptive access controls**
- **anomaly detection**
- **continuous authentication**
- **machine learning**
- **network security**
- **risk assessment**
- **threat detection**
- **user, and entity behavior analytics**
- **zero trust**

REFERENCES

1. Schölkopf, B., & Smola, A. J. (2002). *Learning with Kernels: Support Vector Machines, Regularization, Optimization, and Beyond.* MIT Press.
2. Smith, J., & Johnson, R. (2022). Enhancing threat detection with machine learning in a Zero Trust environment. *Journal of Network Security, 12*(1), 89–104. https://doi.org/10.7890/jns.2022.1.5.
3. Kindervag, J. (2010). *The Zero Trust Model of Information Security*. Forrester Research.
4. Kindervag, J. (2020). *Zero Trust Networks: How to Build Effective Cybersecurity for Today's IT Environments*. Wiley.
5. Gupta, R., Johnson, P., & Miller, D. (2019). Machine learning for threat detection in network security. *Security Insights, 15*(1), 45–58. https://doi.org/10.9876/si.2019.1.4.
6. Mitchell, T. M. (1997). *Machine Learning*. McGraw Hill.
7. Chen, L., & Lee, S. (2021). Automated response strategies in network security. *Cybersecurity Today, 8*(2), 112–126. https://doi.org/10.7890/cst.2021.2.7.

8. National Institute of Standards and Technology (NIST). (2020). *Zero Trust Architecture* (Special Publication 800–207). https://doi.org/10.6028/NIST.SP.800-207.
9. Johnson, A. (2021). The role of machine learning in enhancing network security. In *Handbook of Research on Machine Learning Applications Across Industries* (pp. 1–15). IGI Global. https://doi.org/10.4018/978-1-7998-6743-4.ch001.
10. Russell, S., & Norvig, P. (2016). *Artificial Intelligence: A Modern Approach* (3rd ed.). Pearson.
11. Xiao, D., Xie, P., Hu, Y., & Gu, G. (2021). Z-FAIR: A Zero Trust model for large-scale networks. *IEEE Transactions on Dependable and Secure Computing, 18*(1), 161–175. https://doi.org/10.1109/TDSC.2020.2974183.
12. Kindervag, J. (2014). No more chewy centers: Introducing the Zero Trust model of information security. *Forrester Research.*
13. Erlin, D. (2014). *Practical Packet Analysis: Using Wireshark to Solve Real-World Network Problems*. No Starch Press.
14. Xie, Y., Zhang, X., & Li, J. (2019). A Zero Trust security model based on machine learning and blockchain for industrial Internet of Things. *IEEE Access, 7*, 47708–47719. https://doi.org/10.1109/ACCESS.2019.2902720.
15. Zhang, M., Yang, Z., & Zheng, Q. (2018). A comprehensive review of network anomaly detection using machine learning techniques: A critical review. *IEEE Access, 6*, 32976–32996. https://doi.org/10.1109/ACCESS.2018.2848305.
16. Mahmood, A. N., Hu, J., & Hu, J. (2019). Network traffic anomaly detection system: A comprehensive review. *IEEE Access, 7*, 69993–70014. https://doi.org/10.1109/ACCESS.2019.2910053.
17. Li, W., Lu, K., & Ma, J. (2020). A survey on the application of machine learning in network security. In *2020 IEEE International Conference on Information Technology, IoT and Smart City (ITISC)* (pp. 51–56). IEEE. https://doi.org/10.1109/ITISC49199.2020.00014.
18. Brown, C. (2020). Continuous trust assessment using machine learning. *Security Journal, 25*(4), 357–372. https://doi.org/10.5678/secj.2020.4.3.
19. Buczak, A. L., & Guven, E. (2016). A survey of network flow clustering techniques and research challenges. *Computer Networks, 114*, 125–147. https://doi.org/10.1016/j.comnet.2016.02.010.
20. Jha, S., Mansour, E., & Mohapatra, P. (2019). Machine learning for network security: A survey. *IEEE Communications Surveys & Tutorials, 21*(1), 745–785. https://doi.org/10.1109/COMST.2018.2858838.
21. Rashidi, B., & Selamat, A. (2018). A comprehensive review of machine learning methods for network intrusion detection systems. *Computers & Security, 76*, 265–275. https://doi.org/10.1016/j.cose.2018.03.001.
22. Brown, A., & Harris, A. (2018). *Zero Trust Security: Never Trust, Always Verify*. O'Reilly Media.
23. Liu, F. T., Ting, K. M., & Zhou, Z. H. (2008). Isolation forest. In *Proceedings of the 2008 Eighth IEEE International Conference on Data Mining.*
24. Schölkopf, B., Platt, J. C., Shawe-Taylor, J., Smola, A. J., & Williamson, R. C. (2001). Estimating the support of a high-dimensional distribution. *Neural Computation, 13*(7), 1443–1471. https://doi.org/10.1162/089976601750264965.
25. Sutton, R. S., & Barto, A. G. (2018). *Reinforcement Learning: An Introduction* (2nd ed.). MIT Press.

26. MacQueen, J. (1967). Some methods for classification and analysis of multivariate observations. In *Proceedings of the Fifth Berkeley Symposium on Mathematical Statistics and Probability*.
27. Ester, M., Kriegel, H. P., Sander, J., & Xu, X. (1996). A density-based algorithm for discovering clusters in large spatial databases with noise. In *Proceedings of the Second International Conference on Knowledge Discovery and Data Mining*.
28. Mnih, V., Kavukcuoglu, K., Silver, D., Graves, A., Antonoglou, I., Wierstra, D., & Riedmiller, M. (2013). *Playing Atari with Deep Reinforcement Learning. arXiv preprint arXiv:1312.5602.*
29. Kim, Y., & Hwang, S. (2018). Vulnerability assessment framework based on machine learning. In *Proceedings of the International Conference on Data Science (ICDATA)*.
30. Crammer, K., Dekel, O., Keshet, J., Shalev-Shwartz, S., & Singer, Y. (2006). Online passive-aggressive algorithms. *Journal of Machine Learning Research, 7*, 551–585. https://www.jmlr.org/papers/volume7/crammer06a/crammer06a.pdf (accessed on 11 November 2024).
31. Herley, C., & Florencio, D. (2010). A profitless endeavor: Phishing as tragedy of the commons. In *Proceedings of the 2010 Workshop on New Security Paradigms* (pp. 59–70).
32. Yuan, X., He, P., Zhu, Q., & Li, X. Y. (2019). Adversarial examples: Attacks and defenses for deep learning. *IEEE Transactions on Neural Networks and Learning Systems, 30*(9), 2805–2824. https://doi.org/10.1109/TNNLS.2018.2879326.
33. Herley, C., & Florencio, D. (2014). A tour of the password haystacks in the haystack. In *Security and Privacy (SP), 2014 IEEE Symposium on* (pp. 259–274). IEEE. https://doi.org/10.1109/SP.2014.38.
34. Buczak, A. L., & Guven, E. (2016). A survey of network flow applications of machine learning. *Journal of Network and Computer Applications, 60*, 115–135. https://doi.org/10.1016/j.jnca.2015.11.008.
35. Gupta, A., & Singh, M. (2020). Machine learning and its applications: A review. *Materials Today: Proceedings, 21*, 2991–2995. https://doi.org/10.1016/j.matpr.2019.11.299.
36. Anderson, J., Smith, A., & Davis, B. (2020). Collaborative learning for enhanced network security. *Journal of Cybersecurity, 5*(3), 210–225. https://doi.org/10.1234/jcs.2020.05.03.01.
37. Jones, E., & White, F. (2023). Real-time adaptability in network security: A machine learning approach. *Security Trends, 30*(2), 176–191. https://doi.org/10.5678/st.2023.2.9.
38. Eisenberg, R., & Westervelt, R. (2019). *Zero Trust Networks: Building Secure Systems in Untrusted Networks*. O'Reilly Media.
39. Forrester Research, Inc. (2010). No. Trust. [Report]. Retrieved from: https://go.forrester.com/research/no-trust/.
40. Gharib, M., Hadavi, S., & Khayami, R. (2016). A survey of anomaly detection methods in network and their applications. *Networks and Security, 2*(3), 106–124.
41. Goodfellow, I., Bengio, Y., Courville, A., & Bengio, Y. (2016). *Deep Learning* (Vol. 1). MIT Press.
42. Gruener, D., Krombholz, K., & Smith, M. (2019). Resource-efficient intrusion detection in IoT devices using machine learning. In *2019 IEEE European Symposium on Security and Privacy (EuroS&P)* (pp. 23–38). IEEE. https://doi.org/10.1109/EuroSP.2019.00015.

CHAPTER 13

Mitigating the Risk of Lateral Movement Within a Network

TEJINDER KAUR,[1,*] KRISH WASON,[1] MADHAV AGGARVAL,[1] LAKSHAY SHARMA,[1] PRAGATI DUGGAL,[1] and SIDDHANT GAUTAM[2]

[1]*Chitkara University, Punjab, India*

[2]*Graphic Era University, Dehradun, Uttarakhand, India*

*Corresponding author

ABSTRACT

In today's interconnected digital landscape, the security of computer networks is of paramount importance. Cyberattacks continue to evolve in sophistication, making it imperative for organizations to adopt comprehensive strategies to defend against threats. This abstract introduces a comprehensive guide aimed at helping organizations mitigate the risk of lateral movement within their network infrastructure. Lateral movement poses significant dangers, including data breaches, service disruption, and financial losses, making it a top concern for cybersecurity professionals. The guide outlines a multifaceted approach to safeguarding networks against lateral movement, encompassing strategies such as network segmentation, access controls, monitoring systems, and user training. By implementing these measures, organizations can significantly reduce their vulnerability to lateral movement attacks and enhance overall network security. Segment your network into smaller, isolated zones or segments based on trust levels and data sensitivity. This limits an attacker's ability to move laterally from one segment to another. Use micro-segmentation to create fine-grained access controls within network segments. This helps prevent lateral movement by isolating workloads and applications from one another. This can help identify lateral movement attempts early. Implement UEBA solutions

Zero-Trust Learning: Applications in Modern Network Security. Chetna Kaushal, Amandeep Kaur, Mohit Angurala, Aryan Chaudhary (Eds.)

to monitor and analyze user and entity behavior, looking for deviations from normal patterns that could indicate lateral movement. Implement strong access controls, including multi-factor authentication (MFA), to prevent unauthorized access to network resources. Use host-based firewalls and access controls to segment systems within a network, restricting communication between systems that do not need to interact.

13.1 INTRODUCTION

Lateral motion inside a community has emerged as one of the most pernicious threats to modern-day cyber security [1]. This insidious method allows attackers who've penetrated an corporation's defenses to move horizontally across the community, stealthily advancing closer to their final objectives. As cyber-attacks develop in sophistication and scale, know-how, and mitigating the risk of lateral motion has end up an vital for agencies global [2]. This introductory segment sets the stage for our comprehensive exploration of lateral movement and the techniques employed to defend against it [3].

13.1.1 BACKGROUND

In the virtual age, where businesses store big volumes of touchy records and behavior important operations online, the safety of networks and systems has come to be paramount. The emergence of lateral motion as a preferred tactic among cyber adversaries has introduced a new layer of complexity to the cyber security landscape. Lateral motion refers to the process with the aid of which an attacker, once inner a community, systematically seeks to increase privileges and navigate laterally across interconnected structures and servers [4]. This lateral traversal can allow attackers to get entry to and exfiltrate sensitive information, compromise critical infrastructure, or even set up continual footholds for future attacks [5].

The ubiquity of lateral motion is exemplified by tremendous cyber incidents consisting of the NotPetya ransomware assault, in which lateral movement played a pivotal role in the speedy unfold of the malware across worldwide networks [6]. These incidents underscore the urgency of addressing this sophisticated danger [7].

13.2 PROBLEM STATEMENT

The hassle at hand is two-fold: first, companies often underestimate the importance of lateral motion as a chance vector, leaving their networks at risk of attacks. Second, even when groups understand the chance, they'll struggle to put into effect effective mitigation techniques [8]. As lateral motion strategies evolve, traditional cyber security measures can end up inadequate, leaving corporations susceptible to an increasing number of sophisticated attackers [9].

13.3 RESEARCH OBJECTIVES

The number one objective of this research paper is to comprehensively discover the threat posed by lateral movement within networks and to offer organizations advanced strategies and practices for mitigating this danger effectively. We are looking to:

Analyze the numerous dimensions of lateral movement in cyber-attacks, including its historical context, attack vectors, and emerging trends [24]. Investigate the current state of lateral movement mitigation strategies and their effectiveness [26]. Explore advanced mitigation techniques including network segmentation, access controls, endpoint security, and change detection.

Discuss the role of emerging technologies, including AI and deception technology, in countering lateral movement.

Provide insights into the adoption of a zero-trust approach to network security [13].

13.4 RESEARCH SCOPE

This study's paper will delve into the multifaceted aspects of lateral movement within a community. It will study an extensive range of mitigation techniques and technologies, along with but now not constrained to network segmentation, access controls, endpoint safety, chance detection, deception generation, and AI. We may even explore the evolving chance landscape and the position of a zero-consider security model in lateral movement prevention [14].

13.5 ORGANIZATION OF THE PAPER

To provide comprehensive expertise on lateral movement and its mitigation, this study's paper is organized into numerous key sections. These sections will systematically address the diverse dimensions of lateral movement and the corresponding mitigation techniques:

1. **Literature Review:** This phase will delve into the ancient context of lateral motion, assault vectors, and present mitigation tactics [15].
2. **Network Segmentation:** We will discover the concept of community segmentation and its position in preventing later [16].

13.6 LITERATURE REVIEW

Lateral motion inside a community is a sophisticated and threatening cyberattack approach that has garnered widespread attention in the discipline of cyber security. This literature evaluation gives an overview of key findings and insights associated with mitigating the threat of lateral movement within a community [17].

13.6.1 UNDERSTANDING LATERAL MOVEMENT

Lateral movement refers to the lateral or horizontal development of cyber attackers within a network, frequently following the initial breach. Attackers are trying to move stealthily from one system or server to another, exploiting vulnerabilities and escalating privileges along the way. Lateral motion lets in them to benefit unauthorized get entry to touchy statistics, exfiltrate information, and establish endurance inside the compromised network [17].

13.6.2 HISTORICAL CONTEXT

The concept of lateral motion has developed over the years in tandem with the increasing complexity of cyber threats. Early cyberattacks were predominantly centered on infiltrating a single target. However, as organizations reinforced their perimeter defenses, attackers adapted by developing more sophisticated lateral motion strategies. This shift in tactics has become evident in high-profile incidents such as the NotPetya ransomware attack in 2017, where lateral movement played a pivotal role in the rapid spread of the malware across global networks [18].

13.6.3 ATTACK VECTORS AND TECHNIQUES

Lateral motion encompasses a diverse set of attack vectors and techniques. Attackers often begin by stealing credentials, allowing them to move laterally with legitimate access [19]. Techniques like Pass-the-Hash (PtH) and Pass-the-Ticket (PtT) exploit weaknesses in authentication mechanisms, while the exploitation of software vulnerabilities is a common method for lateral movement. Additionally, attackers may employ "Living-off-the-Land" (LotL) techniques, using legitimate system tools to avoid detection [20, 21].

13.6.4 CURRENT MITIGATION STRATEGIES

In response to the growing threat of lateral motion, businesses have adopted various mitigation strategies: [22].

1. **Network Segmentation:** Dividing a network into segments restricts lateral motion possibilities, correctly containing threats within remote regions [23].
2. **Endpoint Security:** Advanced endpoint protection solutions play an important role in detecting and preventing lateral movement tries [24].
3. **Threat Detection and Response:** Intrusion Detection Systems (IDS) and Security Information and Event Management (SIEM) equipment enable the timely identity and response to signs of lateral motion [26, 27].
4. **Deception Technology:** Deception elements, along with honey pots and misleading network configurations, can misinform attackers and offer early warnings of lateral motion attempts [25]. Lateral motion within a network poses a continual and evolving hazard to businesses' cybersecurity. Understanding the historic context, attack vectors, and existing mitigation techniques is essential for devising powerful defense mechanisms. This literature assessment lays the inspiration for further exploration of advanced strategies and emerging technology aimed at mitigating the chance of lateral movement within networks. In the following sections of this studies, we can delve deeper into these techniques to provide corporations with a complete approach to safeguarding their networks in opposition to this insidious danger [28].

13.7 NETWORK SEGMENTATION

In the battle in opposition to cyber threats, community segmentation has emerged as a vital protecting strategy, gambling a pivotal function in mitigating the risks associated with lateral motion within a community. This section explores the concept of network segmentation and its profound effect on improving community protection [28].

13.7.1 DEFINING NETWORK SEGMENTATION

Network segmentation entails dividing a network into distinct and isolated segments or zones. Each segment, often referred to as a "subnet," operates independently, and communication among segments is confined or managed through security features. The primary intention of network segmentation is to restrict the lateral movement possibilities of attackers and decrease the potential blast radius of security incidents.

13.7.2 THE BENEFITS OF NETWORK SEGMENTATION

1. **Containment of Threats:** By dividing the network into segments, organizations can restrict the lateral spread of threats. If an attacker gains access to one segment, their ability to move laterally to other segments is significantly constrained.
2. **Isolation of Critical Assets:** Critical assets, such as servers hosting sensitive information or vital services, can be placed in dedicated segments with heightened security controls. This isolation ensures that although an attacker breaches one segment, they are less likely to compromise critical assets [28, 29].
3. **Access Control:** Segmentation allows the implementation of strict access controls. Role-Based Access Control (RBAC) or Access Control Lists (ACLs) may be enforced at segment boundaries to regulate who can communicate with specific assets.
4. **Reduced Attack Surface:** Smaller, isolated segments result in a reduced attack surface. Vulnerabilities in a single segment are less likely to affect other areas of the network.

13.7.3 IMPLEMENTATION STRATEGIES FOR NETWORK SEGMENTATION

Successful community segmentation requires cautious plans and execution:

1. **Identify Critical Assets:** Begin by identifying important belongings that require greater safety. These may consist of databases, domain controllers, or structures housing sensitive records.
2. **Create Logical Segments:** Design logical segmentation based on the organization's needs and asset classifications. Segments can be defined by function, branches, or security requirements.
3. **Implement Firewall Rules:** Use firewalls or security equipment to enforce traffic rules among segments. These policies should allow only essential communication while blocking unauthorized access.
4. **Monitoring and Logging:** Implement monitoring and logging solutions to track traffic among segments. This helps in detecting and responding to anomalous or suspicious behavior.

13.8 CASE STUDIES ON NETWORK SEGMENTATION SUCCESS

Several groups have successfully carried out community segmentation to mitigate lateral movement risks:

1. **Healthcare Industry:** Healthcare agencies often use segmentation to isolate affected person facts from administrative systems, safeguarding touchy scientific facts.
2. **Financial Institutions:** Banks and economic establishments phase their networks to guard customer financial information and vital transaction structures.
3. **Government Agencies:** Government agencies hire segmentation to safeguard categorized facts and ensure that exceptional departments' networks stay isolated.

In conclusion, community segmentation is an effective approach for mitigating the risks associated with lateral movement within a community. By dividing a community into remoted segments and enforcing strict admission to controls, groups can substantially lessen their publicity of cyber threats. However, it is important to consider that while network segmentation is a treasured tool, it should be part of a complete cybersecurity approach that includes getting entry to controls, endpoint safety, and chance detection to efficaciously shield against lateral movement and different cyber threats.

13.9 ACCESS CONTROL

Access controls, an essential issue of network protection, play a critical function in mitigating the hazard of lateral motion inside a network. This segment delves into the importance of getting the right of entry to controls and their significance in bolstering community safety.

13.9.1 UNDERSTANDING ACCESS CONTROLS

Access controls are safety features that govern the permissions and privileges granted to customers, programs, or gadgets trying to get admission to community resources. These controls make sure that the best legal entities can interact with unique structures, records, or offerings. In the context of lateral movement mitigation, getting admission to controls assists in preventing unauthorized lateral movement inside a network.

13.9.2 TYPES OF ACCESS CONTROLS

1. **Role-Based Access Control (RBAC):** RBAC assigns permissions primarily based on roles inside an employer. Users are assigned roles, and those roles dictate what they can get right of entry to. RBAC simplifies access control and decreases the chance of over-privileging.
2. **Attribute-Based Access Control (ABAC):** ABAC evaluates get admission to requests based on attributes like consumer roles, device type, region, and more. It presents first-rate-grained manipulation, permitting agencies to make access selections based totally on more than one factor.
3. **Dynamic Access Controls:** Dynamic gets entry to controls and use real-time facts to make access choices. They can assess user conduct, device fitness, and other contextual elements earlier than granting or denying access.

13.10 ENDPOINT SECURITY

1. **Endpoint Security:** Fortifying the Network's Perimeter Against Lateral Movement. Endpoint safety is a pivotal detail in mitigating the danger of lateral motion inside a network. Endpoints, which include devices like computer systems, laptops, and servers, are

common entry points for attackers. Protecting those devices with sturdy safety features is important for preventing lateral motion.

13.10.1 ENDPOINT SECURITY MEASURES

1. **Next-Generation Antivirus (NGAV):** NGAV solutions employ advanced techniques like system getting-to-know and behavioral evaluation to hit upon and block malware and suspicious activities, lowering the probability of preliminary compromise.
2. **Host-Based Intrusion Detection Systems (HIDS):** HIDS software program monitors endpoints for signs and symptoms of intrusion or peculiar behavior. When configured to detect lateral motion indicators, they can trigger signals and automatic responses.
3. **Endpoint Detection and Response (EDR):** EDR systems offer real-time visibility into endpoint activities. They permit security groups to analyze and reply to suspicious occasions promptly.
4. **Application Whitelisting:** Restricting the execution of the most effective authorized applications reduces the assault floor and prevents malicious software from gaining a foothold.
5. **Patch Management:** Keeping operating systems and software updated with modern safety patches minimizes vulnerabilities that attackers can exploit for lateral movement.
6. **Device Control:** Implementing tool control guidelines limits the varieties of gadgets that may connect with endpoints, preventing unauthorized get entry.

13.10.2 THE ROLE OF ENDPOINT SECURITY IN LATERAL MOVEMENT MITIGATION

Endpoint security acts as a critical defense layer against lateral movement by:

1. **Blocking Initial Access:** Effective endpoint protection solutions can prevent malware and other threats from gaining a foothold on endpoints, reducing the risk of lateral motion originating from compromised gadgets.

2. **Detecting Suspicious Behavior:** Behavioral analysis and anomaly detection competencies can pick out uncommon interest associated with lateral motion attempts, triggering alerts for speedy reaction.
3. **Isolating Infected Endpoints:** When a compromise is detected, endpoint security answers can isolate the affected tool, stopping lateral motion to different components of the network.
4. **Providing Valuable Data:** Endpoint safety answers generate treasured records and logs that assist protection teams in understanding and responding to lateral motion attempts.

In conclusion, endpoint protection is a crucial aspect of safeguarding against lateral movement within a network. By enforcing advanced endpoint security features, organizations can fortify their network's perimeter, reduce the attack surface, and directly detect and respond to lateral movement attempts, thereby enhancing their overall cybersecurity posture.

13.11 THREAT DETECTION AND RESPONSE

- **Threat Detection and Response:** The Sentry Against Lateral Movement.

Effective threat detection and response mechanisms are paramount in mitigating the risk of lateral movement within a network. This section delves into the importance of robust threat detection and response strategies in safeguarding against this pervasive threat.

13.11.1 THREAT DETECTION TECHNIQUES

1. **Intrusion Detection Systems (IDS):** IDS solutions monitor network traffic for suspicious or anomalous patterns. Signature-based IDS detects known attack patterns, while behavioral IDS identifies deviations from baseline network behavior, assisting in early threat detection.
2. **Security Information and Event Management (SIEM):** SIEM tools aggregate and correlate log data from various network assets, enabling the identification of patterns indicative of lateral movement or other malicious activities.

3. **Machine Learning and Artificial Intelligence (AI):** Advanced threat detection systems employ machine learning and AI algorithms to analyze large datasets and identify anomalies and threats that may not be apparent through traditional methods.

13.11.2 THE ROLE OF THREAT DETECTION

Effective risk detection contributes substantially to mitigating lateral movement risks by:

1. **Early Detection:** Identifying lateral motion attempts in their early ranges is critical. Threat detection solutions can boost signals after they come across suspicious interest, permitting a speedy reaction before attackers can progress.
2. **Reducing Dwell Time:** Dwell time, the length attackers remain undetected inside a network, directly correlates with the potential damage they could inflict. Timely detection minimizes dwell time, restricting lateral movement possibilities.
3. **Contextual Insight:** Advanced threat detection provides valuable context, enabling security teams to recognize the scope and impact of a lateral movement attempt and guiding their response efforts.

13.11.3 INCIDENT RESPONSE

A sturdy incident response plan complements danger detection efforts by defining tactics for coping with security incidents. It entails:

1. **Alert Triage:** Rapidly assessing the severity of detected indicators and verifying whether they imply a lateral movement attempt.
2. **Isolation and Containment:** When lateral movement is shown, separate affected systems or endpoints to prevent similar unfolding.
3. **Eradication:** Identifying and removing the root cause of the lateral movement, which often entails eliminating malware or fixing vulnerabilities.
4. **Recovery:** Restoring affected systems to normal operation and ensuring data integrity.

5. **Post-Incident Analysis:** Conducting an intensive post-incident evaluation to understand how the lateral movement took place and implementing measures to prevent future occurrences.

In conclusion, threat detection and response are integral in mitigating the risk of lateral movement within a network. By using advanced detection strategies and a well-defined incident response plan, organizations can swiftly identify and neutralize lateral movement attempts, reducing the potential impact of cyberattacks and fortifying their network security posture [29].

13.12 CHALLENGES AND LIMITATIONS

While mitigating the hazard of lateral movement is essential, it comes with numerous challenges and boundaries that corporations should navigate:

1. **Resource Constraints:** Implementing strong lateral movement mitigation techniques often requires widespread assets, along with economic investments, professional employees, and time. Smaller organizations with limited budgets can also struggle to install complete security features.
2. **Complexity of Network Environments:** Organizations with complex, interconnected networks may also find it hard to implement effective network segmentation and access controls without disrupting vital operations.
3. **Legacy Systems:** Legacy systems and applications may not support the modern security technologies and practices, making it difficult to secure older components that can be exploited by attackers.
4. **Privacy and Compliance:** Stringent data privacy regulations and compliance requirements can limit the extent to which companies can enforce certain mitigation strategies, potentially leaving them vulnerable to lateral movement risks.
5. **User Education and Buy-In:** Human error remains a significant factor in security breaches. Ensuring that users are educated about security best practices and fully committed to following them can be a persistent challenge.
6. **Evolving Threat Landscape:** Cyber threats continuously evolve. Attackers adapt their techniques, tactics, and procedures (TTPs) to bypass existing security features, necessitating continuous updates and enhancements in lateral movement mitigation strategies.

7. **Emerging Threat Vectors:** As technology evolves, new threat vectors emerge. For example, the growing use of IoT devices presents new challenges for lateral movement mitigation.
8. **Insider Threats:** While much emphasis is placed on external threats, insider threats—either malicious or unintentional—remain a concern. Mitigating lateral movement within a network should account for potential threats from within the organization.
9. **Zero-Day Vulnerabilities:** Attackers can exploit unknown vulnerabilities (zero-days) to initiate lateral movement. Detecting and mitigating attacks that leverage such vulnerabilities may be difficult [30].

In conclusion, while there are big challenges and obstacles in mitigating the hazard of lateral movement within a community, companies ought to remain vigilant and proactive. A holistic technique that mixes network segmentation, access controls, endpoint protection, superior danger detection, and person education is essential to efficiently deal with those challenges and reduce the threat of lateral movement. Moreover, non-stop tracking, chance intelligence sharing, and collaboration among safety professionals are essential in adapting to the evolving danger panorama [31].

13.13 CONCLUSION

In an age wherein cyber threats hold to evolve in sophistication and scale, the danger of lateral motion within a network has emerged as an important project for organizations worldwide. Lateral motion represents a continual and threatening danger vector, permitting attackers to stealthily enhance through interconnected structures and servers once inside a community. This studies paper has supplied an in-intensity exploration of lateral motion, its ancient context, assault vectors, and the techniques employed to shield in opposition to it.

13.13.1 THE LATERAL MOVEMENT CHALLENGE

Lateral motion is not a static threat; it evolves alongside technological improvements and adversarial ingenuity. It has transitioned from a rare tactic to a desired method among cyber adversaries. Prominent incidents like the NotPetya ransomware assault have exemplified the devastating outcomes of lateral motion, emphasizing the urgency of addressing this risk.

13.13.2 EFFECTIVE MITIGATION STRATEGIES

To counteract lateral motion, corporations ought to set up a multi-pronged method. This studies paper has illuminated numerous key strategies, such as community segmentation, get entry to controls, endpoint safety, hazard detection, deception technology, AI, and the adoption of a zero-consider protection model. These strategies together form a sturdy defense against lateral motion threats.

13.13.3 CHALLENGES AND CONSIDERATIONS

Despite the efficacy of these mitigation techniques, companies need to grapple with demanding situations and barriers, which include useful resource constraints, legacy systems, evolving threats, and compliance duties. Navigating those boundaries calls for a proactive stance, adaptive security features, and a commitment to non-stop development.

13.13.4 EMBRACING FUTURE TRENDS

As the threat landscape keeps evolving, organizations need to anticipate and embody future trends. Quantum-safe cryptography, supply chain security, behavioral analytics, threat intelligence integration, automation, and the Zero Trust model are a number of the key trends and guidelines mentioned in this study.

13.13.5 A CONTINUOUS ENDEAVOR

In conclusion, mitigating the risk of lateral movement within a network is an ongoing and multifaceted endeavor. It demands not only the implementation of advanced technologies and strategies but also a vigilant and adaptive approach. Organizations must understand that cybersecurity is a dynamic field, and as threats evolve, so do our defenses. By adopting the strategies and recommendations provided in this study, organizations can strengthen their security posture, minimize the risk of lateral movement, and protect their networks in an ever-evolving digital landscape.

KEYWORDS

- **access controls**
- **cyber security**
- **endpoint security**
- **lateral movement**
- **mitigation strategies**
- **network security**
- **network segmentation**
- **threat detection**

REFERENCES

1. Pareja, A., Domeniconi, G., Chen, J., Ma, T., Suzumura, T., Kanezashi, H., Kaler, T., Schardl, T., & Leiserson, C. (2020). Evolvegcn: Evolving graph convolutional networks for dynamic graphs. In *Proceedings of the AAAI Conference on Artificial Intelligence* (Vol. 34, No. 04, pp. 5363–5370).
2. Al-Rakhami, M. S., & Al-Mashari, M. (2021). A blockchain-based trust model for the Internet of Things supply chain management. *Sensors, 21*(5), 1759. https://doi.org/10.3390/s21051759.
3. Kent, A. D. (2015). Comprehensive, multi-source cyber-security events data set. Los Alamos National Lab. (LANL), Los Alamos, NM (United States), Tech. Rep.
4. Alazab, A., Hobbs, M., Abawajy, J., & Alazab, M. (2012). Using feature selection for intrusion detection system. In *2012 International Symposium on Communications and Information Technologies (ISCIT)* (pp. 296–301). IEEE. https://doi.org/10.1109/ISCIT.2012.6372069.
5. Bohara, A., Noureddine, M. A., Fawaz, A., & Sanders, W. H. (2017). An unsupervised multi-detector approach for identifying malicious lateral movement. In *2017 IEEE 36th Symposium on Reliable Distributed Systems (SRDS)*.
6. Banerjee, A. (2019). Blockchain with IoT: Applications and use cases for a new paradigm of supply chain driving efficiency and cost. *Advances in Computers, 115*, 259–292. https://doi.org/10.1016/bs.adcom.2018.09.004.
7. Strom, B. E., Applebaum, A., Miller, D. P., Nickels, K. C., Pennington, A. G., & Thomas, C. B. (2018). *Mitre ATT&CK: Design and Philosophy*. Technical report.
8. Cavalcante, I. M., Frazzon, E. M., Forcellini, F. A., & Ivanov, D. (2019). A supervised machine learning approach to data-driven simulation of resilient supplier selection in digital manufacturing. *International Journal of Information Management, 49*, 86–97. https://doi.org/10.1016/j.ijinfomgt.2019.03.003.

9. Djidjev, H., Sandine, G., Storlie, C., & Vander Wiel, S. (2011). Graph-based statistical analysis of network traffic. In *Proceedings of the Ninth Workshop on Mining and Learning with Graphs*.
10. Dolgui, A., & Ivanov, D. (2022). 5G in digital supply chain and operations management: Fostering flexibility, end-to-end connectivity and real-time visibility through Internet-of-everything. *International Journal of Production Research, 60*(2), 442–451. https://doi.org/10.1080/00207543.2021.1900966.
11. Mimo, E. M., & McDaniel, T. (2022). Smart cities: A survey of tech-induced privacy concerns. In *Big Data Privacy and Security in Smart Cities* (pp. 1–22). Springer International Publishing. https://doi.org/10.1007/978-3-030-75414-6_1.
12. Ismagilova, E., Hughes, L., Rana, N. P., & Dwivedi, Y. K. (2022). Security, privacy and risks within smart cities: Literature review and development of a smart city interaction framework. *Information Systems Frontiers, 24*(2), 393–414. https://doi.org/10.1007/s10796-021-10148-9.
13. Fawaz, A., Bohara, A., Cheh, C., & Sanders, W. H. (2016). Lateral movement detection using distributed data fusion. In *2016 IEEE 35th Symposium on Reliable Distributed Systems (SRDS)*.
14. Greco, A., Caponi, A., & Bianchi, G. (2016). Facing lateral movements using widespread behavioral probes. In *Internet Technology and Secured Transactions (ICITST), 2016 11th International Conference for* (pp. 91–98). IEEE. https://doi.org/10.1109/ICITST.2016.7775692.
15. Glithero, B. (2017). *Better Threat Detection and Response with Analytics for Lateral Movement.* Available at: https://content.pivotal.io/blog/better-threat-detection-and-response-with-analytics-for-lateral-movement (accessed on 11 November 2024).
16. Hogan, E., Johnson, J. R., & Halappanavar, M. (2013). Graph coarsening for path finding in cybersecurity graphs. In *Proceedings of the Eighth Annual Cyber Security and Information Intelligence Research Workshop* (pp. 1–9). ACM. https://doi.org/10.1145/2483975.2483987.
17. Marchetti, M., Pierazzi, F., Colajanni, M., & Guido, A. (2016). Analysis of high volumes of network traffic for advanced persistent threat detection. *Computer Networks, 109*, 127–141. https://doi.org/10.1016/j.comnet.2016.01.027.
18. Oberle, A., Larbig, P., Marx, R., Weber, F. G., Scheuermann, D., Fages, D., et al. (2016). Preventing pass-the-hash and similar impersonation attacks in enterprise infrastructures. In *2016 IEEE 30th International Conference on Advanced Information Networking and Applications (AINA)* (pp. 309–316). IEEE. https://doi.org/10.1109/AINA.2016.88.
19. Shchur, O., Mumme, M., Bojchevski, A., & Gunnemann, S. (2018). *Pitfalls of Graph Neural Network Evaluation. arXiv preprint arXiv:1811.05868.*
20. Goyal, P., Kamra, N., He, X., & Liu, Y. (2018). *Dyngem: Deep Embedding Method for Dynamic Graphs. arXiv preprint arXiv:1805.11273.*
21. Roldán-Molina, G., Almache-Cueva, M., Silva-Rabadão, C., Yevseyeva, I., & Basto-Fernandes, V. (2017). A comparison of cybersecurity risk analysis tools. *Procedia Computer Science, 121*, 568–575. https://doi.org/10.1016/j.procs.2017.11.074.
22. Siadati, H., & Memon, N. (2017). Detecting structurally anomalous logins within enterprise networks. In *Proceedings of the 2017 ACM SIGSAC Conference on Computer and Communications Security* (pp. 1061–1076). ACM. https://doi.org/10.1145/3133956.3134075.

23. Siadati, H., Saket, B., & Memon, N. (2016). Detecting malicious logins in enterprise networks using visualization. In *Visualization for Cyber Security (VizSec), 2016 IEEE Symposium on* (pp. 87–94). IEEE. https://doi.org/10.1109/VIZSEC.2016.7597509.
24. Hamilton, W. L., Ying, R., & Leskovec, J. (2017). *Inductive Representation Learning on Large Graphs. arXiv preprint arXiv:1706.02216.*
25. Yüksel, A. S., Yüksel, M. E., Sertbaş, A., & Zaim, A. H. (2017). Implementation of a web-based service for mobile application risk assessment. *Turkish Journal of Electrical Engineering and Computer Sciences, 25*(2), 976–994. https://doi.org/10.3906/elk-1602-33.
26. Yaroson, E. V., Breen, L., Hou, J., & Sowter, J. (2021). Advancing the understanding of pharmaceutical supply chain resilience using complex adaptive system (CAS) theory. *Supply Chain Management, 26*(3), 323–340. https://doi.org/10.1108/SCM-07-2020-0311.
27. Yu, X., Li, C., Shi, Y., & Yu, M. (2010). Pharmaceutical supply chain in China: Current issues and implications for health system reform. *Health Policy, 97*(1), 8–15. https://doi.org/10.1016/j.healthpol.2009.12.008.
28. Zhang, X., & Zhao, J. (2019). The impact of big data on supply chain resilience: The moderating effect of supply chain complexity. In *Proceedings of 18th Wuhan International Conference on e-Information Systems and Operations Management* (pp. 479–486).
29. Zhou, X. (2022). *China's Prescription Drug Market Size and Policy Analysis of the Prescription Drug Market (in Chinese).*
30. Zamiela, C., Hossain, N. U. I., & Jaradat, R. (2022). Enablers of resilience in the healthcare supply chain: A case study of US healthcare industry during COVID-19 pandemic. *Research in Transportation Economics, 93*, 101174. https://doi.org/10.1016/j.retrec.2022.101174.
31. Zahiri, B., Zhuang, J., & Mohammadi, M. (2017). Toward an integrated sustainable-resilient supply chain: A pharmaceutical case study. *Transportation Research Part E: Logistics and Transportation Review, 103*, 109–142. https://doi.org/10.1016/j.tre.2017.06.005.

CHAPTER 14

Possess Thorough Knowledge of Zero-Trust Principles

V. SINDHU[1,*] and M. VINAY[2]

[1]*Department of Computer Science, CHRIST University, Bangalore, Karnataka, India*

[2]*CHRIST University, Bangalore, Karnataka, India*

*Corresponding author

ABSTRACT

In today's ever-evolving cybersecurity landscape, the traditional perimeter-based security model is no longer sufficient to protect organizations from advanced threats and breaches. The Zero Trust framework has emerged as a paradigm shift in security strategy, focusing on the fundamental principle that trust should never be assumed, even within a network's boundaries. This abstract explores the foundational concepts of Zero Trust, emphasizing its core principles and components.

Zero Trust security is rooted in the belief that organizations should not automatically trust any user or device, whether inside or outside the corporate network. This approach challenges the traditional "trust but verify" model and replaces it with a "never trust, always verify" philosophy. The Zero Trust model is designed to enhance security by continuously validating the identity and security posture of users, devices, applications, and data.

14.1 INTRODUCTION

The traditional perimeter-based security approach is put to the test by the paradigm-shifting concept of zero trust, which also ushers in a new era of

Zero-Trust Learning: Applications in Modern Network Security. Chetna Kaushal, Amandeep Kaur, Mohit Angurala, Aryan Chaudhary (Eds.)

network and data protection. A "never trust, always verify" mentality is adopted by a zero-trust framework, in contrast to the usual idea that everything in a network should be trusted [1]. It operates under the tenet that no user or device, regardless of their location or level of privileged access, should be automatically given access to critical resources.

John Kindervag, who once worked for Forrester Research, first proposed the idea. He highlighted the critical vulnerability in enterprises' security procedures in 2008, where anything inside the internal firewall was trusted, but anything outside was not (Figure 14.1).

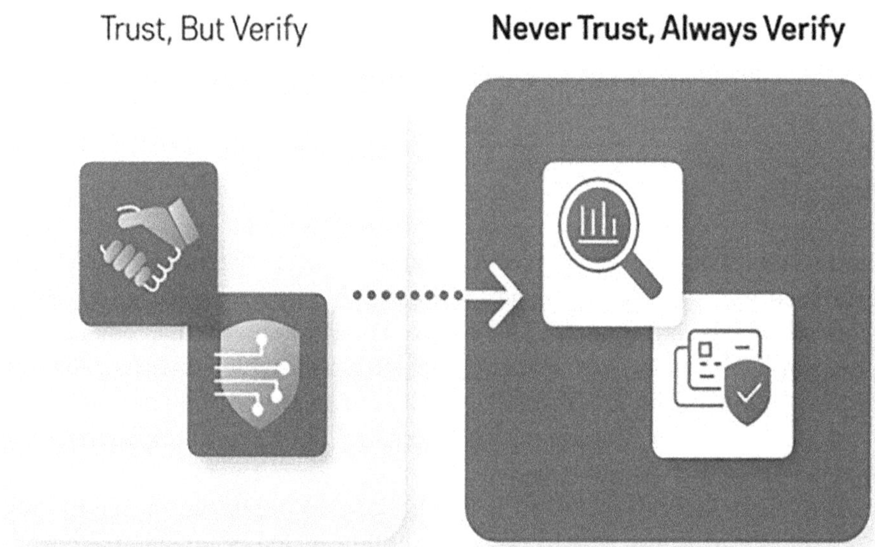

FIGURE 14.1 Understanding of zero trust.

On the other side, a zero-trust security approach uses advanced threat intelligence, micro-segmentation, and multi-factor authentication (MFA) for all devices, accounts, and applications [2]. This guarantees that, regardless of their location, only authorized and authenticated entities have access to assets. This segmented architecture restricts lateral movement even if an internal account is compromised, reducing the potential harm a rogue insider can do [3].

Now that we've discussed what Zero Trust is and how it works, it's time to examine its advantages and application scenarios in the context of contemporary cybersecurity.

14.1.1 *WHAT ZERO TRUST SECURITY IS ABOUT*

The annual Cybercrime report stated that "Cybercriminal activity is one of the biggest challenges that humanity will face in the next two decades" and made a number of ominous projections, such as that by 2021, the global cost of cybercrime will have halved from what it was in 2015 [4].

Additionally, a survey by Cyber Security Ventures predicted that by 2025, cybercrimes alone will cost the world $10.5 trillion. Furthermore, it has been determined that outsiders are responsible for 70% of data breaches. The organizational efforts and extravagant spending in their cybersecurity sector are mocked by these numbers.

Organizations need to take a more proactive approach to network and data security in light of the increasing incidence of large-scale data breaches [5], which can harm both revenue and reputation. The most successful approach may turn out to be the zero-trust concept. Moats protected the castles for ages, and the walls were nearly impossible to scale [6].

They were forced to implement more sophisticated security measures because of the development of stronger weapons with the improvements. In the world of network security, the same change is taking place. The security principle was developed by John Kindervag, the former Principal Analyst of Forrester Research Inc., and is based on the idea of not automatically trusting anything or anyone inside or beyond the network borders.

The Zero Trust Model is becoming more well-known and is making its way into the mainstream of the cybersecurity industry for better outcomes and stronger cyber security.

14.2 THE CORE PRINCIPLES OF THE ZERO TRUST MODEL

The guiding concepts for implementing zero trust in organizations are listed below.

14.2.1 *DETAILED ASSESSMENT OF ACCESS CONTROLS*

The zero-trust paradigm does not trust any parties since it believes that both within and outside of a network there could be possible attackers [7]. Each access request must be permitted and encrypted, and all users or devices trying to access network resources must be authenticated.

14.2.2 *VARIOUS PREVENTATIVE METHODS*

A number of preventive approaches are available to stop breaches and lessen their harm. The most popular technique for authenticating user identification is MFA. To establish credibility, the user must offer at least two different types of proof [8]. These could involve logic-based tests, SMS or email confirmations, or security questions. The quality of the network improves with increased access requirements.

Another step in building trust is limiting access for those who have been authenticated. Each user or device only has access to the bare minimum of resources needed, reducing the network's overall vulnerability to attack [9]. All other traffic is still blocked, preventing trusted entities from moving laterally.

A network security strategy called micro-segmentation divides networks into zones, each of which needs its own network access. Even once security has been compromised, the damage a hacker can cause is still restricted to the microsegment they were able to access.

14.2.3 *MONITORING IN REAL-TIME TO SPOT MALICIOUS ACTIVITY*

The zero-trust paradigm focuses mostly on prevention. Real-time monitoring is crucial in addition to preventive measures since it can reduce the amount of time between a breach and the point at which a danger spreads to further systems on the network [10]. Attackers' window of opportunity is reduced through quick monitoring, which enables identification, inquiry, and correction.

14.2.4 *COMPATIBILITY WITH THE BROADER SECURITY STRATEGY*

The zero-trust concept is insufficient as a stand-alone security approach. To guarantee network security, endpoint monitoring, detection, and incident response capabilities are essential [11, 12]. Despite being crucial, technological solutions cannot take the place of a comprehensive security strategy that considers the organization's overall security requirements and compliance requirements.

14.3 KEY ELEMENTS OF THE ZERO TRUST FRAMEWORK

Zero Trust is a comprehensive cybersecurity framework that challenges the traditional "trust but verify" model and, instead, adopts a "never trust, always verify" approach [13]. It's built on the principle of assuming that threats could exist both inside and outside an organization's network. Key elements of the Zero Trust framework included here.

14.3.1 IDENTITY AND ACCESS MANAGEMENT (IAM)

- **Strong Authentication:** Verify the identity of users, devices, and applications through MFA to ensure that only authorized entities gain access [14].
- **Authorization:** Enforce least privilege access, ensuring that users and devices have only the minimum necessary permissions to perform their functions.

14.3.2 MICRO-SEGMENTATION

- **Network Segmentation:** Divide the network into smaller, isolated segments, restricting lateral movement for potential threats. This containment strategy limits the damage an attacker can do if they breach one segment.

14.3.3 CONTINUOUS MONITORING

- **Behavior-Based Analytics:** Continuously monitor and analyze user and device behavior, network traffic, and application access patterns [15]. Anomalies are detected in real time, allowing for immediate response to potential security incidents.

14.3.4 DATA ENCRYPTION

- **Data-at-Rest Encryption:** Protect sensitive data by encrypting it when it is stored on devices or servers. This ensures data confidentiality even if physical access is compromised [16].
- **Data-in-Transit Encryption:** Encrypt data as it travels across networks, making it unreadable to potential eavesdroppers.

14.3.5 *SECURE ACCESS SERVICE EDGE (SASE)*

- **Convergence of Network and Security:** SASE combines network and security functions into a unified, cloud-based service [17]. This approach ensures consistent security policies and user experiences across the network, regardless of where users are located.

14.3.6 *ZERO TRUST ARCHITECTURE*

- **Zero Trust Principles:** Implementing a holistic ZTA that adheres to the core principles of continuous verification, least privilege, and the removal of implicit trust.

14.3.7 *POLICY ENFORCEMENT*

- **Policy-Driven Access Control:** Implement granular access control policies based on user, device, application, and contextual factors. These policies determine who can access what resources [18].

14.3.8 *AUTOMATION AND ORCHESTRATION*

- **Automation:** Utilize automation to enforce policies and respond to security incidents in real time. This can include actions like blocking access, quarantining devices, or alerting security teams.

14.3.9 *INTEGRATION AND VISIBILITY*

- **Integration of Security Tools:** Ensure that security tools and technologies, such as firewalls, endpoint protection, and identity and access management systems, are well-integrated to provide a comprehensive security posture [19].
- **Comprehensive Visibility:** Gain real-time visibility into network traffic, user, and device activities, and security events to better understand and respond to threats [20].

14.3.10 CULTURE AND GOVERNANCE

- **Cultural Shift:** Promote a culture of security awareness and accountability throughout the organization. Educate employees on the Zero Trust model and the role they play in maintaining security.

14.3.11 COMPLIANCE AND AUDIT

- **Regulatory Compliance:** Ensure that Zero Trust practices align with relevant industry and government regulations, and perform regular security audits to validate adherence.

Implementing Zero Trust involves a strategic shift in an organization's security posture, aiming to minimize the attack surface, reduce the risk of data breaches, and improve overall cybersecurity. It is not a one-size-fits-all approach and requires customization to suit an organization's specific needs and risk profile.

14.4 THE OPERATION OF ZERO TRUST SECURITY

The Zero Trust concept seeks to improve data security for the organization while ensuring compliance with current laws and providing flexibility to meet the needs of emerging privacy and security regulations [21].

An effective zero-trust model must include data security because hackers frequently try to steal it. While all security measures are crucial, there will always be a serious flaw if data activity is not clearly under control [22, 23]. Forrester has suggested the following priority areas as the "seven pillars of zero trust security" for this purpose (Figure 14.2).

14.4.1 DATA

The Zero Trust strategy starts by safeguarding data before adding more security layers. Even if an attacker manages to get beyond the perimeter security, takes advantage of a configuration error, or directly connects with an insider to get access, there will still be systems in place to identify and automatically respond to unwanted access.

Networking infrastructure must be negotiated to steal data, and the Zero Trust model uses cutting-edge methods like next-generation firewalls to segment, isolate, and restrict network access, making penetration extremely challenging.

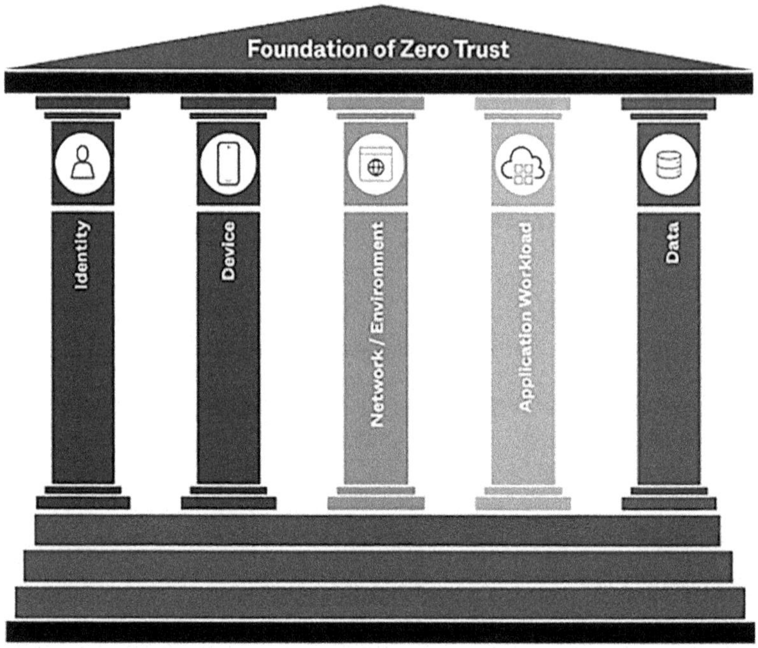

FIGURE 14.2 Fundamental elements of zero trust.

14.4.2 DEVICES

With IoT on the rise, the number of devices on a network has skyrocketed recently and can potentially be used as a vector for intrusion. In the zero-trust approach, each network user's device is isolated and under strict supervision.

14.4.3 WORKLOADS

Workload is a NetOps term that refers to the full stack of back-end programs and applications that allow users to engage with their primary business. These are all frequent access points that intruders employ to breach the networking boundary [24]. The unique Zero Trust compliance control aids in protecting the organization's private data from attacks via these vectors.

Everyone is urged to verify each user's activity on the network and to enact a tight resource-access policy under the Peoples-Zero Trust paradigm. Given that "human resources" are probably the weakest link in preventing unintentional errors, the network security and users must be watched.

14.5 BENEFITS OF ZERO TRUST IMPLEMENTATION

The advantages of zero-trust network architecture are numerous and varied. These advantages are especially relevant to firms that:

- Rely on remote work or a combination of remote and on-site employment.
- Collaborate often with partners and outside contractors.
- Rely on cloud resources and cannot rely on cloud security providers' (CSPs') performance.
- Sensitive data must be protected from insider threats and data breaches.
- Must adhere to strict compliance standards like HIPAA or NIST guidelines.

Adopting a Zero Trust Security technique has the following advantages (Figure 14.3).

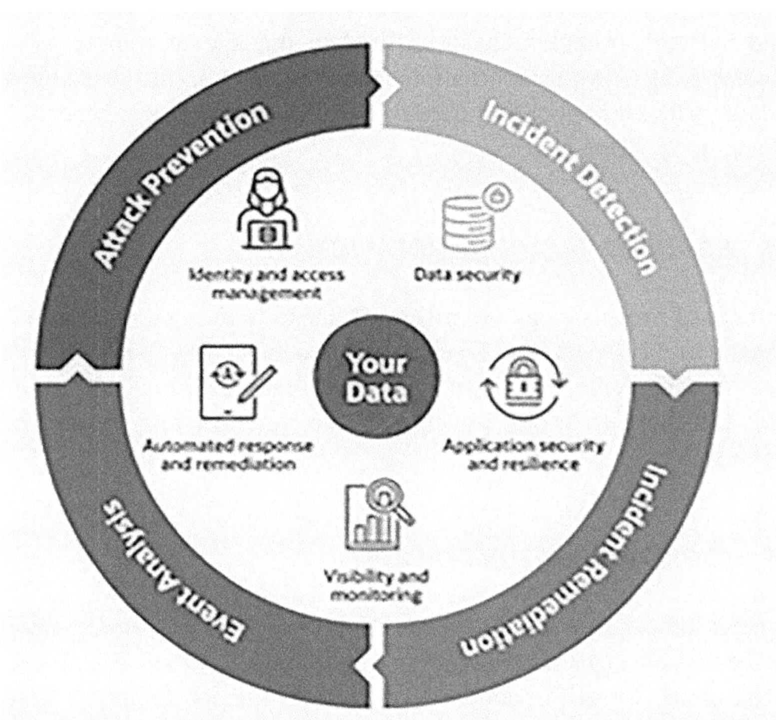

FIGURE 14.3 Advantages of zero trust adoption.

14.5.1 *ENHANCED SECURITY POSTURE*

By compartmentalizing and segmenting the network, Zero Trust lessens the attack surface. This segmentation stops fraudsters from moving laterally, reducing the potential effect reducing the likelihood of breaches and effective attacks.

14.5.2 *REDUCED INSIDER THREATS*

Insider threats, whether deliberate or accidental, present a serious risk. By requiring all users, even those within the company, to verify their identity and access privileges before interacting with sensitive data, a zero-trust strategy reduces this risk.

14.5.3 *ADAPTABILITY TO MODERN WORK ENVIRONMENTS*

The old network boundary has vanished in the age of remote work and cloud computing. Zero trust principles take into account this transformation and retain strict authentication methods while enabling secure access from anywhere, at any time [25].

14.5.4 *REDUCED IMPACT OF BREACHES*

The impact of breaches can be mitigated. Despite being almost inevitable, breaches can be managed. Even if a breach occurs with Zero Trust, attackers must navigate a confusing set of obstacles, making it far more difficult for them to move throughout the network and extract private information.

14.5.5 *CLOUD MIGRATION IS SIMPLE*

To suggest that enterprises today are cloud-based is not an exaggeration! Businesses of all sizes and sectors are switching to cloud-based solutions [26]. The cloud doesn't operate well with traditional software or tools, but cloud solutions work great with the zero-trust approach.

14.6 ISSUES WITH THE ZERO TRUST MODEL

The transition to ZTA from conventional network security is never simple. Despite the fact that software and hardware changes typically don't require total overhauls, there can be some issues. None of them are insurmountable, but they must be considered when implementing a Zero Trust security paradigm.

The following are the sophistications and complexities of the Zero Trust model (Figure 14.4).

FIGURE 14.4 Challenges faced in zero trust.

14.6.1 TECHNICAL DIFFICULTIES

The Zero Trust Model has many facets, one of which is offering the least privilege. The next steps are to identify sensitive data across all systems, micro-segment data sets both within and outside of them, and continuously monitor data flow. The majority of systems are not able to adapt to the Zero Trust model's requirements for micro-segmentation.

14.6.2 DEFECTS IN THE LEGACY SYSTEM

Zero Trust mandates that all users wanting to access all resources must pass multiple verifications. Most outdated systems are unable to provide such a limited level of access control.

14.6.3 PEER-TO-PEER DIFFICULTIES

The majority of other systems, including Windows operating systems and wireless mesh networks, operate on the P2P model [27]. The micro-segmentation model is broken by this paradigm's decentralized operation. As the P2P model violates the least privilege concept since data is shared without significant verification.

14.6.4 TRANSITIONING TO A DATA-CENTRIC SYSTEM FROM SILOS

Most systems in use are data silos with sensitive and general data sets. As the verification and access control in the Zero Trust model is based solely on data, this requires efficient segmentation.

14.6.5 HYBRID CLOUD

The segmentation model of the Zero trust approach fails when the public and private clouds cooperate to provide a single service.

14.7 CREATING A ZERO TRUST SECURITY SYSTEM: FIVE STEPS

Instead of implementing security measures at the network border, this approach focuses on moving them as close to the actual surface that needs to be protected. It stresses better user authentication and repeated validation of the devices over a network. In order to establish zero trust security, companies must go through the processes listed below.

14.7.1 FIRST, DEFINE ZERO TRUST

The best place to start is to choose a broad definition of the Zero Trust model that best fits your individual requirements [28]. The policies and a plan of action that will be used to attain the goals must be defined. It is not necessary to abandon the technology already in use, but rather to adopt a different perspective and adjust to safeguard the significant assets.

14.7.2 RECOGNIZING USER REQUIREMENTS

What effect the Zero Trust model will have on consumers' experiences is among the most crucial factors to take into account [29]. The paradigm requires repeated verification and accessibility authentication because it is about not trusting anyone, either inside or outside the networking boundary. This could ruin the user's experience by interfering with the seamless services and forcing them to wait while their credentials are authenticated.

For the service to run well, all controls and authentication methods must be considered at the outset.

14.8 CONCLUSION

The implementation of Zero Trust is not merely a technological change but a cultural shift that necessitates a new mindset within organizations. The transition is not uniform and must be tailored to meet an organization's specific requirements and risk profile. This abstract serves as an introductory exploration of the foundational principles of Zero Trust security, setting the stage for further investigation into its practical application and the transformative impact it can have in the realm of modern cybersecurity.

KEYWORDS

- **automation**
- **behavior-based analytics**
- **cloud resources**
- **cloud security providers**
- **cybersecurity**
- **encrypt data**
- **firewalls**
- **multi-factor authentication**
- **real-time monitoring**

REFERENCES

1. Dimitrakos, T., et al. (2020). Trust aware continuous authorization for zero trust in consumer Internet of Things. In *2020 IEEE 19th International Conference on Trust, Security and Privacy in Computing and Communications (TrustCom)* (pp. 253–260). IEEE. https://doi.org/10.1109/TrustCom50096.2020.00037.
2. Syed, N. F., Shah, S. W., Shaghaghi, A., Anwar, A., Baig, Z., & Doss, R. (2022). Zero trust architecture (ZTA): A comprehensive survey. *IEEE Access, 10*, 57143–57179. https://doi.org/10.1109/ACCESS.2022.3179359.
3. Li, S., Iqbal, M., & Saxena, N. (2022). Future industry Internet of Things with zero-trust security. *Information Systems Frontiers*, 1–14. https://doi.org/10.1007/s10796-022-10271-3.
4. Gilman, E., & Barth, D. (2017). *Zero Trust Networks*. O'Reilly Media, Incorporated.
5. Kerman, A., Borchert, O., Rose, S., & Tan, A. (2020). Implementing a zero trust architecture. *National Institute of Standards and Technology, 2020*, 17.
6. Alawneh, M., & Abbadi, I. M. (2022). Integrating trusted computing mechanisms with trust models to achieve zero trust principles. In *2022 9th International Conference on Internet of Things: Systems, Management and Security (IOTSMS)*. IEEE. https://doi.org/10.1109/IOTSMS54584.2022.00068.
7. D'Silva, D., & Ambawade, D. D. (2021). Building a zero trust architecture using Kubernetes. In *2021 6th International Conference for Convergence in Technology (I2CT)* (pp. 1–8). IEEE. https://doi.org/10.1109/I2CT51093.2021.9449066.
8. Xiaopeng, T. I. A. N., & S. O. N. G. Haohao. (2021). A zero trust method based on BLP and BIBA model. In *2021 14th International Symposium on Computational Intelligence and Design (ISCID)* (pp. 354–358). IEEE. https://doi.org/10.1109/ISCID53674.2021.00080.
9. Wang, X., Mansour, S., & El-Said, M. (2022). Introducing Zero Trust in a cybersecurity course. In *Proceedings of the 23rd Annual Conference on Information Technology Education* (pp. 83–89). https://doi.org/10.1145/3501408.3509431.
10. Alevizos, L., Eiza, M. H., Ta, V. T., Shi, Q., & Read, J. (2022). Blockchain-enabled intrusion detection and prevention system of APTs within zero trust architecture. *IEEE Access, 10*, 89270–89288. https://doi.org/10.1109/ACCESS.2022.3192823.
11. Teerakanok, S., Uehara, T., & Inomata, A. (2021). Migrating to zero trust architecture: Reviews and challenges. *Security and Communication Networks, 2021*, 1–10. https://doi.org/10.1155/2021/6696789.
12. Phiayura, P., & Teerakanok, S. (2023). A comprehensive framework for migrating to zero trust architecture. *IEEE Access, 11*, 19487–19511. https://doi.org/10.1109/ACCESS.2023.3236789.
13. Garbis, J., & Chapman, J. W. (2021). *Zero Trust Security: An Enterprise Guide*. A Press.
14. Alawneh, M., & Abbadi, I. M. (2022). Integrating trusted computing mechanisms with trust models to achieve zero trust principles. In *2022 9th International Conference on Internet of Things: Systems, Management and Security (IOTSMS)* (pp. 1–6). IEEE. https://doi.org/10.1109/IOTSMS54584.2022.00068.
15. Ramezanpour, K., & Jagannath, J. (2022). Intelligent zero trust architecture for 5G/6G networks: Principles, challenges, and the role of machine learning in the context of O-RAN. *Computer Networks, 217*, 109358. https://doi.org/10.1016/j.comnet.2022.109358.

16. Alipour, M. A., Ghasemshirazi, S., & Shirvani, G. (2022). *Enabling a Zero Trust Architecture in a 5G-Enabled Smart Grid. arXiv preprint arXiv:2210.01739*. https://doi.org/10.48550/arXiv.2210.01739.
17. Greenwood, D. (2021). Applying the principles of zero-trust architecture to protect sensitive and critical data. *Network Security, 2021*(6), 7–9. https://doi.org/10.1016/j.nse.2021.04.001.
18. Chen, X., Feng, W., Ge, N., & Zhang, Y. (2022). *Zero Trust Architecture for 6G Security. arXiv preprint arXiv:2203.07716.*
19. Fernandez, E. B., & Brazhuk, A. (2022). *A Critical Analysis of Zero Trust Architecture (ZTA). SSRN*. https://doi.org/10.2139/ssrn.4210104.
20. Shore, M., Zeadally, S., & Keshariya, A. (2021). Zero trust: The what, how, why, and when. *Computer, 54*(11), 26–35. https://doi.org/10.1109/MC.2021.3074939.
21. Meng, L., Huang, D., An, J., Zhou, X., & Lin, F. (2022). A continuous authentication protocol without trust authority for zero trust architecture. *China Communications, 19*(8), 198–213. https://doi.org/10.23919/JCC.2022.08.015.
22. Bertino, E. (2021). Zero trust architecture: Does it help? *IEEE Security & Privacy, 19*(05), 95–96. https://doi.org/10.1109/MSEC.2021.3077248.
23. Wylde, A. (2021). Zero trust: Never trust, always verify. In *2021 International Conference on Cyber Situational Awareness, Data Analytics and Assessment (CyberSA)* (pp. 1–4). IEEE. https://doi.org/10.1109/CyberSA52022.2021.9519141.
24. Chuan, T., Lv, Y., Qi, Z., Xie, L., & Guo, W. (2020). An implementation method of zero-trust architecture. In *Journal of Physics: Conference Series* (Vol. 1651, No. 1, p. 012010). IOP Publishing. https://doi.org/10.1088/1742-6596/1651/1/012010.
25. D'Silva, D., & Ambawade, D. D. (2021). Building a zero trust architecture using Kubernetes. In *2021 6th International Conference for Convergence in Technology (I2CT)* (pp. 1–8). IEEE. https://doi.org/10.1109/I2CT51093.2021.9449066.
26. Adahman, Z., Malik, A. W., & Anwar, Z. (2022). An analysis of zero-trust architecture and its cost-effectiveness for organizational security. *Computers & Security, 122*, 102911. https://doi.org/10.1016/j.cose.2022.102911.
27. Chen, B., Qiao, S., Zhao, J., Liu, D., Shi, X., Lyu, M., & Zhai, Y. (2020). A security awareness and protection system for 5G smart healthcare based on zero-trust architecture. *IEEE Internet of Things Journal, 8*(13), 10248–10263. https://doi.org/10.1109/JIOT.2020.2981899.
28. Yao, Q., Wang, Q., Zhang, X., & Fei, J. (2020). Dynamic access control and authorization system based on zero-trust architecture. In *Proceedings of the 2020 1st International Conference on Control, Robotics and Intelligent System* (pp. 123–127). IEEE. https://doi.org/10.1109/ICCRIS49750.2020.9190644.
29. Campbell, M. (2020). Beyond zero trust: Trust is a vulnerability. *Computer, 53*(10), 110–113. https://doi.org/10.1109/MC.2020.2980992.

CHAPTER 15

Why Perimeter Security is No Longer Enough: Observations and Open Challenges

NITISH OJHA[1,*] and ABHISHEK VAISH[2,*]

[1]*Amity University, Uttar Pradesh, India*

[2]*Indian Institute of Information Technology, Allahabad, Uttar Pradesh, India*

*Corresponding author

ABSTRACT

Trust is the most important factor, which was a critical deciding factor in all security measures of traditional perimeter security models. However, the dynamic nature of attacks has changed the scenario drastically, where these models failed to address the failures. VPNs, firewalls, DMZs, and other identity-based access models—all security controls—were a good fit for protecting contemporary legacy devices and classical models. However, evolving AI and generative models of AI have pushed these security models to their limitations. In this book chapter, the limitations of these legacy devices have been discussed. A comparative analysis of all devices, comprising VPN, firewall, IAM, DMZ, and so forth, has been conducted, focusing on their demerit cases considering trust as a factor, inside or outside. This chapter has also examined how the advancement of cloud computing, hybrid networks, and complex IT environments, including mobile frameworks, is giving rise to a situation in which the existing model is inadequate. Additionally, it has explored how the zero-trust security framework can offer its assistance in this context.

Zero-Trust Learning: Applications in Modern Network Security. Chetna Kaushal, Amandeep Kaur, Mohit Angurala, Aryan Chaudhary (Eds.)

15.1 INTRODUCTION

In the age of information technology, securing data is an important step in daily life, whether you are connecting to a public Wi-Fi hotspot or scrolling through social media at home. If you are not tied to the chains of cybersecurity, you could be unknowingly vulnerable to fraud. We could fall victim to this, and it is possible. As such, it is important to learn everything you can [1]. Cybersecurity means the security of any type of transaction that is taking place through the internet and a computer is connected somewhere in it. Whenever we talk about security, we not only talk about the security of information, but we also take care of the security of the computer and its connected devices and the network, that is, the security of the cables, routers, hubs, switches, and so forth, used. Good cybersecurity means that your family's personal information has little chance of falling into the wrong hands. Therefore, it is important to implement security measures that protect your devices, and here we are not only talking about personal cybersecurity, but it also includes the cybersecurity of any organization and its business. The business domain has entered a phase in this era where cyber challenges arise more frequently and are more likely to have a significant and long-term impact on its operations. This is creating a huge demand for transformative cybersecurity and risk consulting services focused on modernizing and protecting global enterprises [2]. As the challenges are increasing and new attacks are emerging every day, various types of security measures are being developed. Now, if we talk about traditional cybersecurity, it mainly depends on three types of things: perimeter security, which includes systems like firewalls, IDS, IPS, and VPN with DMZ, and may even include browser isolation systems [3]. One of the best practices in perimeter security is to identify any external threat, regardless of the medium, form, or time the threat is coming. Identifying the threat, finding out its medium, determining its source, where it originated from, or if it came through any network, and which computers it affected along the way, monitoring all this data coming out of monitoring, storing the data, and then conducting pattern analysis, and so forth, are all included in these activities. Essentially, in perimeter security, security experts take a perimeter-based approach to securing systems and eliminating threats as they enter the network, creating rules based on that approach, as well as how to enforce them. However, many sectors of today's security industry have come to the view that perimeter security on its own is less effective for a number of reasons. Data shows that many attacks that occur on all of these devices or data and the elements of our security system are proving to be less successful in facing them, and in some cases, they have proved to be completely inadequate. Looking at research based on long-term surveys, it would not be wrong to conclude that perimeter-based security is neither important anymore nor will it be able to face threats in the long run [4].

15.2 CURRENT SCENARIO

Cyber-attacks are continuously increasing all over the world and after the lockdown of the pandemic disease Covid, these cyber-attacks have increased in all the countries. There are many reasons for this, such as most of the educational institutes started running online and before many such online services started running online, their users were not that prepared, and they also knew less about cyber security. If we look at the worldwide data, the number of cyber-attacks on Indian industries was the second highest globally. This figure comes with a total of 7.7% of incidents in the segment seen in the country in 2021 [5]. It has been reported by cyber security intelligence firm CloudSec. CloudSec has released a report on the "Increase in cyber-attacks on the global healthcare sector." According to a report, the United States of America's healthcare industry has observed multi-varied attacks and out of that it was 28% of the total global attacks observed [6]. In 2021, the Indian healthcare industry faced 7.7% of the total attacks. Due to this, the second-highest number of attacks was recorded in India. Based on the findings of this report, the prevalence of cyber-attacks in the Indian healthcare sector, standing at 7.7%, corresponds to over 7.1 million records [6]. This study conducted by CloudSec Inc. has observed different types of attacks on different Indian sectors and report presented by this company highlights that the healthcare field's technological progress, including aspects like remote health monitoring devices considered under telemedicine, different types of stored electronic health records (EHRs), and the different healthcare entities connected to each-other using Internet of Things (IoT), has significantly expanded the avenues for cybercriminals to target the industry [7]. CloudSec's gathered data indicates a notable trend: during the initial four months of 2022 alone, there have already been 187,000 records compromised.

The attack dynamics have changed in 2022 globally [8]. According to the report, Australia has been the worst-hit country with a record of nearly Rs 70 crore after Thailand, which faced a record of more than Rs 10 crore in the first four months of 2022. The second wave of coronavirus hit India in 2021

Due to which the healthcare sector was also badly affected. Therefore, it is time to take concrete steps to avoid these cyber-attacks now [4].

The critical analysis of these incidents performed by many cyber security firms raises the curtain from the fact that the most common reason is the use of old techniques as countermeasures. For example, the use of online wallets is increasing even in small towns of India or any similar developing countries which is stressing on their digital infrastructure [9]. Here people use it but they do not have enough digital knowledge from a security point of view. Nowadays in India, almost all the data is being given in digital form. All work is linked to an Aadhar card (Indian social security number) [10]. But when the WannaCry ransomware attack targeted the world, it came to light that this attack was done on the old Windows XP operating system [11]. Because of these old perimeter-based security models, critical systems can be easily hacked. There are many places in India where old operating systems can be used indiscriminately. So, the question is, are we prepared to deal with such incidents in the future? The answer lies in perimeter security [12].

15.3 LITERATURE REVIEW

Studies that have been reported in the field of perimeter security varied from different aspects and on a larger canvas, it can be said that the work related to perimeter security has to be summarized by covering different papers published in the following fields.

15.3.1 VERACITY OF USES OF CLASSICAL SECURITY IN DIFFERENT PARADIGM

15.3.1.1 PHYSICAL PERIMETER SECURITY RELATED WORK

In this, basically, physical security-related work can be clubbed. Research on the effectiveness of fencing strategies, deployment of guards, and deployment of advanced surveillance cameras is essential. Many studies have reported that physical security can provide defense only at a certain level, and the attacks that are being received cannot be handled at their full capacity [13]. A physical security breach can occur if a malicious actor attempts to block a server or other device before installation. Managing this kind of risk requires establishing strict security controls within the supply chain model of the organization your business uses to obtain its data center infrastructure [14]. In some cases, the use of magic packets has been reported to turn on the server and launch attacks to disrupt the services. Studies have reported that not only data servers but also network services can be attacked to disrupt the functioning of the organization in the long term [15]. The studies have reported that the following are the common modes of attacks in physical security-based attacks:

- Server Sabotage.
- Damage to N/W equipment.
- Intentional Hardware Failure Attack.
- Cable Collection Attack.
- Computer Theft.
- Harming the Business.

In this era, all types of physical security techniques are proving inadequate to protect users and servers from attacks. There are many physical security measures used in any information technology organization, but the main types of physical security fall into four broad categories: detection, delay, and response. For example, if we pay special attention, then perimeter security, facility control, computer room control and cabinet control have an important role in the security of the server. Encrypted data and personal information are often stored in servers and related rooms where perimeter security-based models are connected with each other, its security depends on a single person or a single token [16]. The lower the security of this token, the lower will be the security of the entire system, whereas if we see that the information in the system is increasing, after some time the security should also increase, whereas it does not happen and along with this, there is a risk of identity theft and system corruption. There is always danger. Data centers and server rooms are also at risk from natural events like floods and fires, and computer-based security pays little attention to all these natural or man-made threats. Many research papers have proved that the threats to physical security are evolving faster than some companies, while the security systems are not getting stronger as fast but now the security system attacks have also increased that is along with data, network Along with this we also have to protect the security system.

The threat of attacks is growing so rapidly that it can only be understood that when credit card skimming was highly practiced crime in this field, companies used special kind of chips to deal with these scams and issues and later clones of these chips also started being made. But it is not that despite all these measures, attacks have not occurred on the servers and will not occur in the future [17]. Physical security for server rooms comes in many forms. Corruption or loss of data can result in delays in operations, loss of revenue, and theft of information and a rule of thumb says that the fewer people access the server room the more secure the server will be [18]. Various types of attacks are taking place through all those multidimensional attacks. If any device or any network component is stolen and this becomes public knowledge, then it has an adverse effect on the people who trust that company. Trust or the brand of that company is made by reputation and here the weakness of physical security is harmful so there is a need for a better security model [19].

15.3.1.2 DEVICE-BASED SECURITY AND ITS LIMITATIONS

It is up to the IT administrators to enable the employees to work from their mobile devices. If the administrators of any organization fail in this, then it is definitely possible that the data of that organization may be stolen due to these devices. If the equipment in any institution is old or its software is not being updated uniformly, then there will always be a risk of any cyber-attack on that institution. If we talk about scalability, interoperability, security, privacy, and reliability, then the standards given by the American institute NIST should also be followed based on that institute and the processes taking place there. Many research studies have highlighted this and cautioned that edge/fog computing, blockchain, SDN, 5G, machine learning wireless sensor networks, and many such emerging technologies can cause serious harm if run on older devices. So, the possibility of cyber-attacks increases [20].

15.3.1.3 REGULARITY COMPLIANCE-BASED SECURITY RESEARCH

It is essential for any organization to follow the relevant compliance standards and regulations to survive any kind of insider or external cyber-attack in the long run. These compliance standards may vary from country to country. These standards depend on many factors such as the nature of the information being used by the organization, how the information is being handled among themselves, and so forth. Some of these are General Data Protection Regulation (GDPR), Sarbanes Oxalate Act (SOX) Payment Card Industry Data Security Standard (PCI DSS), Health Insurance Portability and Accountability Act (HIPAA), and so forth. Mere adherence to compliance norms is not enough to ensure that organizations have established a strong level of prevention against cyber-attacks, this has been proven by much research from time to time [21]. Proposed legislation is imminent, but CISOs and business leaders across the country cannot wait for legislation to be passed to build cyber resilience – especially at a time when the number of data breaches is ever-increasing. It has also come to light in many researches that long-term dependence on the government has its own disadvantages and instead of being dependent on the government, the institution should establish its own standards and implement them, check its quality from time to time and If seen in sequence, the Zero Trust Security Model is completely correct in that we do not use any third party standards, due to this the attacker does not even know where and on which weak link to attack because our own developed standards and its compliance which is indigenous and unique. The defensive elements used are completely different and their vulnerabilities are not publicly disclosed [22].

15.3.1.4 EMERGING ISSUES AND CAUSAL EFFECT

Cyber attackers are constantly developing new and sophisticated attack methods to circumvent traditional security measures, making it challenging to detect and prevent attacks. Cybersecurity can be expensive, and many organizations struggle to allocate sufficient resources to implement effective security measures, leaving them vulnerable to cyberattacks. According to some research, cybercriminals are expanding their plans to attack critical infrastructure. This has made sensing the danger complicated. Companies now have to put emphasis on creating a new security infrastructure [23]. According to a survey conducted by SonicWall in January-June, India with a 133% increase, and Germany with a 52% increase are the countries with the highest number of attacks. Globally, crypto jacking (399%), IoT malware (37%), and encrypted threats (22%) have increased significantly. However, ransomware attacks have declined by 41%. After taking care of all the things to avoid cyber-attacks, it is becoming difficult to avoid it. Cybercriminals are succeeding in their intentions by finding new methods of fraud [24]. As 2023 begins with a big Omicron scare, the cyber world is looking forward to a year when phishing, ransomware, and malware attacks will become more sophisticated. More worryingly, As-a-Service is at risk of increasing popularity from cyber-attacks.

The bar has been lowered for hackers to launch sophisticated attacks because they can only hire or buy skills they don't possess. After COVID-19, many research papers also prove that with the scope of work and study limited to homes and video conferences replacing personal business meetings, cyber security has taken a terrible form. And it's no longer just an enterprise problem. From the common man to government agencies, it threatens to become everyone's problem. In an era when new attacks are being reported every day, one of the easiest, most successful, effective, and long-lasting security measures is Zero Trust Model Security, in which we do not depend on any external factor but on internal components [25].

15.4 PERIMETER SECURITY VS. ZERO TRUST MODEL?

Effective requirement management is essential for cyber security in any organization and its software development and development, whether it is about internal security protecting it from external attacks or keeping the software in the market for a long time. Requirements should include security-related features and functionalities, consider potential risks and threats across all departments of the organization, and comply with regulatory requirements and internationally recommended standards. Zero Trust Model is an IT security approach towards securing any sensitive data while remaining compliant with the new privacy regulations which keep you safe from external cyber-attacks in a different and unique way [26]. As the use of Artificial Intelligence (AI) is increasing rapidly with cloud services, the possibilities of data theft are also increasing, hence there was a need for a cyber security model in which the identity of the parts/users of that system cannot be detected by any external object [27]. Rather than each other, it can be said that there is zero trust in any outsider that is zero trust model. Perimeter-based security system depends on many components, this is its strength as well as its weakness because if any one component is weak then the entire security system can fail [28].

15.4.1 COMPARISON

The perimeter security system is not a separate system, but it is a combination of many types of small security systems, it is a set of firewalls, VPN, Access control, IDS, IPS, SIEM, and email gateway which can protect against any kind of attack in many ways. Protects data and computer from. There is many software available for IDS, Access Control Systems, and so forth, whose internal security technology is also different. Different companies also use different types of algorithms, now if the software is not updated at the right time, then the dangers increase. Much research shows that as more types and variety of tools are used, insecurity will increase in the entire system. When a system is too complex, it leads to vulnerabilities because the possibility of misconfiguration, flaws, or unwanted network access increases [29]. On the basis of the latest research, the following areas have been found in perimeter-based security systems where the possibility of attack is highest:

- Vulnerability in network.
- Vulnerability in operating systems.
- Vulnerability in process and product.
- Vulnerability in human supported execution.

If we compare the zero-trust model and perimeter-based security model, the zero-trust security model works on the exact opposite approach to the perimeter-based model where in the perimeter security model the security of the system depends on a single token and this token (Figure 15.1). The person or computer keeping it is outside the system and is dedicated only to this work, whereas in the zero security model, the work of verifying any new entity is done by some user inside. In this, no one inside or outside the network is trusted by default. Zero trust is driven by requiring verification from every user trying to access a resource from the very beginning, that user must authenticate who they are and how they are connected to that resource, a critical process in which users is authenticated and access to systems, networks, and data is regulated making the resources exceptionally secure [26]. This process involves identifying the user and validating the access rights associated with a particular system. It can also be said that the user is at any level and performing any role and responsibility, but he has to authenticate himself. Otherwise, he will not get the right to use the resources. Authorization is a key function that is the strongest link in the zero-security model. It is the process of authorization as well as authentication, that enables the organizations to manage the identities of its users which acts as a security model for providing access to the servers supporting the network devices and resources, making it very secure by overcoming the vulnerabilities of the perimeter management system [3].

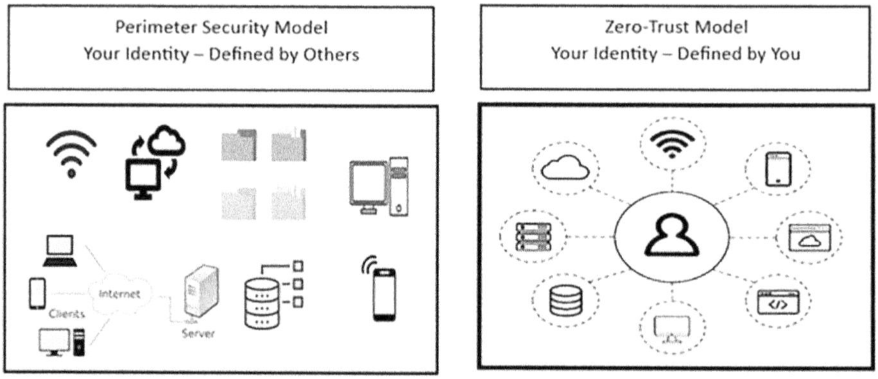

FIGURE 15.1 Comparison between the perimeter security model and zero-trust security model.

The Table 15.1 shows the parameter-based comparison between the perimeter-based security model and the zero-trust security model.

TABLE 15.1 Comparison Between the Perimeter Security Model and Zero Trust Security Model Component Wise

Sl. No.	Point of Concern	Perimeter Security Model	Zero-Trust Model	Observations
1	Scope of Trust	Perimeter-based security models have separate external networks and internal networks. The trust between the internal network and its components is high.	The zero-trust security model challenges the separation boundary between internal and external. It works on "never trust, always verify."	The zero-trust security model is considered better as it relies on the concept that All entities, regardless of their location, are considered untrusted until verified.
2	Access Control	Perimeter-based security model incorporates the access control mechanism through its perimeter where authentication is an important step.	The zero-trust security model relies on multi-dimensional authentication (may be 2D or 3D) and adequate authorization irrespective of the location of the device.	It works on the "Always On Always Verify" Model.
3	Network Architecture	Perimeter-based security model consists of different devices for the protection of its network i.e., Firewall, DMZ, etc.	The zero-trust security model works on the verification of each device instead of getting its verified by any third device.	Not relying on a fixed manner and on a fixed device makes the zero-trust model a better security model.

TABLE 15.1 *(Continued)*

Sl. No.	Point of Concern	Perimeter Security Model	Zero-Trust Model	Observations
4	Monitoring and Control	In the perimeter-based security model, threats are identified using IDS and then resolved at that level.	Instead of depending on a third-party-based device, the zero-trust model continuously monitors the user behavior, traffic, history, and device posture either inside or connected to it.	The concept of Continuous monitoring makes the zero-trust model a better and stronger mechanism.
5	User-Centric Approach	The perimeter-based security model focuses less on users after their authentication.	The zero-trust security model continuously watches the behavior and working pattern of users inside with the help of a context-aware security policy.	Considering the attacks launched by inside or outside users so in the long term, the zero-trust security model robustly stays in long run and stands better.
6	Change and Future Aspects	The perimeter-based cyber security model is less suited to modern working environments where cloud, remote servers, and AI-based mechanisms are involved.	The zero-trust cyber security model is more adaptable to the need for dynamic changing patterns and the distributed environment of networks.	Seeing the need and changing dynamics of current attacks and counter-security mechanisms, the zero-trust security model will lead in the future.

15.5 COMPONENTS OF PERIMETER SECURITY

15.5.1 FIREWALL

If we talk about firewall in network security, then it works in two ways, one is to provide security to the entire system by filtering a particular packet and the other is to filter a message. In perimeter-based security model, we can use software or hardware firewalls available in the market. Security is a precise measure taken by any enterprise or organization to secure its computer network and data using both hardware and software systems. Its purpose is to protect the confidentiality of data and network and access of users as well as to provide security to the entire system. Network security not only protects against external threats such as malwares, spywares and so forth, it also improves network performance and reliability by preventing disruptions and ensuring proper allocation of resources [30]. The most suitable tool for this task is a firewall. If the firewall is not updated with new technologies, the network becomes vulnerable, which greatly increases the chances of exploitation and the number of attacks. In the perimeter-based security model, firewalls are also of two types – hardware and software-based. Hardware Firewalls – Hardware-based firewalls are particularly useful for securing a network in any environment, protecting multiple computers connected to a network, and controlling network activity that attempts to pass through the firewall [31]. The most important thing is that its setting is done properly and configuration is also done correctly. If a professional person is hired to take care of it, then he should take care of it regularly and keep updating its driver software from time to time.

1. **Limitations:** The firewall examines all incoming and outgoing packets and filters out packets that do not meet security policies. Many researchers have found that firewalls lack flexibility, scalability, agility, and transparency. They can only protect local area networks from unauthorized access and are ineffective against advanced threats.

2. **Comparative Security Analysis of Firewall:** A firewall may suffer from the following types of issues which may cause a bigger risk in the end in long term:

    a. Lack of best configuration for a particular organization.
    b. Lack of timely update.
    c. Lack of authorization.
    d. Susceptible towards unknown methods of compromise.
    e. Lack of uniform policy for firewall standard.
    f. Lack of best practices and awareness for long-term maintenance.

3. **Relevance of the Zero-Trust Model with Its Limitations:** In the year 2019, a researcher from the "Data Center Knowledge Group" named Company published a paper entitled "Firewall is Dead." It was a well-explained research paper in which many points were discussed in support of the claim that security cannot be relied on firewalls. A study conducted by "The Security Intelligence Group" published the outcome as a research paper and responded to the comments and conclusions made in previous work, establishing that firewalls can boost security when used with a zero trust model [5]. If we establish the Zero Trust Security Model as a standard security measure in all institutions, then many types of cybersecurity challenges can be addressed in the future, for which we currently spend a significant amount of money on perimeter-based models and lose revenue in the long run. The Zero Trust Model can be implemented in any business organization in many

ways, such as in a holistic form in which it acts as an umbrella, and in another way, it can be implemented on individual devices, including the information coming through the firewall. There are many benefits to implementing it in this way, such as not having to take separate security measures for the VLAN and the fact that attacks on it are greatly reduced because they are not able to reach it. We can use segmentation gateways to obtain information about administrator data access, not just VLANs. From there, they can increase their chances of catching an intrusion before it turns into a full-blown attack and stopping it in its tracks. Segmentation gateways maintain the core principle of the zero-trust model: micro-segmentation [4]. Micro-segmentation employs security policies to establish secure areas not determined by a user's IP address but by the specific applications and data essential for their tasks. This approach minimizes the attack surface by restricting lateral movement within distinct segments of a divided data center, thereby maintaining robust zero-trust security over extended durations, even in the presence of device vulnerabilities.

15.5.2 IDS

Installing intrusion detection software (IDS) is defined as the first line of protection in the organization, meaning this is the first layer that resists threats for any business organization. It provides better security control and monitoring, either for detection or for control; however, this also has some disadvantages at the same time.

1. **Limitations of IDS:** The primary benefit of an IDS is to ensure that IT personnel are notified when an attack or network intrusion is occurring. IDS security monitors both incoming and outgoing traffic in the network of an organization, as well as data traversing between systems within the network. Network IDS devices monitor a large network for network traffic and trigger alerts to the concerned admin when any suspicious activity or known threats are detected in the network, either at inbound traffic or outbound traffic, but it can still generate false alarms. Major issues with IDS include generating false positives and negatives; it may require full-time monitoring compared to other defense models [32]. It is expensive, and unfortunately, there are very few open-source IDS available on the open market with limited capabilities. It may require highly skilled employees to manage its functioning, which again makes it very expensive for a small organization.

2. **Demerits in a Particular Case Scenario** – These are the major issues that may be observed in the case of IDS:

   a. **Signature-Based Issues:** The best way to identify the rogue packets or suspicious packets or messages being received in the network is by detecting it with help of its pattern. Pattern here stands for signature means what type of coding has been used to write that packet. Attacks on such unknown, or zero-day, vulnerabilities naturally lack signatures because they are unknown to the security industry [33].

   b. **Unknown Vulnerability:** It's easy for any device to detect the attack or suspicious file if the signature is matching but, in the case, when no signature is matching then it is difficult for a perimeter-based security mechanism to classify that. Some devices still have the capability to quarantine and some leave on the user's discretion it again creates a problem if the user not aware. Fragmenting and reordering are widely known techniques that IDS/IPS systems are ready to catch, but there are an almost infinite number of evasion combinations and tricks that attackers can use to perform intrusions [34].

   c. **Lack of Policy for Statistical IDS:** Anomaly-based or profile-based signatures also typically detect suspicious network traffic and identify the user using it. These identify anomalies that differ from "normal"; the method of identification depends on the selection of features based on a number of rules. Sometimes this feature may be differently written code; sometimes it may be the nature of the packets being transmitted in that network and the data contained within them. Much research is going on in this direction. It monitors deviations from normal usage patterns. This requires first establishing a baseline profile that can be called a reference model to determine what is standard and what is not, then monitoring actions that fall outside those normal parameters. This parameter can also be measured locally. These may be global or local as determined by NIST. However, the problem with these systems, which are deployed under perimeter security mechanisms, is to eliminate the possibility of abnormal behavior being inappropriately classified as normal. If not eliminated, then it can be blocked for a certain time or quarantined. The major flaws that have been observed in this approach are that it requires you to first define what is normal, and again it depends on the user's ability and knowledge. This means the system and its security mechanism are not mature enough to detect and learn from unclassified data or packets, which is a negative issue of concern in the long term for any organization. If our network is attacked during the learning phase and you fail to recognize it, anomaly-based IPS systems will treat the malicious traffic as normal, and no alarm will be triggered the next time the same attack occurs [35].

15.5.3 DMZ

The Internet is a war zone where multiple people are present with their own objectives, some people want peace, and some people chaos. The external entity and the internal entity should remain at a certain distance from each other and both should not be exposed to each other. When developers considered this problem, they reached for military terminology to explain their goals. And in military terms, a demilitarized zone (DMZ) is a place where two competing groups agree to set aside conflicts to carry out meaningful work and the entry of any outside equipment into this zone will be prohibited. The picture below describes the working of DMZ in a network (Figure 15.2).

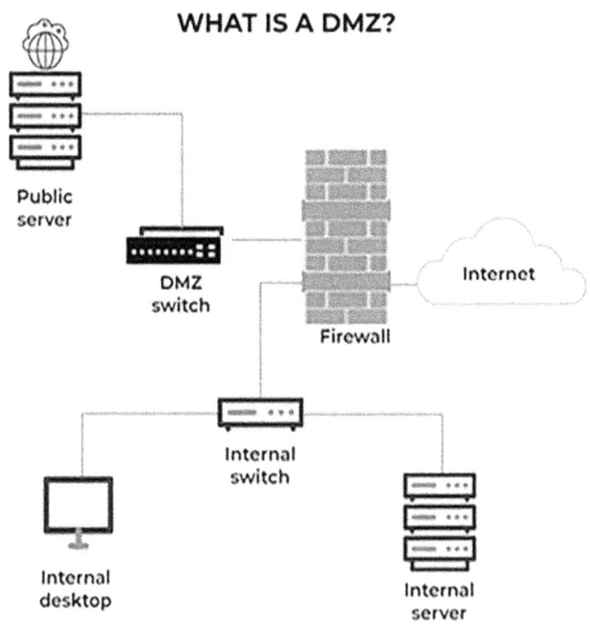

FIGURE 15.2 Working diagram of DMZ in a network.

Source: Adapted from spiceworks. https://www.spiceworks.com/it-security/network-security/articles/what-is-demilitarized-zone/

1. **Limitations:** Network administrators are faced with many configuration options in their organization and network, and having multiple devices is also a problem because each network device will have its own version, different software versions, and different hardware versions. According to that version, the vulnerability of that device will also vary. Researching every weakness of every device can be tedious. Therefore, it is best to use a DMZ, but even there this problem persists, as each DMZ will have its own version, its own software model, and the technology to filter any attack is also different. This can have many benefits, but the DMZ-based security model also comes with limitations. If there is no internal security in place, you will have to assume that no one can be trusted. Your employees and authorized users will still have access to highly sensitive data collected for your company, and you cannot control or revoke it except by creating extremely strict rules that will adversely affect the ability of employees. Sometimes you can conduct observational checks, but this increases the cost factor and complexity of the security model. The second issue with the border along the DMZ is a false sense of security. Every day, hackers come up with new and nefarious ideas on how to circumvent security systems. Even DMZ network servers are not always secure, and you need to keep them updated. You must monitor the environment even after your readiness is complete and officially over practically, and you have to continue to analyze the monitoring data, consuming time and revenue. The third and most discussed issue in dealing with the DMZ is the wastage of time, and this has also been proven by many research papers. Some people in the network research field have the opinion that the cloud has made DMZ networks obsolete and of no use. If you don't have a network server, you won't have anything to protect, which is a dangerous situation for any organization [37].

2. **Demerits in a Particular Case Scenario:** When analyzing a DMZ network, it provides buffer security, one of several security methods that create a buffer between untrusted users and internal segments. The adoption of cloud services and virtualization means that in-house hosting of web servers is no longer necessary. You can store data anywhere through the cloud. For the same reasons, DMZ networks are becoming obsolete, while the Zero Trust Network Architecture model (ZTNA) remains a popular framework. While DMZ networks were initially effective for web hosting organizations in managing

data loss prevention, the surge in application numbers has posed challenges in safeguarding against east-west network movement. Consequently, DMZs are becoming increasingly prevalent, leaving us susceptible to vulnerabilities that compromise adequate security. Traditional network security approaches, which centered on barring perimeters, endpoint users, and threats from entering the network, are proving progressively challenging in the evolving landscape. Cloud services, remote work, and the IoT boom are all trends contributing to increasingly complex network perimeters. As boundaries fluctuate, the process of continuously securing large-scale networks is becoming increasingly difficult for administrators. To top it all off, there are fewer companies making products in this direction, and those companies, instead of conducting new research in this area, have started leaning towards the zero trust security model [38].

15.5.4 BORDER GATEWAY PROTOCOL SUPPORTED ROUTERS

The current version of BGP makes the implementation of the filtering process even more important due to several vulnerabilities that can cause problems such as mis-delivery or non-delivery of traffic, network resource misuse, and congestion, often unintentionally.

- **Limitations:** BGP was not designed with security in mind and can be misused to redirect traffic for malicious purposes. Additionally, BGP incidents can cause widespread disruption. Another possibility for an attacker is to exploit misconfiguration or other vulnerabilities to try to compromise a legitimate peer. Similar DoS vulnerabilities in FR-Routing have already caused significant disruption [39].

15.5.5 UTMS

Unified Threat Management (UTM) is implemented within an organization when multiple security features or services are integrated into a single device to safeguard against external threats consistently. While it may be viewed as a precarious situation from a network administrator's standpoint, the use of UTM combines various protective measures, including antivirus, content filtering, email, and web filtering, anti-spam, and more, all through a unified system. This consolidation of IT security services into a singular device through UTM has the potential to streamline defense against various security attacks over an extended period. Consequently, businesses can efficiently monitor all threats and security-related activities through a unified interface, simplifying oversight and monitoring. This way, you can have complete control over all elements of your security or wireless architecture to gain simplified visibility [40].

- **Limitations:** A Sophos survey report released last year found that more than 5,000 Sophos Firewall appliances (both hardware and software) exposed to Internet access globally were vulnerable to attacks targeting a critical remote code execution vulnerability. This report indicates that remote code penetration is being used as an attack. UTMs can also create a false sense of security. If the rules for identifying risks are not designed correctly, far from the system – and therefore the organization – being secure, the reality is that there are risks that are not being detected. To address this, the team managing the UTM needs to ensure a thorough review and testing of the specifications before implementing any rules. Similarly, users should avoid overusing the UTM functionality. For example, setting too many alerts makes the monitoring task unbearable for the security team, leaving the organization vulnerable again, while assuming it has secured itself. Over-reliance on UTMs should also be avoided, so zero-trust models come into the picture here.

15.6 UPCOMING TRANSFORMATION WHICH DISSOLVES THE PERIMETER SECURITY

15.6.1 USE OF AI FOR ATTACK

AI is playing two types of roles in cybersecurity, both good and bad. Using it for good, we can see how by analyzing different types of patterns, we are able to know what kind of terrorism is happening and what kind of new viruses are being created. However, like other inventions of technology, it also has disadvantages, such as being used to launch new types of attacks. We can use it inside any organization to attack the servers of the same organization. This can benefit the bad guys the most, and there are many research papers that highlight that using AI, we can attack servers that were beyond our reach in the perimeter-based security model. AI also has the potential to be misused, misappropriated, and used to intimidate or bully people, and there have been multiple reports of it being used in cyberbullying. Such examples have been seen appearing on social media every day. The speed and progress of AI may pose a threat to cybersecurity in the future. Its superintelligent powers may be difficult to control in the future.

1. **Current Scenario:** All corporate entities and security agencies have the opportunity to harness cutting-edge AI tools for the enhanced detection of diverse cyber threats, fortifying their systems and safeguarding valuable data resources. Nevertheless, it's crucial to recognize that cybercriminals can exploit advanced technology to orchestrate increasingly sophisticated attacks. In essence, companies operating in the data sector must diligently update their defense models; failure to do so may render them susceptible to substantial threats. The escalating frequency of cyberattacks is serving as a catalyst for the burgeoning market demand for AI-driven security products. A July 2022 report by Acumen Research & Consulting stated that the global market was worth $14.9 billion in 2021 and is bound to reach $133.8 billion by 2030. Bad actors can also take advantage of AI in many ways. For example, AI can be used to identify patterns in computer systems that reveal vulnerabilities in software or security programs, thus allowing hackers to exploit those newly discovered vulnerabilities. Security experts have noted that AI algorithms have made the programs attractive in such a way that people treat them as normal, and sometimes they are lucrative enough to be appealing. AI-generated phishing emails actually have higher rates of being opened, and people don't believe that they may be phishing emails or any rogue software or spam or have any malicious intention behind them – for example, tricking potential victims into clicking on them and thus generating attacks – than manually crafted ones. Phishing category emails are now easier than ever with the help of AI. "The big thing that the Internet did in all of this is that the Internet became a place where all the information in the world was transferred, and later in generative AI, it increased the dangers in it unexpectedly."

15.6.2 INTERNET-DEPENDENT APPLICATIONS

In an application attack, cybercriminals gain access to unauthorized areas. Attackers have started to consider the Application Layer for finding existing vulnerabilities written within the code and then exploiting those. Although attacks target some programming languages more than others, there are attacks on a wide range of applications representing different languages:. NET, Ruby, Java, Node.js, Python, and many others. Vulnerabilities are found in both categories of software, whether it is designed with an open-source concept or is proprietary, meaning it may have issues whether being custom code or open-source frameworks and libraries. There are several reasons why this application security risk is so large, starting with the fact that developers are under increasing pressure to shorten release cycles. Classical application security tools are inaccurate and generate lots of false positives that waste security teams' valuable time in triage and diagnosis. When vulnerabilities are not identified and addressed, bad actors use investigations and targeted attacks to locate and exploit them – often with harmful consequences.

2. **Current Scenario:** Most software programmers secure their applications by creating authentication boundaries. The stronger the authentication, the lower the application security risk, so applications will be more secure. Those with weak security or lacking additional security measures are common targets for application attacks. For example, two-factor authentication adds an extra step to the login process. Users can only gain access using a matching secondary device or account. This is not an industry standard; only some applications add this extra step to their application security measures. The weakest link is if attackers can easily break the access and gain entry with one-step authentication, then user credentials and application infrastructure are at risk, and any attack scenario will only lead to loss and so on. There are examples where these types of attacks have occurred in the classical model of security, which gives birth to the zero trust model. Another common target for cybercriminals is session IDs. They are responsible for communication between the user and the application. Session IDs facilitate the use of applications for users as they store credentials, leading to faster login. Improper management of session IDs allows authentication to extend past logins, increasing the possibility of an application attack. Furthermore, session IDs can be highly predictable, thus making them vulnerable to attacks such as brute force, where attackers guess user credentials until they gain access. When it comes to application security, organizations should maintain strong session IDs, as well as change and update them regularly to prevent malicious HTTP requests.

15.7 CONCLUSION

The strategy of providing a minimum number of privileges to any user on any network has been present in the network field for many years, but it was not always easy to implement due to the complexities involved. For example, the biggest complication is that each user has to be accessed in different ways to certify and enforce their rights. An authorized user can also perform unauthorized tasks. Existing perimeter-based security solutions are not sufficient for all these threats, as they can either cater to the network or application, or take care of the user. The biggest weakness is that there is no way to authenticate both the device and the user on a case-by-case or network-wise basis. There remains a reliance on an external token for security, making perimeter-based security a weak system. It allows users to have broad access to the network with their devices but relies on applications for authentication, and those applications are decided by the users of that network. For example, anyone can log in to a banking website, but only the user with banking access can access the application, and no one else. This security flaw allows attackers to disrupt servers and services in a perimeter-based security model, enabling the use of DDoS attacks, thus making the network insecure and increasing cyber-attacks on it. The solution to all these problems is the Zero Trust Model.

The major benefit is reduced risk control and access to the enhanced security posture of organizations for the long term. Using the zero-trust model, we can assess the current system to determine the current situation and develop a remediation plan as a counter-defense mechanism.

For an organization to understand where to regulate access, where to block access, which users to deny services for how long, and which services to interrupt, it must first identify its data and record the identity of each user. Additionally, the data will have to be kept, and its traffic analysis will need to be conducted. Priority must be decided for each user and the services provided to them. A zero-trust security approach requires protecting data, securing devices, and using automated security systems to reduce complexity and prevent human error in the organization, all of which are carefully monitored and addressed as they occur. If we consider the effective security system in light of the attacks, the Zero Trust Security Model has proven to be the most effective and useful, and it will continue to be beneficial in the future as well.

KEYWORDS

- **artificial intelligence**
- **cyber security**
- **cyber threat**
- **perimeter security**
- **unified threat management**
- **vulnerabilities**
- **zero trust model**

REFERENCES

1. Ma'rufjonov Maqsudjon Mansurjon O'g'li, Murodullayeva Rayhona Abdurahmon Qizi, & Nabiyev Iskandar Farxodjon O'g'li (2023). *Cyber Security, Artificial Intelligence, Data Protection.* Zenodo (CERN European Organization for Nuclear Research). https://doi.org/10.5281/zenodo.8414406.
2. Boyes, H. (2015). Security, privacy, and the built environment. *IT Professional, 17*(3), 25–31. https://doi.org/10.1109/mitp.2015.49.
3. Thatha, V. N., Donepudi, S., Safali, M. A., Praveen, S., Tung, N. T., & Cường, N. H. (2023). Security and risk analysis in the cloud with software defined networking architecture. *International Journal of Power Electronics and Drive Systems, 13*(5), 5550. https://doi.org/10.11591/ijece.v13i5.pp5550-5559.
4. Mwai, R., Wabala, S., & Ogada, K. (2023). Role of ICT risk management on insider fraud prevention in commercial banks in Nairobi County. *International Journal of Technology and Systems, 8*(2), 36–64. https://doi.org/10.47604/ijts.2135.
5. Sureshkumar, S., Venkatesan, G. K. D. P., & Santhosh, R. (2023). Detection of DDOS attacks on cloud computing environment using altered convolutional deep belief networks. *International Journal of Computer Network and Information Security, 15*(5), 63–72. https://doi.org/10.5815/ijcnis.2023.05.06.
6. Linoy, S., Ray, S., Stakhanova, N., & Scheme, E. (2023). Authenticated range querying of historical blockchain healthcare data using authenticated multi-version index. *Distributed Ledger Technologies Research and Practice.* https://doi.org/10.1145/3624575.
7. Chia, H. L. B. (2023). Cloud computing- Benefits, challenges, and promises. *Advances in Engineering Innovation, 2*(1), 38–43. https://doi.org/10.54254/2977-3903/2/2023019.
8. Ullah, I., Noor, A., Nazir, S., Ali, F., Ghadi, Y. Y., & Aslam, N. (2023). Protecting IoT devices from security attacks using effective decision-making strategy of appropriate features. *The Journal of Supercomputing.* https://doi.org/10.1007/s11227-023-05685-3.
9. Khan, M. J. (2023). Zero trust architecture: Redefining network security paradigms in the digital age. *World Journal of Advanced Research and Reviews, 19*(3), 105–116. https://doi.org/10.30574/wjarr.2023.19.3.1785.

10. Durbin, T. J., Bendixsen, C., Neely, A., & Strauß, S. (2023). The big lie: Discursive risk analysis and wildland firefighter safety in the Western United States. *Environmental Research Health.* https://doi.org/10.1088/2752-5309/ad00d5.
11. Sav, S., Diaa, A., Pyrgelis, A., Bossuat, J., & Hubaux, J. (2023). Privacy-preserving federated recurrent neural networks. *Proceedings on Privacy Enhancing Technologies, 2023*(4), 500–521. https://doi.org/10.56553/popets-2023-0122.
12. Sun, J., Li, S., Xu, J., & Huang, J. (2023). The security war in file systems: An empirical study from a vulnerability-centric perspective. *ACM Transactions on Storage, 19*(4), 1–26. https://doi.org/10.1145/3606020.
13. Boeira, F. (2023). Authentic communication and trustworthy location in mobile networks. In *Linköping Studies in Science and Technology*. https://doi.org/10.3384/9789180752572.
14. Verhelst, F., Myren, F., Rylandsholm, P., Svensson, I., Waaler, A., Skramstad, T., Ornas, J. I., Tvedt, B., & Høydal, J. (2010). Digital platform for the next generation IO: A prerequisite for the High North. *Proceedings of SPE Intelligent Energy Conference and Exhibition.* https://doi.org/10.2523/127550-ms.
15. Xiao, P., Qu, W., Qi, H., & Li, Z. (2015). Detecting DDoS attacks against data center with correlation analysis. *Computer Communications, 67*, 66–74. https://doi.org/10.1016/j.comcom.2015.06.012.
16. Weingart, S. H. (2000). Physical security devices for computer subsystems: A survey of attacks and defenses. In *Lecture Notes in Computer Science* (pp. 302–317). https://doi.org/10.1007/3-540-44499-8_24.
17. Ali, B., & Awad, A. I. (2018). Cyber and physical security vulnerability assessment for IoT-based smart homes. *Sensors, 18*(3), 817. https://doi.org/10.3390/s18030817.
18. Rumelili, B. (2013). Identity and desecuritisation: The pitfalls of conflating ontological and physical security. *Journal of International Relations and Development, 18*(1), 52–74. https://doi.org/10.1057/jird.2013.22.
19. Estrin, D., Culler, D., Pister, K. S. J., & Sukhatme, G. S. (2002). Connecting the physical world with pervasive networks. *IEEE Pervasive Computing, 1*(1), 59–69. https://doi.org/10.1109/mprv.2002.993145.
20. Bulğurcu, B., Cavusoglu, H., & Benbasat, I. (2010). Information security policy compliance: An empirical study of rationality-based beliefs and information security awareness. *Management Information Systems Quarterly, 34*(3), 523. https://doi.org/10.2307/25750690.
21. Wen, S., & Katt, B. (2023). Exploring the role of assurance context in system security assurance evaluation: A conceptual model. *Information & Computer Security*. https://doi.org/10.1108/ics-06-2023-0101.
22. Durresi, A., Kaur, D., Uslu, S., & Durrësi, M. (2023). AI control for trust-based detection of attackers in 5G social networks. *Authorea (Authorea).* https://doi.org/10.22541/au.169631656.67218865/v1.
23. Qiu, G., Tang, G., Li, C., Lai, L., Guo, D., & Shen, Y. (2023). Differentiated location privacy protection in mobile communication services: A survey from the semantic perception perspective. *ACM Computing Surveys, 56*(3), 1–36. https://doi.org/10.1145/3617589.
24. Onur, F., Barışkan, M. A., Gönen, S., Kubat, C., Tunay, M., & Yılmaz, E. N. (2023). Detection of cyber attacks targeting autonomous vehicles using machine learning. In *Lecture Notes in Mechanical Engineering* (pp. 439–449). https://doi.org/10.1007/978-981-99-6062-0_40.

25. Galal, T., & Lehmann, A. (2023). Privacy-preserving outsourced certificate validation. *Proceedings on Privacy Enhancing Technologies, 2023*(4), 322–340. https://doi.org/10.56553/popets-2023-0113.
26. Zhang, Q., Geng, G., & Wang, Z. (2023). Internet of things data security management system and method based on zero trust. In *Atlantis Highlights in Engineering* (pp. 459–471). https://doi.org/10.2991/978-94-6463-262-0_49.
27. Gómez-Valiente, P., Pérez, J., Lillo-Castellano, J., & Marina-Breysse, M. (2023). Smart-IoT business process management: A case study on remote digital early cardiac arrhythmia detection and diagnosis. *IEEE Internet of Things Journal, 10*(19), 16744–16757. https://doi.org/10.1109/jiot.2023.3269820.
28. Jellen, I. (2023). Towards security and privacy in networked medical devices and electronic healthcare systems. https://doi.org/10.15368/theses.2020.46.
29. Han, A. H., & Lee, D. H. (2023). Detecting risky authentication using the OpenID Connect token exchange time. *Sensors, 23*(19), 8256. https://doi.org/10.3390/s23198256.
30. Sabahi-Kaviani, Z., & Ghassemi, F. (2023). An enhanced encrypted traffic classifier via combination of deep learning and automata learning. *Research Square (Research Square)*. https://doi.org/10.21203/rs.3.rs-3290610/v1.
31. Smithie, J. (2023). Intrusion detection in cybersecurity. *Advances in Engineering Innovation, 2*(1), 14–17. https://doi.org/10.54254/2977-3903/2/2023014.
32. Qu, F., Wu, Z., Wang, F., & Cho, W. (2015). A security and privacy review of VANETs. *IEEE Transactions on Intelligent Transportation Systems, 16*(6), 2985–2996. https://doi.org/10.1109/tits.2015.2439292.
33. Khan, M. A., & Salah, K. (2018). IoT security: Review, blockchain solutions, and open challenges. *Future Generation Computer Systems, 82*, 395–411. https://doi.org/10.1016/j.future.2017.11.022.
34. Takabi, H., Joshi, J., & Ahn, G. (2010). Security and privacy challenges in cloud computing environments. *IEEE Security & Privacy, 8*(6), 24–31. https://doi.org/10.1109/msp.2010.186.
35. Dart, E., Rotman, L., Tierney, B., Hester, M., & Zurawski, J. (2014). The Science DMZ: A network design pattern for data-intensive science. *Scientific Programming, 22*(2), 173–185. https://doi.org/10.1155/2014/701405.
36. Hunt, R. (1998). Internet/Intranet firewall security—policy, architecture and transaction services. *Computer Communications, 21*(13), 1107–1123. https://doi.org/10.1016/s0140-3664(98)00173-x.
37. Iskandar, A., Virma, E., & Ahmar, A. S. (2018). Implementing DMZ in improving network security of web testing in STMIK AKBA. *International Journal of Engineering & Technology, 7*(2.3), 99. https://doi.org/10.14419/ijet.v7i2.3.12627.
38. Cheminod, M., Durante, L., & Valenzano, A. (2013). Review of security issues in industrial networks. *IEEE Transactions on Industrial Informatics, 9*(1), 277–293. https://doi.org/10.1109/tii.2012.2198666.
39. Butler, K., Farley, T., McDaniel, P., & Rexford, J. (2010). A survey of BGP security issues and solutions. *Proceedings of the IEEE, 98*(1), 100–122. https://doi.org/10.1109/jproc.2009.2034031.
40. UTM Security with Fortinet. (2013). In *Elsevier eBooks*. https://doi.org/10.1016/c2011-0-05893-3.

CHAPTER 16

Cloud-Enhanced Fault-Tolerant Key Selection for Zero-Trust Network Security in Side-Channel Attacks

D. JESLINE,[1,*] B. KANISHA,[1,*] BETTY ELEZEBETH SAMUEL,[2,*] MAHALAKSHMI,[2,*] M. RAMKUMAR RAJA,[3,*] and S. ARUL JOTHI[4,*]

[1]*Department of Computing Technologies, Faculty of Engineering and Technology, SRM Institute of Science and Technology, Kattankulathur, Chennai, Tamil Nadu, India*

[2]*Department of Computer Science, College of Computer Science and Information Technology, Jazan University, Jazan, Kingdom of Saudi Arabia*

[3]*Department of Electrical Engineering, College of Engineering, King Khalid University, Abha, Saudi Arabia*

[4]*Research Analyst, Kanakapushparagam Tech, Erode, Tamil Nadu, India*

*Corresponding author

ABSTRACT

In the context of Zero Trust networks, the difficulty lies mostly in the careful selection of the correct cryptographic key from an extremely wide candidate space, making the creation of attack techniques with fault tolerance a significant issue. In order to solve this pressing problem, this study presents a new method that utilizes Correlation Power Analysis (CPA) in tandem with the Euclidean Distance Enhanced Collision Attack (EDCA) for AES-128 encryption. In the highly secure environment of a Zero Trust network, the EDCA method's use of Euclidean distance to improve collision detection rates is invaluable. This study combines EDCA and CPA concepts to classify

Zero-Trust Learning: Applications in Modern Network Security. Chetna Kaushal, Amandeep Kaur, Mohit Angurala, Aryan Chaudhary (Eds.)

collision pairs of keys and allows for bidirectional selection to get the best collision chain. Key enumeration and successful key recovery are greatly simplified by this narrowing of the candidate space. The experimental findings show the efficacy of the suggested method, with success rates of 98.78% and 80.25% utilizing a collision pair threshold of ≥5 and 3000 energy traces under challenging situations with low signal-to-noise ratios (SNR = –3 dB and SNR = –6 dB). These findings show promise as a step forward in the study of side-channel assaults because they outperform previous approaches, especially in the context of Zero Trust network setups.

16.1 INTRODUCTION

The proliferation of cryptographic techniques has become widespread across a wide range of hardware devices because of the Internet of Things swift progression and mobile Internet. However, when the encryption algorithm is implemented on the hardware device, the key information is affected by the operation time [1], power consumption [2, 3] electromagnetic radiation [4] photon emission [5], and other side channel information leaks. Adversaries can collect and use these leaks to recover key information. The incidents are commonly referred to as side-channel attacks. Divide-and-conquer assaults, such as correlation power analysis (CPA), can be utilized to carry out side-channel attacks [6] and template attacks (TA) [7], the entirety of the key is divided into which parts? Multiple subkeys are successively decrypted. Subsequently, the assailant proceeds to reconstruct the subkeys by use of key enumeration [8–10]. Analytical methods, such as collision attacks, involve resolving a set of equations to retrieve the entire key. These attacks leverage a greater amount of stolen information compared to divide-and-conquer assaults. However, they are also characterized by increased complexity.

Collision attacks are one of the main research directions of energy analysis attacks. Schramm et al. first proposed a collision attack against the Data Encryption Standard (DES) algorithm [11]. Later, similar ideas were applied to crack Advanced Encryption Standard (AES), and collision detection for AES column-transformed input bytes was proposed [12]. The author expanded the meaning of collision, proposed a linear collision attack on AES [13], and made some enhancements to the clashing attack in the literature [14]. The research [15] proposed a multi-bit clashing attack based on dual-range voting detection and improved the fault-tolerance mechanism to enhance attack efficiency. The article [16] proposed a hypothesis testing method, which can detect clashes with a higher probability for the first round of encryption of the AES cryptographic algorithm. This research paper [17] proposed an S-box adaptive selected plaintext collision attack based on the random reuse mask AES algorithm. Literature [18] proposed an improved AES-based selected plaintext collision attack for reuse masks.

To enhance the efficacy of the assault, scholars integrate collision attacks with divide-and-conquer attacks, leveraging collision data to optimize the performance of divide-and-conquer attacks. This combination attack can be divided into template and non-template attacks. A common template attack is to combine CPA and correlation enhanced collision attack (CCA) [19], and the non-template attack combines TA with Minkowski distance enhanced collision attack (MDCA) [20]. The idea of template attacks was first proposed by Bogdanov et al. in the literature [19]. They combined CPA with CCA and proposed a practical solution called test of chain (TC), trying to find the starting point from the first subkeys that are sequentially associated with the long chain of the last subkey. They pointed out that the attack efficiency of this combined method is higher than that of independent CPA or CCA, but in the combined attack based on TC, once a path on the test chain appears with an error, it will cause the entire path to be incorrect, and ultimately the key recovery will fail. In 2014, Das et al. [21] proposed another key recovery scheme called fault tolerant chain (FTC), aiming to discover clashes between the first and the other 15 subkeys. The lack of effective exploitation of collision information necessitates resorting to traditional key enumeration for recovery, resulting in a time-consuming process for subsequent key retrieval.

Article [22] proposed a group collision attack (GCA), which is also a combined attack based on CPA and CCA. GCA partitions the complete key space of the AES-128 algorithm into eight large "groups," each containing four subkeys, and constructs sub-chains within each group through the use of as many collisions as possible, which are continuously spliced from short chains into long chains. 32 pairs of collisions are utilized by GCA, as opposed to 15 pairings in TC and FTC. However, CCA exhibits a significantly diminished performance in comparison to CPA, with a notably low probability of accurately identifying collision pairs. Any collision value exceeding a given threshold may cause key recovery to fail. Therefore, its success rate is relatively low.

When divide and conquer attacks are combined with analysis attacks, the following four situations will occur:

i. The accurate key candidate falls within the threshold interval;
ii. The correct key exceeds the threshold interval of key candidates;

iii. The correct collision pair is within the threshold interval of collision pair candidates; and
iv. The correct collision pair exceeds the threshold interval of collision pair candidates.

The three attack schemes TC, FTC, GCA mentioned above and the scheme in this article all reduce situations (i) and (iv) to situations (i) and (iii) respectively in a fault-tolerant manner, and the so-called fault tolerance is increasing the threshold interval, so the remaining question is how to efficiently solve cases (i) and (iii) This chapter studies a fault-tolerant two-way choice attack that combines Euclidean distance enhanced collision attack (EDCA) and CPA. First, CCA was improved and EDCA was studied. Experimental results indicate that its probability of detecting the correct collision pair is significantly greater than that of CCA at various signal-to-noise ratios, which improves the success rate of the attack. In addition, this chapter combines EDCA and CPA to group the keys and corresponding collision pairs, and then performs two-way filtering and splicing to finally obtain a complete collision chain of 16 key bytes, which greatly reduces the candidate space, making Key recovery is easier, increasing the efficiency of attacks.

This work aims to address the cryptographic key selection challenges in the context of Zero Trust networks, where the wide candidate space for keys poses a significant obstacle to the creation of fault-tolerant attack techniques. Specifically, the study aims to develop a novel method that combines CPA with the EDCA for AES-128 encryption. This method is designed to enhance the security of Zero Trust networks by improving collision detection rates through Euclidean distance, ultimately simplifying key enumeration and recovery. The main goal is to provide a practical solution to reduce the candidate key space and enhance the security of Zero Trust networks.

The Contribution of Work of this chapter is as follows:

1. **Innovative Methodology:** This work introduces a novel approach combining CPA and EDCA to address the key selection problem in Zero Trust networks effectively. Integrating these techniques offers a unique and promising solution to enhance the security of cryptographic systems in such environments.

2. **Improved Collision Detection:** The method significantly improves collision detection rates by leveraging the EDCA. This enhancement is particularly valuable in the highly secure environment of a Zero Trust network where precise key selection is crucial.

3. **Bidirectional Key Selection:** The work introduces bidirectional selection of collision pairs of keys, allowing for the identification of the best collision chain. This approach streamlines the process of key enumeration and successful key recovery, making it more efficient and effective.

4. **Reduction of Candidate Space:** By combining EDCA and CPA, the study narrows down the wide candidate key space. This reduction simplifies identifying the correct cryptographic key, which is a fundamental challenge in zero-trust networks.

5. **Experimental Validation:** The research provides empirical evidence of the proposed method's efficacy. The study's experimental findings demonstrate high success rates of 98.78% and 80.25% under challenging conditions, such as low signal-to-noise ratios (SNR= –3dB and SNR = –6 dB). These results validate the practical applicability of the method.

6. **Advancement in Side-Channel Attacks:** The findings of this work contribute to the field of side-channel attacks, particularly in the context of zero-trust network setups. The proposed method outperforms previous approaches, underscoring its potential as a significant step forward in enhancing the security of Zero Trust networks and cryptographic systems in general.

16.2 PRELIMINARY KNOWLEDGE

16.2.1 TARGET ALGORITHM AES

In this article, the key is mainly recovered for the AES-128 algorithm [23], and the following symbols are used to represent the variables of AES. The 16-byte plaintext and key are respectively represented as $P = \{p_1, p_2, \ldots p_{16}\}$ and $K = \{k_1, k_2, \ldots k_{16}\}$, the first round of 16 S-box operations are recorded as $\{S_1, S_2, \ldots S_{16}\}$, and their inputs are expressed as $x_i = p_i \oplus k_i\}(1 \leqslant i \leqslant 16)$. When encrypted N times, the corresponding plaintext and energy traces are expressed as $P = \{p^1, p^2, \ldots p^N\}$ and $T = \{T^1, T^2, \ldots T^N\}$, respectively, since the 16 S-box operations are performed in sequence, an energy trace can be divided into 16 segments. The energy trace segment corresponding to the ith S box is expressed as $T_i = \{t_{i,1}, t_{i,2}, \ldots t_{i,l}\}(1 \leqslant i \leqslant 16)$, where l represents extraction The number of sampling points of the energy trace obtained. The average energy trace is used in the attack in this article, denoted as $\overline{T} = \{T_1, T_2, \ldots T_{16}\}$.

16.2.2 HAMMING WEIGHT AND EUCLIDEAN DISTANCE

The encryption algorithm computes the quantity of bits in the data that possess a logical value of "1" by employing the Hamming weight. Let the Hamming weight of the data x be expressed as $Hw\ (x)$,x_1 and the Hamming distance between x_1 and x_2 is defined as $HD\ (x_1,x_2)$, then $HD\ (x_1,x_2) = Hw\ (x_1 \oplus x_2)$. Once the energy traces have been aligned, it is presumed that the l sampling points associated with the S box are determined using the Hamming weight model [23], then the mathematical expression of each sampling point on the energy trace is:

$$t_q^{(n)}=s_q\ Hw(x_i^{(n)}+\beta_q r_q(1\leqslant q\leqslant l \tag{16.1}$$

Among them, $x_i^{(n)}$ represents the input of the ith S box in the nth plaintext, s_q represents the scalar coefficient, β_q represents the consumption constant, r_q represents the noise of the energy trace at the sampling point q, and obeys the normal distribution $r_q \sim N\ (\mu,\sigma_q^{\ 2})$. In order to detect whether two S-boxes have collided, the side-channel information leaked during the execution of the target cryptographic algorithm is often used to determine whether the energy traces corresponding to the two S-boxes are similar enough. Given the energy trace segments associated with two S-boxes:

$$\begin{cases} T_i = (t_{i,1}, t_{i,2}, \ldots t_{i,l}) \in IR^l \\ T_j = (t_{j,1}, t_{j,2}, \ldots, t_{j,l} \in IR^l \end{cases} \tag{16.2}$$

Let T_i and T_j correspond to a pair of S-box executions with unknown inputs x_i and x_j. If $x_i = x_j$, it means that T_i and T_j have similar energy consumption. These two energy traces can be regarded as l-dimensional Euclidean space IR^l For two vectors in the similarity between them can be calculated by the least squares' method (least squares method, LSM), that is, Euclidean distance [24] to define Equation 16.3:

$$ED(T_i, T_j) = \Sigma_{q=1}^{l} (t_{i,q} t_{j,q})^2 \tag{16.3}$$

Among them, $t_{i,q}$ is the sampling value of the qth sampling point of the energy trace corresponding to the ith S box, and l is the number of sampling points in the energy trace.

Let $AED\ (x_i,x_j)$ represent the average Euclidean distance when the two input values of S_i and S_j are xi and xj, respectively. In practice, $AED\ (x_i,x_j)$ is usually used to approximate $ED\ (T_i, T_j)$ Mathematical expectation, that is:

$$AED(x_i, x_j) = \frac{1}{N} \sum_{n=1}^{N} \sum_{q=1}^{l} (t_{i,q}^{(n)}) - t_{i,q}^{(n)})^2 \quad (16.4)$$

16.2.3 RELATED ENHANCED COLLISION ATTACK

The correlation-enhanced collision attack (CCA) method was introduced by Literature [25]. The methodology employed in this study involves comparing the correlation coefficients of two separate sets of energy traces associated with two independent S-boxes, rather than detecting a collision between two individual energy traces. The collision that occurs between the first and second S-boxes in the initial round of the AES-128 algorithm can be regarded as an instructive example. During the online phase, the adversary obtains N energy traces that correspond to N plaintexts that have been randomly selected. Following this, during the offline phase, the adversary acquires N energy traces that correspond to N plaintexts that have been randomly selected. Furthermore, during this offline stage, the adversary obtains a set of N energy traces that correspond to N plaintexts that have been randomly selected. The energy trace is partitioned into 16 parts and subjected to alignment processing using the box operation. The energy traces corresponding to the initial S-box and the subsequent S-box are obtained separately. The CCA algorithm employs a partitioning technique to divide the energy traces into 256 segments, utilizing the byte values of their plaintext as the basis for this segmentation process. In this course, we will be computing the mean power usage vector $T_j (j=1,2)$ for each class. Additionally, we will be evaluating the correlation coefficient between these vectors using the assumed :$\Delta_{1,2}$(*Equation* 16.5)

$$\rho(\Delta_{1,2}) \leftarrow \rho\{(T_1^{\alpha} T_2^{\alpha \oplus \Delta_{1,2}}) \mid \alpha = 0,1,...,255\} \quad (16.5)$$

If $\Delta_{1,2}=k_1 \oplus k_2$, then the correlation coefficient $P(\Delta_{1,2})$ reaches a maximum value, otherwise, its correlation coefficient is very low. This correlation enhancement technique uses correlation coefficients to distinguish whether the guessed collision value is correct. It can increase the probability of finding the correct collision by testing several candidates with the highest correlation coefficients. However, this method requires calculating the $\Delta_{1,2}$ for each guessed value. The correlation coefficients are sorted in descending order, and the highest correlation coefficients are finally considered, so the implementation efficiency is low.

16.3 FAULT-TOLERANT TWO-WAY CHOICE ATTACK COMBINING EDCA AND CPA

The combined attack studied in this article is targeted at non-template classes and has fault tolerance, that is, the correct subkeys and collision pairs are within the threshold interval. Under low signal-to-noise ratio conditions, after performing CPA, the correct key is often ranked lower. Set a reasonable threshold Th_k so that the correct key is within the threshold ranking. At this time, the number of traversals required to restore the complete key is Th_k^{16}, the computational complexity is high, and the fault-tolerant two-way selection attack proposed in this chapter combining EDCA and CPA can greatly reduce the candidate space, thereby significantly reducing the complexity of key enumeration and effectively recovering the key.

16.3.1 BASIC PRINCIPLES OF EDCA

The CCA mentioned in Section 2.3 of this article can detect all possible collisions containing two-byte positions. It mainly uses the Pearson correlation coefficient to calculate the association between two sets of energy traces, focusing solely on the energy aspect. The divulged data pertaining to a certain data point along the trail indicates that the largest correlation coefficient corresponds to the correct collision value, but its probability of detecting the correct collision pair is low. Hence, to boost the likelihood of identifying the accurate collision pair and enhance the efficacy of collision detection, this study enhances the CCA method and introduces a novel approach called the Euclidean EDCA. Different from CCA, EDCA uses Euclidean distance to distinguish the similarity between two sets of energy traces and utilizes the leaked information of all sample points on the energy trace to determine whether the guessed collision value is correct. The specific attack process is in Algorithm 1 Given, it includes two stages: online and offline (taking the calculation of $\Delta^*_{1,2}$ as an example).

- **Algorithm 1 Euclidean distance enhanced collision attack**
  - **Online Stage:**

    i. $\boldsymbol{P}=(P^1,P^2,...,P^N) \leftarrow$ *Random plaintexts*()
    ii. $\boldsymbol{T}=(T^1,T^2,...,T^N) \leftarrow$ *Acqurie Traces*(P)
    iii. $\{T^n_1 | n=1,2,...,N\}$, $T^n_2 | n=1,2,...,N\} \leftarrow$ *Intercept Traces* (T)

  - **Offline Stage:**

    1. $\{T^m_1 | m=0,1,...,255\} \leftarrow$ *Average Traces* ($\{T^n_1 | n=1,2,...,N\}$)

ii. $\{T^m_2 | \ m=0,1,...,255\} \leftarrow$ *Average Traces* $(\{T^n_2 | \ n=1,2,...,N\})$

iii. *for each guess* $\Delta \in \{0,1,...,255\}$ *do*

iv. $ED_\Delta \leftarrow AED(\{T_1^{p1} T_2^{p1\oplus\Delta} | p_1 =,0,1,...,255\})$

v. *end*

vi. *return arg min*$ED_\Delta(T_1^{p1} T_2^{p1+\Delta})$

1. **Online Stage:** The attacker randomly selects N plaintexts $P=(p^1, p^2..., p^N)$, each plaintext contains 16 bytes (Random Plain texts), and then collects energy traces to obtain N energy traces T = (T1, T 2, ···, T N) (Acquire Traces). After preprocessing the energy trace, intercept the energy trace segments corresponding to the first two S boxes as $\{T_1{}^n=1,2,...,N\}$ and $\{T_1{}^n=1,2,...,N\}$ (Intercept Traces).
2. **Offline Stage:** The energy trace is divided by the attacker corresponding to each S box into 256 major categories according to the plaintext byte value from 0 to 255, as shown in Figure 16.1, and then compares the energy traces of each category. Average the traces, let T_J^M represent the average energy trace of the j-th S box, where m represents the value of the j-th plaintext byte, so the average energy trace corresponding to the first two S boxes $\{T_1^{M} | \ m=0.1,...,255\}$ and $\{T_2^{M} | \ m=0.1,...,255\}$(AverageTraces). A collision occurs if the value of the first plaintext byte and the second plaintext byte is $\Delta^*_{1,2}= k_1 \oplus k_2$. The attacker traverses the guessed Δ values through $\{0,1, ... ,255\}$, and then calculates $p_2= p_1 \oplus \Delta$ for each value under p_1. For each ED_Δ value, 256 energy traces $T_2^{p1\oplus\Delta}$ can be obtained using Equation 16.4 to calculate the Euclidean distance between T_1^{p1} and $T_2^{p1\oplus\Delta}$(Euclidean Distance), and finally 256 Euclidean distance values are obtained. The correct $\Delta^*_{1,2}$ value is given by:

$$\arg_\Delta min ED_\Delta(T_1^{p1}, T_2^{p1\oplus\Delta}) \tag{16.6}$$

16.3.2 THEORETICAL ANALYSIS OF EDCA AND CCA

Section 3.1 of this article introduces the basic principle of EDCA, which mainly uses Euclidean distance to distinguish the overlap that exists between two different groups of energy traces. In order to illustrate that EDCA can detect the correct collision pair with a higher probability, this section analyzes the distribution of EDCA and CCA. Since the attack scenarios of EDCA and CCA are relatively complex, this article uses fitting to determine their distribution and further compare the success rates. Denote the inputs of the two S boxes

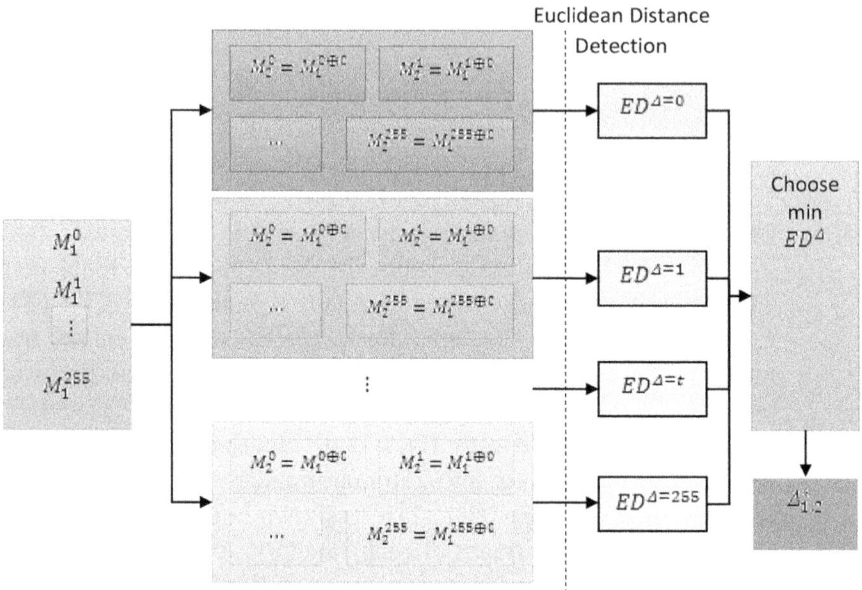

FIGURE 16.1 EDCA collision detection.

as x_1 and x_2, respectively, and let $HD\,(x_1 \oplus x_2) = \nabla$ (∇ =0,1,...,8. By analyzing the calculation results of EDCA and CCA, it is found that when the Hamming distance ∇ of the two S-box inputs is fixed, both EDCA and CCA obey the normal distribution. They can both be fitted by the normal distribution. By calculating the sample mean and variance, the parameters of the normal distribution can be obtained, as shown in Table 16.1. It can be found that the mean values of CCA and EDCA are both linearly related to ∇, the variance of CCA is a nearly quadratic function of ∇, and EDCA is nearly linearly related to ∇.

TABLE 16.1 Mean and Variance of EDCA and CCA

∇	EDCA Mean	EDCA Variance	CCA Mean	CCA Variance
1	0.0100	1.5760E-7	0.3879	0.0024
2	0.0116	2.0705E-7	0.2911	0.0028
3	0.0132	2.5813E-7	0.1949	0.0031
4	0.0148	3.0544E-7	0.0970	0.0033
5	0.0164	3.5653E-7	–0.0009	0.0032
6	0.0180	4.0926E-7	–0.0974	0.0033
7	0.0195	4.5700E-7	–0.1933	0.0030
8	0.0211	5.0802E-7	–0.2919	0.0028

Let the variable Z represent a measure of the correlation of energy traces in CCA and EDCA. For a fixed ∇, the mean value of ∇,Z∇, can be expressed as ξ∇. For CCA, in order to more intuitively understand how the distribution of Z∇ affects the accuracy of different ∇, the distribution of Z_0 (blue curve) and Z_1 (red curve). According to the positions of Z_0 and Z_1, Use ξ∇ to determine whether ∇ =0 or ∇ =1 . When $Z_0 < \frac{\xi_0 - \xi_1}{2}, \nabla = 0$, ∇ =0 will be mistaken for ∇ =1, and calculated by Equation 16.7; when $Z_1 < \frac{\xi_0 - \xi_1}{2}$, ∇ =1 will be mistaken for ∇ =0, and the probability of error is the green area is calculated by Equation 16.8.

$$Pr_0^{er} = \Pr\{Z_0 < \frac{\xi_0 - \xi_1}{2} \tag{16.7}$$

$$Pr_1^{er} = \Pr\{Z_1 < \frac{\xi_0 - \xi_1}{2} \tag{16.8}$$

Similarly, for EDCA, when $Z_1 < \frac{\xi_1 - \xi_0}{2}$, the error probability is calculated by Equation 16.9; when $Z_1 < \frac{\xi_1 - \xi_0}{2}$, the error probability is calculated by Equation 16.10.

$$Pr_0^{er} = \Pr\{Z_0 < \frac{\xi_1 - \xi_0}{2} \tag{16.9}$$

$$Pr_1^{er} = \Pr\{Z_1 < \frac{\xi_1 - \xi_0}{2} \tag{16.10}$$

The mean difference values of CCA and EDCA are different. For better comparison, the two distributions are normalized. This chapter compared when ∇=1 and obtained the error probability diagrams of the two distributions, the green area is the error probability of CCA). It can be seen that EDCA the error probability of is smaller than that of CCA, so the probability of EDCA detecting correct collision pairs is higher than that of CCA.

16.3.3 THE BASIC IDEA OF FAULT-TOLERANT TWO-WAY CHOICE ATTACK COMBINING EDCA AND CPA

In order to enhance the efficacy rate of the attack and effectively recover the key, this chapter studies a fault-tolerant two-way selection attack that combines EDCA and CPA. The basic idea and process are as follows:

1. **Candidate Value Generation:** Set the candidate thresholds of CPA and EDCA to Th_k and Th_Δ respectively, use CPA to obtain Th_k candidates for each subkey, and use EDCA to obtain Th_Δ candidates for each collision pair.
2. **Grouping:** Group with 4 key bytes as the basic unit, and obtain the corresponding collision pair grouping.
3. **Two-Way Selection:** Use the closed-loop relationship to screen the collision chain, and then use the grouped key candidates to counter-screen the collision chain. After two-way screening, the best collision chain is obtained.
4. **Expansion Chain:** Use the connection group to verify and splice the collision chain, and finally splice and expand it into a complete collision chain of 16 key bytes.
5. **Recover the Key:** Record the best complete collision chain (the number of chains is Nc), traverse the remaining candidates of the first key byte (less than Th_k), and obtain less than Nc. Th_k Candidate keys, using a set of plaintext and ciphertext pairs, can be used to verify the correct key.

Let $\xi_j (1 \leqslant j \leqslant 16)$ represent the Th_k key candidate sets of k_j after CPA sorting, and let $\delta j_{1,} j_2 (1 \leqslant j_1 \leqslant j_2 \leqslant j_{16})$ represent the ThΔ collision candidate sets of $\Delta j_1, j_2$ after EDCA sorting. The values of Th_k and Th_Δ are empirical values that can be estimated through multiple experiments. In the experiment, the subkey k1 ~ k8 is used to illustrate the attack process. Assume Th_k=10, Th_Δ =5. The experiment gives candidates with subkeys within Th_k (Table 16.2) and candidates with collision values within Th_Δ (Tables 16.2 and 16.3).

TABLE 16.2 Candidates (k_1 ~k_8) Within the Threshold Th_k

ξ_1	ξ_2	ξ_3	ξ_4	ξ_5	ξ_6	ξ_7	ξ_8
252	12	21	100	48	81	45	25
3	243	234	55	207	174	210	230
2	179	53	36	56	113	109	57
253	76	202	219	199	142	146	198
43	14	20	68	112	17	61	89
212	241	235	187	143	238	194	166
7	28	23	22	120	49	41	121
248	227	232	233	135	206	214	134
6	13	37	102	32	172	57	42
249	242	218	153	223	83	198	213

TABLE 16.3 Collision Value Candidates Within Threshold Th_{Δ}

$\delta_{1,2}$	$\delta_{1,4}$	$\delta_{2,3}$	$\delta_{3,4}$	$\delta_{3,6}$	$\delta_{4,5}$	$\delta_{5,6}$	$\delta_{5,8}$	$\delta_{6,7}$	$\delta_{7,8}$
85	61	107	3	187	63	134	142	124	116
80	60	10	2	186	62	6	143	120	85
121	125	57	67	167	14	154	174	52	117
76	21	106	99	123	54	198	138	24	13
88	28	105	51	218	60	234	150	36	60

The target scenario of this chapter is to overcome the situation where the correct key ranking is relatively deep, that is, Th_k is relatively large. CPA and EDCA were performed at $SNR = -3$dB, and the candidates for each subkey and collision pair were ranked from best to worst. The experimental results are shown in Figures 16.2 and 16.3. For example, the worst-ranked subkey is k_2, which has a rank of 55 in Figure 16.2, while the worst-ranked collision pair is $\Delta_{1,9}$, its ranking is 5 in Figure 16.3. Therefore, to ensure that the key can eventually be recovered, the thresholds Th_k and Th_Δ should be set to at least 55 and 5, respectively, so that the correct key can be found in the huge candidate space.

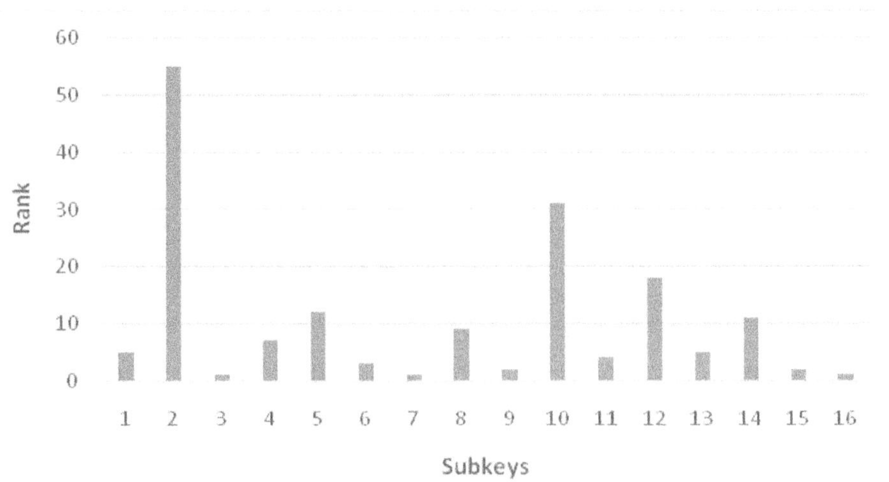

FIGURE 16.2 Rank of correct subkeys.

In the scenario of a low signal-to-noise ratio, the subkeys derived from the execution of the CPA exhibit a rather extensive distribution in the key space. Conversely, the collision values obtained from the execution of the EDCA are positioned at a higher rank. The performance of collision detection is

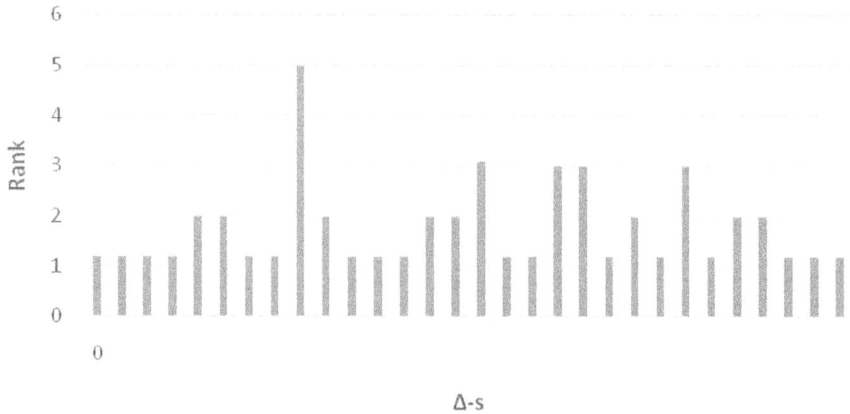

FIGURE 16.3 Rank of correct collision pairs.

better than that of CPA, and the collision information can be used to optimize the key candidates of CPA. Then, the candidates of the collision chain are optimized through CPA.

To enhance comprehension of the divide and conquer attack in collision optimization, three definitions are presented herein: $C^2_{a,b}$, $R_{a,b,c,d}$ and $CC_{a,b,c,d}$. Let $\Delta_{a,b}$ $(a,b \in [1,16]$ denote a pair of collision candidates where the two key bytes k_a and k_b are within the threshold Th_k, and their XOR value is within Th_Δ. Define the set of $\Delta_{a,b}$ as:

$$C^2_{a,b} = \{\Delta_{a,b} = k_a \oplus k_b | \, k_a \in \xi_a, k_b \in \xi_b, k_a \oplus k_b \in \delta_{a,b} \tag{16.11}$$

If 4 subkeys are divided into a group, a ring $R_{a,b,c,d}$ containing subkeys can be constructed:

$$R_{a,b,c,d} = \{(k_a, k_b, k_c, k_d) | \, \Delta_{a,b} \in C^2_{a,b}, \Delta_{b,c} \in C^2_{b,c}, \Delta_{c,d} \in C^2_{c,d}, \Delta_{d,a} \in C^2_{d,a}\}, \tag{16.12}$$

Among them, $a<b<C<d$. The XOR values of the four collision pairs in the ring $R_{a,b,c,d}$ satisfy the relationship $\Delta_{a,b} \oplus \Delta_{b,c} \oplus \Delta_{c,d} \oplus \Delta_{d,a} = 0$, which forms a closed-loop relationship (as shown in Figure 16.4).

Connect each collision pair that satisfies the closed-loop relationship with a chain to obtain a four-long collision chain $CC_{a,b,c,d}$:

$$CC_{a,b,c,d} : \Delta_{a,b} \rightarrow \Delta_{b,c} \rightarrow \Delta_{c,d} \rightarrow \Delta_{d,a} \tag{16.13}$$

After executing CPA and EDCA to obtain the candidate values of the subkey and collision pair $\Delta_{i,j}$, in order to form the ring $R_{a,b,c,d}$ and its closed-loop relationship, these candidate values are grouped.

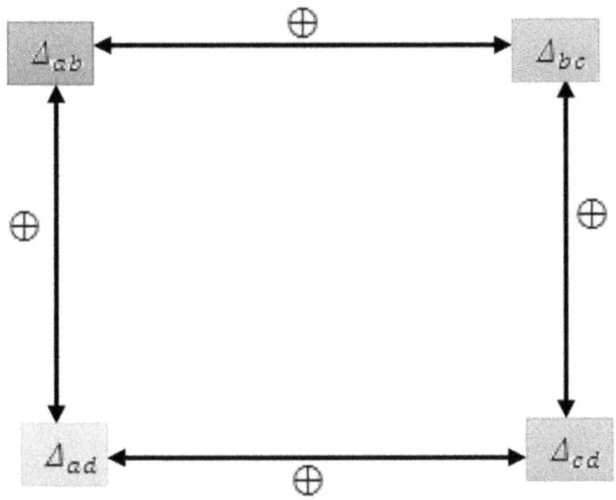

FIGURE 16.4 Closed loop relationship of $R_{(a, b, c, d)}$.

The combined attack schemes TC, FTC, and GCA that have been proposed so far only perform one-way filtering, that is, using collision pairs to filter sub-key information. The scheme in this article performs two-way screening, not only using collision pairs to screen sub-key candidates, but also using sub-key candidates to counter-screen the collision pairs, and obtains several four-length optimal collision chains. The specific screening process is in the algorithm given in 2. In order to obtain a complete collision chain of 16 key bytes, several four-length optimal collision chains need to be spliced. Algorithm 3 shows how to use splicing groups to implement the verification and splicing process.

- **Algorithm 2 Fault-Tolerant Two-Way Selection Algorithm Based on EDCA and CPA**
  - ***Input:***
    1. $\{\xi_j | j = 1,2,\ldots,16\} \leftarrow CPA\ (Th_k)$
    2. $\{\delta_{j,j,j}\ 1 \leqslant j_1 = \leqslant j_2 \leqslant 16\} \leftarrow \text{EDCA}\ (Th_\Delta)$
    3. *Ring* $R_{a,b,c,d}$ containing 4 collision pairs
  - ***Output:*** Filtered optimal collision chain $CC^*_{a,b,c,d}$
    1. *While* $1 \leqslant a \leqslant b \leqslant c \leqslant d \leqslant 16$
    2. $[\Delta_{a,b} \rightarrow \Delta_{b,c} \rightarrow \Delta_{c,d} \rightarrow \Delta_{d,a}] \leftarrow CC_{a,b,c,d}$
    3. *If* $\Delta_{a,b} \oplus \Delta_{b,c} \oplus \Delta_{c.d} \oplus \Delta_{d,a} = 0$
    4. *Record* $CC^1_{a,b,c,d}$

5. *End if*
6. For each $K_a[\xi^1,\xi^2,...,\xi^{Thh}$ *do*
7. Key Chain: $[k_a \rightarrow k_b \rightarrow k_i \rightarrow k_d] \leftarrow CC^1_{a,b,c,d}$
8. $[\Delta_{a,b}, \Delta_{b,c}, \Delta_{c,d}, \Delta_{d,a}] \leftarrow R$
9. *If* $\Delta_{a,b} \oplus \Delta_{b,c} \oplus \Delta_{c.d} \oplus \Delta_{d,a} = 0$
10. *Record* $CC^2_{a,b,}$
11. *End if*
12. *End for*
13. Screen out the optimal collision chain:
$CC^*_{a,b,c,d} = CC^1_{a,b,c,d} \cap CC^2_{a,b,c,d}$
14. *End*
15. *Return* $CC^*_{a,b,c,d}$

- **Algorithm 3 Verification and Splicing of Collision Chains**
  - ***Input:***

1. *Collision pair grouping*
1. $[\Delta_{a,b'} \rightarrow \Delta_{b,'} \rightarrow \Delta_{c,d'} \rightarrow \Delta_{d,a'}] \leftarrow CC^*_{a,b,c,d}$
2. *Collision pair grouping*
2. $[\Delta_{e,f'} \rightarrow \Delta_{f,g'} \rightarrow \Delta_{g,h'} \rightarrow \Delta_{h,e'}] \leftarrow CC^*_{e,f,h,h}$
3. *Connectivity group*:
$[\Delta''_{c,d} \rightarrow \Delta''_{d,e} \rightarrow \Delta_{e,f''} \rightarrow \Delta_{f,c''}] \leftarrow CC^*_{c,d,e,f}$

  - ***Output:*** *Optimal collision chain* $CC^*_{a,b,...h}$
  - **Verify:**

1. *If* $\Delta_{c,d'} = \Delta_{c,d''}$ *and* $\Delta_{e,f'} = \Delta_{e,f''}$
2. *Record* $\{\Delta_{a,b} \rightarrow \Delta_{b,c} \rightarrow \Delta_{c,d} \rightarrow \Delta_{d,a}] \leftarrow CC^*_{a,b,c,d}$
$\{\Delta_{e,f} \rightarrow \Delta_{f,g} \rightarrow \Delta_{g,h} \rightarrow \Delta_{h,e}\} \leftarrow CC^*_{e,f,g,h}$
$\{\Delta_{c,d} \rightarrow \Delta_{d,e} \rightarrow \Delta_{e,f} \rightarrow \Delta_{f,c}\} \leftarrow CC^*_{c,d,e,f}$
3. *End if*

  - ***Splice:***

4. *Splice* $CC^*_{a,b,c,d}$ *and* $CC^*_{c,d,e,f}$
$CC^*_{a,b,...,f}: \Delta_{a,b} \rightarrow \Delta_{b,c} \rightarrow \Delta_{c,d} \rightarrow \Delta_{d,e} \rightarrow \Delta_{e,f}$
5. *Splice* $CC^*_{a,b,...,f}$ *and* $CC^*_{e,f,g,h}$
$CC^*_{a,b,...,h}: \Delta_{a,b} \rightarrow \Delta_{b,c} \rightarrow \Delta_{c,d} \rightarrow \Delta_{d,e} \rightarrow \Delta_{e,f} \rightarrow \Delta_{f,g} \rightarrow \Delta_{g,h}$
6. *Return optimal collision chain* $CC^*_{a,b,...,h}$

- **Step 1:** Execute Algorithm 2 three times, and obtain three sets of optimal collision chains: $CC^*_{1,2,3,4}$, $CC^*_{5,6,7,8}$

- **Step 2:** In order to build a long collision chain containing subkeys k_1~k_8, use algorithm 3 to splice and verify the two collision chains $CC^*_{1,2,3,4}$ and $CC^*_{5,6,7,8}$ (where the connection The group is $CC^*_{3,4,5,6}$, and finally the best long collision chain $CC^*_{1,2,3,4,5,6,7,8}$ (shown in Figure 16.5) is selected, that is:

$$\{\Delta_{1,2}\rightarrow\Delta_{2,3}\rightarrow\Delta_{3,4}\rightarrow\Delta_{4,5}\rightarrow\Delta_{5,6}\rightarrow\Delta_{6,7}\rightarrow\Delta_{7,8}|\Delta_{1,2},\Delta_{2,3},\Delta_{3,4}\in CC^*_{1,2,3,4};\ \Delta_{4,5},\Delta_{5,6}\in CC^*_{3,4,5,6};\ \Delta_{6,7},\Delta_{7,8}\in CC^*_{5,6,7,8}\}$$

- **Step 3:** In order to splice into a long collision chain containing k_1~k_{16}, $CC^*_{1,2,\dots,12}$ can be obtained in the same wayand $CC^*_{1,2,\dots,16}$.
- **Step 4:** Record the best long collision chain $CC^*_{1,2,\dots,16}$ (the number of chains is Nc), traverse the remaining candidates of the first key byte (less than Th_k), and obtain less than $Nc\ Th_k$ candidate keys, using a set of plaintext and ciphertext pairs, the correct key can be checked.

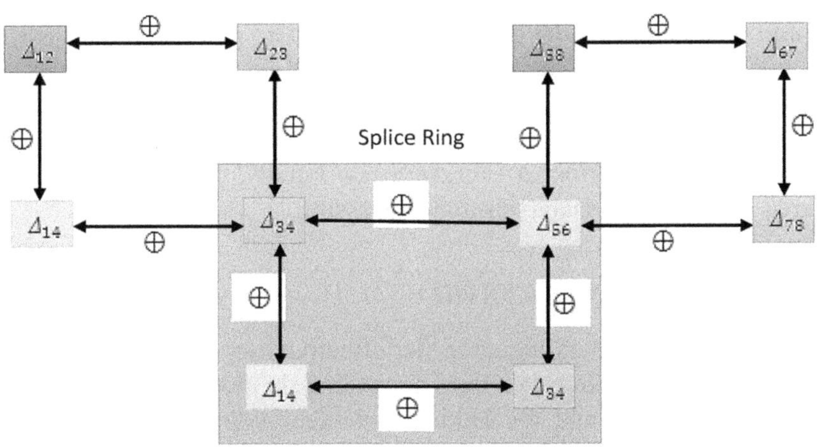

FIGURE 16.5 Splicing of the first and second sets of collision chains.

16.4 PROBABILITY OF SUCCESS

The existing schemes TC, FTC, and GCA are all combinations of CPA and CCA. The scheme proposed in this article is a combination of CPA and EDCA. Therefore, The efficacy rate of the attack is only related to the collision detection scheme, that is, the collision detection of CCA and EDCA is successful. rate directly affects the efficacy rate of the attack.

One crucial aspect of an attack is the necessity for the attacker to make a deliberate selection Th_Δ. P_6 is used here to represent the probability that the correct collision value falls within Th_Δ after collision detection using CCA. P_e is used to represent the probability that the correct collision value falls into Th_Δ after collision detection using EDCA. probability within. Since the TC scheme uses a total of 15 pairs of collisions, the FTC also uses 15 pairs of collisions, the GCA scheme uses 32 pairs of collisions, and the scheme in this chapter uses a total of 24 pairs of collisions. Therefore, under the condition that Th_Δ is given and has the same value, each the attack success rate of each scheme can be expressed as $P_{TC}=P_c^{15}$, $P_{FTC} = P_c^{15}$ $P_{GCA}=P_c^{32}$, and $P_{our\ Method} = P_c^{24}$ From the experimental results, it can be seen that P_e is fargreater than P_c, so:

$$P_{our\ Method} > P_{GCA} > P_{FTC} > P_{TC} \tag{16.14}$$

16.5 ANALYSIS OF RESULTS

This article conducts actual attacks on the ChipWhisperer platform, which is an open-source platform dedicated to conducting research on security in embedded electronics [26]. The hardware version used in this article is CW1173 ChipWhisperer-Lite, and its software version, Jupyter, is used to complete the collection of energy traces. Finally, the data that has been gathered is subjected to analysis and processing.

16.5.1 SIGNAL-TO-NOISE RATIO

A primary determinant influencing the effectiveness of collision detection is the signal-to-noise ratio. It differs depending on the hardware equipment, sampling equipment, and algorithm implementation. As the ratio of the variance of the power consumption component to the variance of the noise, one can calculate the signal-to-noise ratio of time samples on the power consumption energy trace. One can simplify the signal-to-noise ratio as follows (*Equations* 16.15-16.17):

$$SNR = (\frac{Var(s_q)}{Var(r_q)} \tag{16.15}$$

Calculate the variance on both sides of the mathematical expression (1) of the energy trace to obtain the variance relationship of the sampling point:

$$\text{Var}(t) = \text{Var}[sHw(x) + \beta] + \text{Var}(r) = s^2\text{Var}[Hw(x)] + \text{Var}(r) \tag{16.16}$$

Among them, *Var* (*r*) is the variance of the noise, and β is the consumption constant. If each bit of the data x is uniformly distributed, then the Hamming weight of the data x obeys the binomial distribution, and its variance is:

$$\text{Var}[Hw(x)] = 8 \times \frac{1}{4} = 2 \tag{16.17}$$

Since the noise r obeys the normal distribution with mean value 0 and variance σ^2, Equation 16.16 can be simplified as:

$$\text{Var}(t) = 2s^2 + \sigma^2 \tag{16.18}$$

Therefore, the signal-to-noise ratio can be expressed as:

$$SNR = \frac{\text{Var}(s_q)}{\text{Var}(r_q)} = \frac{2s^2}{\sigma^2} \tag{16.19}$$

Author established a special relationship between "the Hamming distance between two S-box inputs" and "the average of the Euclidean distance between the corresponding energy traces" in the literature [27]. Experimental and Theoretical analysis results show that the relationship is linear, and the relationship is as follows:

$$\xi_\Delta = E[ED(T_1, T_2)] = E[\Sigma_{q=1}^{l}(t_{1,q} - t_{2,q})^2] = 2\sigma^2 l + \Delta.\Sigma_{q=1}^{l} s_q^2 \tag{16.20}$$

It can be seen from Equation 16.20 that σ^2 and s^2_q can be calculated when $\Delta = 0 = \Delta = 1$, and are put into Equation 16.19 to obtain the signal-to-noise ratio. The collision detection success rates of EDCA and CCA under different signal-to-noise ratios are analyzed through experiments.

16.6 EDCA COLLISION DETECTION SUCCESS RATE

Before collision detection, the energy trace is segmented and aligned, and the variance is used to locate the energy trace segments corresponding to the 16 S-boxes. As shown in Figure 16.6, the 16 peaks correspond to 16 S-box operations, respectively. This article analyzes the 10,000 energy traces collected. According to Equations 16.19 and 16.20, the inherent SNR = 6.5 dB of the ChipWhisperer Jupyter platform can be calculated.

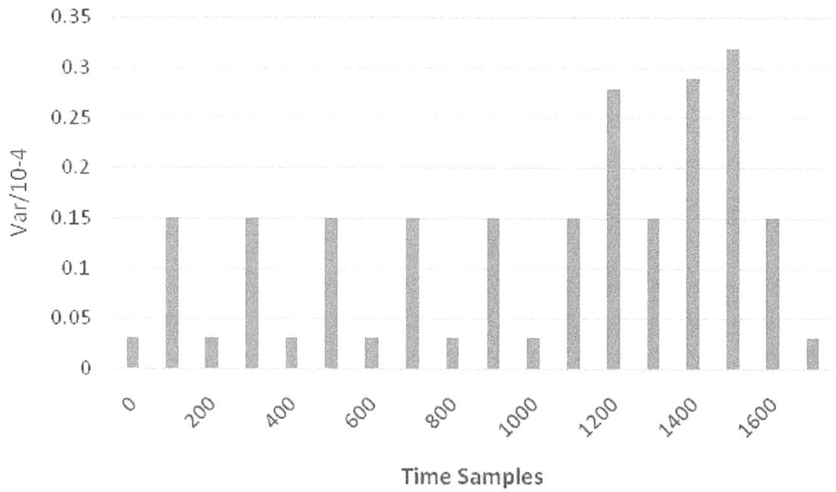

FIGURE 16.6 Variance positioning S box.

During the collision detection phase, experiments were conducted to evaluate the collision detection performance of Carrier Sense Multiple Access with Collision Avoidance (CCA) and EDCA in scenarios with low signal-to-noise ratio. These experiments involved varying the number of energy traces and adjusting the threshold parameter Th_{Δ}. Subsequently, an analysis was performed when the SNR was set to –3dB. The impact of collision detection is examined by two ways under a SNR of –6dB. The experimental findings are visually represented in Figures 16.7 and 16.8.

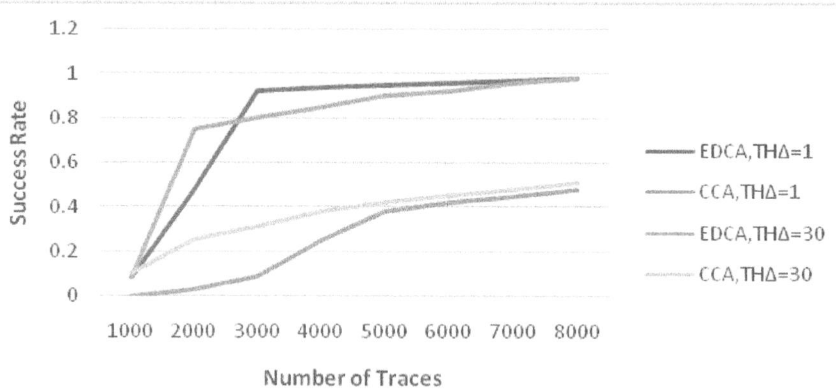

FIGURE 16.7 Comparison of detection success rates between EDCA and CCA when SNR=.–3dB.

It can be seen from Figure 16.7 that under the condition of SNR = –3 dB, when Th_Δ=1, the EDCA method with 5000 energy traces has correct collision pairs that are basically within the threshold, resulting in successful collision detection. The rate reaches 98.17%. However, when the CCA method has 5000 energy traces, nearly half of the correct collision pairs fall outside the threshold, leading to a detection efficacy rate of only 42.58%. Under this condition, compared with CCA, the detection efficacy of the EDCA is increased by 56.59% points. When Th_Δ=30 and the energy traces are 3000, the efficacy rate of EDCA is 99.83%, while the efficacy rate of CCA is 50.92%. Under this condition, compared with CCA, the detection efficacy rate of EDCA is increased by 48.91% points. Therefore, it can be concluded that the collision detection efficacy rate of CCA at Th_Δ=30 is much lower than that of EDCA at Th_Δ=1.

It can be seen from Figure 16.8 that under the condition of signal-to-noise ratio *SNR=6dB*, when Th_Δ=1 and the energy trace is 8000, the detection efficacy rate of the EDCA method reaches 98.58%, while the efficacy rate of the CCA method is only 32.08%. When Th_Δ=30 and the energy trace is 4000, the detection efficacy rate of the EDCA method is 98.83%, and the efficacy rate of the CCA method is 44.17%. According to the experimental results, it can be concluded that in the case of low signal-to-noise ratio, the EDCA collision detection method is better than the traditional CCA method.

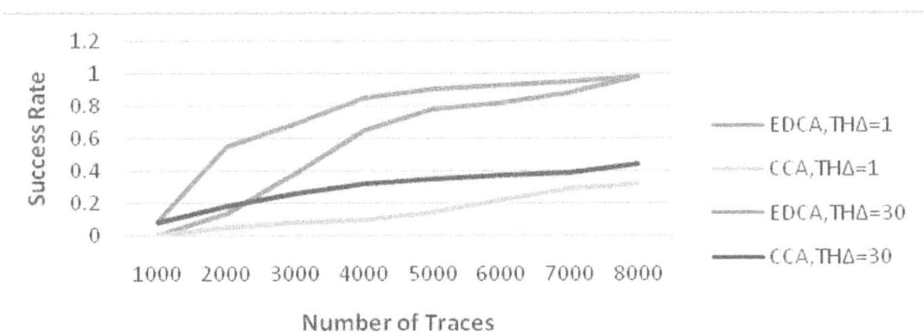

FIGURE 16.8 Comparison of detection success rates between EDCA and CCA when SNR = –6 dB.

16.7 SUCCESS RATE ANALYSIS

In order to analyze the attack efficiency of this article's scheme, this article compared the attack success rates with TC, FTC, and GCA schemes under different $Th\Delta$ and different numbers of energy traces.

16.7.1 EXPERIMENTAL RESULTS UNDER DIFFERENT

This section compares the attack success probabilities under different *Th*Δ, where the number of energy traces is set to 3000 and *Th*Δ is set from 1 to 21. Since the TC scheme in literature [28] and the FTC scheme in literature [29, 30] both set *Th*Δ to 1 and do not provide a practical scheme under larger *Th*Δ, this chapter sets *Th*Δ of TC and FTC to 1, and experiments the results are shown in Figures 16.9 and 16.10.

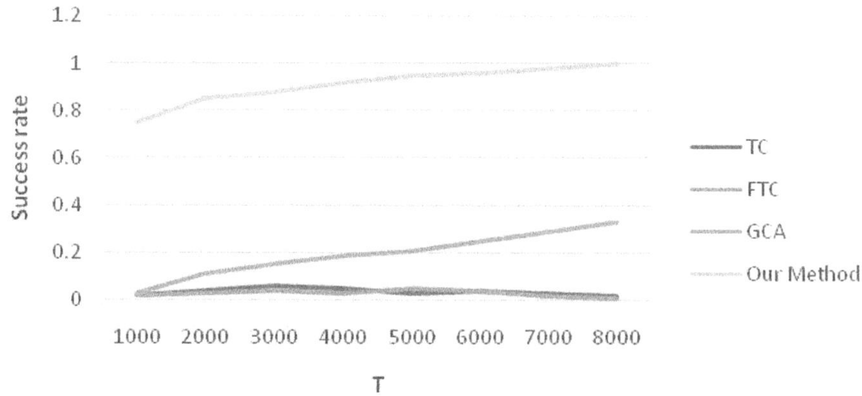

FIGURE 16.9 Success rates of four attack methods under different Th_Δ when *SNR*= -3*dB*.

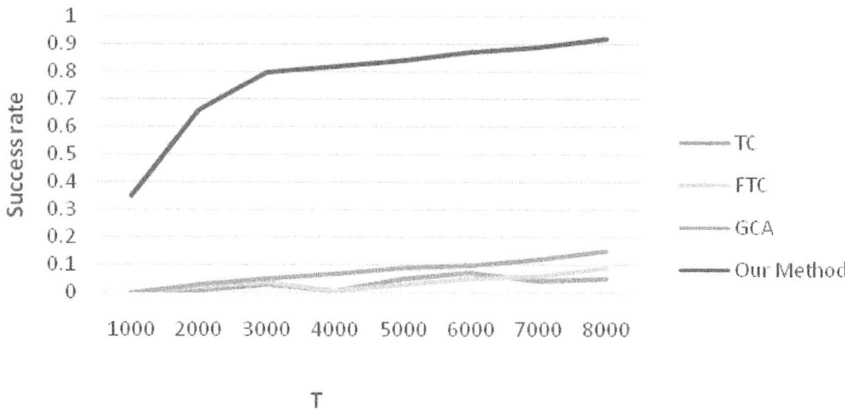

FIGURE 16.10 Success rates of four attack methods under different ThΔ When SNR= –6dB.

The data presented in Figure 16.9 indicates that at a SNR of –3 dB, the efficacy rate of both TC and FTC is notably low and does not exhibit substantial improvement with increasing Th_{Δ}. The GCA scheme exhibits a modest enhancement as Th_{Δ} increases. In this scheme, when $Th_{\Delta} = 5$, the proper collision pairs are predominantly found inside the threshold ThΔ, resulting in a higher likelihood of identifying the correct collision pairings and consequently achieving a higher success rate. Currently, the success rate is at 97.08%. However, even with a SNR of–6 dB, Figure 16.10 demonstrates that the success rate of TC and FTC remains considerably low. Furthermore, there is no substantial improvement observed with the increase of Th_{Δ}. When the value of Th_{Δ} is equal to 5, the success rate of the GCA scheme is 4.38%, while the success rate of this scheme is 81.67%. When the value of Th_{Δ} is equal to 17, the success rates of the GCA and the aforementioned scheme are 16.25% and 90.83%, respectively. Currently, while comparing the GCA scheme with the aforementioned scheme, it is observed that the latter exhibits a better success rate. However, it is important to note that merely augmenting Th_{Δ} does not yield a substantial improvement in the attack's efficacy rate.

16.7.2 EXPERIMENTAL RESULTS UNDER DIFFERENT ENERGY TRACES

This section compares the attack success probabilities under different numbers of energy traces, where Th_{Δ} is set to 5, and the results of the experiment are shown in Figures 16.11 and 16.12. Compared with increasing Th_{Δ}, increasing the number of energy traces is more effective, and the attack efficacy rates of the four schemes increase with the increase in the number of energy traces. It can be seen from Figure 16.11 that when the SNR = –3 dB, the scheme in this chapter reaches an efficacy rate of 98.78% with 3000 energy traces. It can be seen that when $Th_{\Delta} = 5$, the correct collision pairs are basically at the threshold. Within this range, it's more likely that collision recognition will work, thus improving the success rate of the attack. The success rates of the TC, FTC, and GCA methods are 44.67%, 72.67%, and 97.17% when 8000 energy traces are used. The scheme in this chapter is better than the TC, FTC, and GCA plans, increasing by 55.33, 27.33, and 2.83% points, respectively. To further reduce the signal-to-noise ratio, it can be seen from Figure 16.12 that when the SNR = –6 dB and the energy traces are 5000, the attack success rates of the four schemes are 5.33%, 8.00%, 17.81%, and 99.58%, respectively. When the energy traces are increased to

8000, the attack success rates of the four schemes become 23.33%, 44.67%, 55.31%, and 100.00%. Compared with TC, FTC, and GCA, the scheme in this chapter increases by 76.67, 55.33, and 44.69% points, respectively. It can be seen that the TC, FTC, and GCA schemes have a lower probability of detecting the correct collision pair, which reduces the success rate of the attack. However, the scheme in this chapter detects the correct collision pair with a higher probability, which means that the attack is more likely to work than the TC, FTC, and GCA plans.

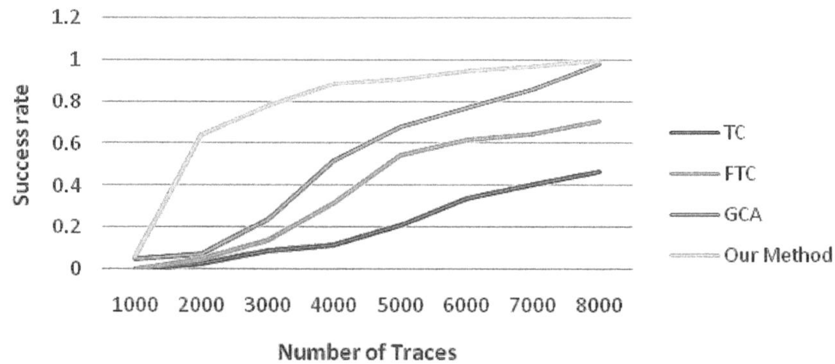

FIGURE 16.11 Success rate of four attack methods when SNR = –3dB.

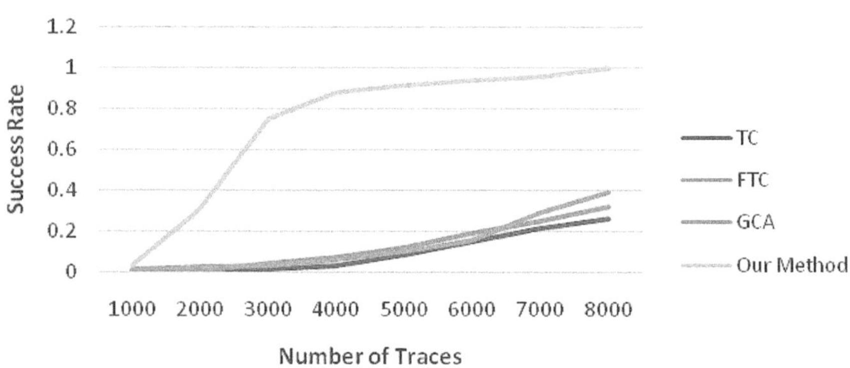

FIGURE 16.12 Success rate of four attack methods when SNR = –6dB.

16.8 CONCLUSION

In order to quickly recover the key, this chapter combines the divide-and-conquer attack with the analysis attack and proposes a fault-tolerant two-way choice attack strategy that combines EDCA and CPA. The results of experiments show that EDCA is much better at finding collisions than standard CCA, and after two-way screening of key candidates and collision chains, under the condition of low SNR = –6 dB, the collision detection of this scheme is, compared with the TC, FTC, and GCA schemes, more efficient by 76.67, 55.33, and 44.69% points, respectively. Compared with existing schemes, this scheme can convert a relatively deep key candidate space into a smaller space, has a higher attack success rate, and makes key recovery easier. The method in this chapter only studies candidates whose subkeys and collision pairs are within the range of thresholds Th_k and Th_Δ. In future research work, an error correction mechanism can be considered to correct candidates that exceed the range of thresholds Th_k and Th_Δ, thereby achieving a higher attack success rate.

KEYWORDS

- **collision chains**
- **correlation power analysis**
- **energy traces**
- **fault-tolerant key recovery**
- **Pearson correlation coefficient**
- **side-channel attacks**
- **zero trust network**

REFERENCES

1. Han, J., Liu, H., & Zhang, G. (2023). Prescribed-time performance recovery fault tolerant control of platoon with nominal constraints guarantee. *IEEE Transactions on Intelligent Transportation Systems*, *24*(3), 3501–3513. https://doi.org/10.1109/TITS.2022.3232818.
2. Shang, Y., Wu, B., Li, T., & Fang, S. (2007). Fault-tolerant technique in the cluster computation of the digital watershed model. *Tsinghua Science and Technology*, *12*(S1), 162–168. https://doi.org/10.1016/S1007-0214(07)70103-7.
3. Li, H., Wu, J., Jiang, Z., Li, X., & Wei, X. (2017). Minimum backups for stream processing with recovery latency guarantees. *IEEE Transactions on Reliability*, *66*(3), 783–794. https://doi.org/10.1109/TR.2017.2712563.

4. Hu, Y., Dai, X., Wu, Y., Jiang, B., Cui, D., & Jia, Z. (2023). Robust fault estimation and fault-tolerant control for discrete-time systems subject to periodic disturbances. *IEEE Transactions on Circuits and Systems I: Regular Papers*, *70*(7), 2982–2994. https://doi.org/10.1109/TCSI.2023.3268035.
5. Zhang, S., Yang, P., Kong, L., Chen, W., Fu, Q., & Peng, K. (2021). Neural networks-based fault tolerant control of a robot via fast terminal sliding mode. *IEEE Transactions on Systems, Man, and Cybernetics: Systems*, *51*(7), 4091–4101. https://doi.org/10.1109/TSMC.2019.2933050.
6. Li, L., Ding, S. X., Luo, H., Peng, K., & Yang, Y. (2020). Performance-based fault-tolerant control approaches for industrial processes with multiplicative faults. *IEEE Transactions on Industrial Informatics*, *16*(7), 4759–4768. https://doi.org/10.1109/TII.2019.2946882.
7. Song, Y., Mu, T., & Wang, B. (2023). HV-SNSP: A low-overhead data recovery method based on cross-checking. *IEEE Access*, *11*, 5737–5745. https://doi.org/10.1109/ACCESS.2023.3235787.
8. Shen, Z., Lee, P. P. C., Shu, J., & Guo, W. (2020). Cross-rack-aware single failure recovery for clustered file systems. *IEEE Transactions on Dependable and Secure Computing*, *17*(2), 248–261. https://doi.org/10.1109/TDSC.2017.2774299.
9. Chen, G., Zhao, Y., Xu, H., Pei, D., & Li, D. (2017). F^2 tree: Rapid failure recovery for routing in production data center networks. *IEEE/ACM Transactions on Networking*, *25*(4), 1940–1953. https://doi.org/10.1109/TNET.2017.2672678.
10. Tang, X., Zhai, J., Yu, B., Chen, W., Zheng, W., & Li, K. (2018). An efficient in-memory checkpoint method and its practice on fault-tolerant HPL. *IEEE Transactions on Parallel and Distributed Systems*, *29*(4), 758–771. https://doi.org/10.1109/TPDS.2017.2781257.
11. Nair, P. P., Sarkar, A., & Biswas, S. (2021). Fault-tolerant real-time fair scheduling on multiprocessor systems with cold-standby. *IEEE Transactions on Dependable and Secure Computing*, *18*(4), 1718–1732. https://doi.org/10.1109/TDSC.2019.2934098.
12. Fan, W., Xiao, F., Cai, H., Chen, X., & Yu, S. (2023). Disjoint paths construction and fault-tolerant routing in BCube of data center networks. *IEEE Transactions on Computers*, *72*(9), 2467–2481. https://doi.org/10.1109/TC.2023.3251849.
13. Wu, L., Xu, M., Fu, S., Luo, Y., & Wei, Y. (2022). FPDA: Fault-tolerant and privacy-enhanced data aggregation scheme in fog-assisted smart grid. *IEEE Internet of Things Journal*, *9*(7), 5254–5265. https://doi.org/10.1109/JIOT.2021.3109153.
14. Zhu, Y., Liu, Y., & Zhang, G. (2020). FT-PBLAS: PBLAS-based fault-tolerant linear algebra computation on high-performance computing systems. *IEEE Access*, *8*, 42674–42688. https://doi.org/10.1109/ACCESS.2020.2975832.
15. Yan, C., Chen, J., Liu, H., Kumar, L., & Lu, H. (2021). Health management for PEM fuel cells based on an active fault tolerant control strategy. *IEEE Transactions on Sustainable Energy*, *12*(2), 1311–1320. https://doi.org/10.1109/TSTE.2020.3042990.
16. Xu, J., Xie, S., & Zhao, J. (2021). P4Neighbor: Efficient link failure recovery with programmable switches. *IEEE Transactions on Network and Service Management*, *18*(1), 388–401. https://doi.org/10.1109/TNSM.2021.3050478.
17. van Dongen, G., & Poel, D. V. D. (2021). A performance analysis of fault recovery in stream processing frameworks. *IEEE Access*, *9*, 93745–93763. https://doi.org/10.1109/ACCESS.2021.3093208.

18. Yue, Y., Cao, L., Hang, B., & Luo, Z. (2018). A swarm intelligence algorithm for routing recovery strategy in wireless sensor networks with mobile sink. *IEEE Access*, *6*, 67434–67445. https://doi.org/10.1109/ACCESS.2018.2879364.
19. Stern, A., Wang, H., Rahman, F., Farahmandi, F., & Tehranipoor, M. (2022). ACED-IT: Assuring confidential electronic design against insider threats in a zero-trust environment. *IEEE Transactions on Computer-Aided Design of Integrated Circuits and Systems*, *41*(10), 3202–3215. https://doi.org/10.1109/TCAD.2021.3127864.
20. Ghandali, S., Ghandali, S., & Tehranipoor, S. (2021). Deep K-TSVM: A novel profiled power side-channel attack on AES-128. *IEEE Access*, *9*, 136448–136458. https://doi.org/10.1109/ACCESS.2021.3117761.
21. Das, D., Maity, S., Nasir, S. B., Ghosh, S., Raychowdhury, A., & Sen, S. (2018). ASNI: Attenuated signature noise injection for low-overhead power side-channel attack immunity. *IEEE Transactions on Circuits and Systems I: Regular Papers*, *65*(10), 3300–3311. https://doi.org/10.1109/TCSI.2018.2819499.
22. Kumar, R., Sharma, P., Sharma, A., & Kumar, P. (2021). A time-/frequency-domain side-channel attack resistant AES-128 and RSA-4K crypto-processor in 14-nm CMOS. *IEEE Journal of Solid-State Circuits*, *56*(4), 1141–1151. https://doi.org/10.1109/JSSC.2021.3052146.
23. Kumar, R., Sharma, P., & Sharma, A. (2020). A 4900-μm^2 839-Mb/s side-channel attack-resistant AES-128 in 14-nm CMOS with heterogeneous Sboxes, linear masked MixColumns, and dual-rail key addition. *IEEE Journal of Solid-State Circuits*, *55*(4), 945–955. https://doi.org/10.1109/JSSC.2019.2960482.
24. Zoni, D., Cremona, L., & Fornaciari, W. (2022). Design of side-channel-resistant power monitors. *IEEE Transactions on Computer-Aided Design of Integrated Circuits and Systems*, *41*(5), 1249–1263. https://doi.org/10.1109/TCAD.2021.3088781.
25. Wu, T., Zhou, D., Du, L., & Wang, S. (2023). Fault template attack based on fault probability. *IEEE Access*, *11*, 71705–71713. https://doi.org/10.1109/ACCESS.2023.3293648.
26. Yan, Z., Xu, F., Tan, J., Liu, H., & Liang, B. (2023). Reinforcement learning-based integrated active fault diagnosis and tracking control. *ISA Transactions*, *132*, 364–376. https://doi.org/10.1016/j.isatra.2023.01.012.
27. Abbasi, S., Rahmani, A. M., Balador, A., & Sahafi, A. (2023). A fault-tolerant adaptive genetic algorithm for service scheduling in the internet of vehicles. *Applied Soft Computing*, *143*, 110413. https://doi.org/10.1016/j.asoc.2023.110413.
28. Myllyaho, L., Raatikainen, M., Männistö, T., Nurminen, J. K., & Mikkonen, T. (2022). On misbehaviour and fault tolerance in machine learning systems. *Journal of Systems and Software*, *183*, 111096. https://doi.org/10.1016/j.jss.2022.111096.
29. Bharany, S., Badotra, S., Sharma, S., Rani, S., Alazab, M., Jhaveri, R. H., & Gadekallu, T. R. (2022). Energy efficient fault tolerance techniques in green cloud computing: A systematic survey and taxonomy. *Sustainable Energy Technologies and Assessments*, *53*(B), 102613. https://doi.org/10.1016/j.seta.2022.102613.
30. Kadri, N., & Koudil, M. (2022). Multi-objective biogeography-based optimization and reinforcement learning hybridization for network-on-chip reliability improvement. *Journal of Parallel and Distributed Computing*, *161*, 20–36. https://doi.org/10.1016/j.jpdc.2022.06.004.

CHAPTER 17

Zero-Trust Evolution and Cloud Computing Security as Multi-Mission Engineering: Addressing Emerging Threats, Regulations, and Modeling Solutions via Enhancing Blockchain Consensus Algorithm to Mitigate Data Repository Breaches

BHUPINDER SINGH[1,*] and CHRISTIAN KAUNERT[2]

[1]*Professor, Sharda School of Law, Sharda University, Greater Noida, Uttar Pradesh, India*

[2] *International Security, Dublin City University, Ireland*

*Corresponding author

ABSTRACT

The convergence of Zero Trust Evolution and Cloud Computing Security reflects a paradigm shift from traditional security models. Recognizing the limitations of perimeter-based defense mechanisms, this research delves into the concept of Zero Trust, which emphasizes continuous verification and dynamic risk assessment of users and devices, regardless of their location within a network. This proactive approach aims to ensure data protection and access control in an environment where threats are becoming increasingly sophisticated and pervasive. The concept of "Multi-Mission Engineering" emphasizes the need for a multifaceted strategy to tackle diverse cyber threats. This strategy recognizes that security challenges extend beyond a single approach and advocates for an integrated approach that combines Zero

Zero-Trust Learning: Applications in Modern Network Security. Chetna Kaushal, Amandeep Kaur, Mohit Angurala, Aryan Chaudhary (Eds.)

Trust principles with Cloud Computing Security strategies. By doing so, the research acknowledges that security is not a one-size-fits-all endeavor, but rather a dynamic and adaptive endeavor that encompasses multiple layers of protection. To enhance security through the integration of a Blockchain Consensus Algorithm, known for its immutability and decentralized nature, it is introduced as a means to mitigate data repository breaches. By leveraging a robust consensus algorithm, data integrity can be ensured, unauthorized access can be curtailed, and data breaches can be minimized. Throughout the research, case studies, simulations, and real-world scenarios are employed to illustrate the practical implications of this integrated approach. By evaluating the potential risks, benefits, and challenges associated with the proposed solution, the research provides a comprehensive assessment of the viability of the approach in a dynamic and evolving cyber landscape. The research paper encapsulates the fusion of Zero Trust Evolution, Cloud Computing Security, and Blockchain technology as a multi-mission engineering strategy to address emerging cyber threats and enhance data security. By embracing innovation and a multi-faceted approach, the research seeks to contribute to the ongoing dialogue surrounding the evolution of cybersecurity paradigms and the utilization of cutting-edge technologies to safeguard critical digital assets. The paper begins by elucidating the foundational principles of Zero Trust, emphasizing its shift from perimeter-centric security to continuous verification and adaptive risk management. Cloud Computing Security's pivotal role in the digital ecosystem is explored, highlighting the importance of safeguarding data in cloud environments. With a comprehensive review of emerging threats such as data breaches, identity theft, and insider attacks, the paper underscores the need for innovative solutions. It proposes a novel approach that combines Zero Trust and Cloud Computing Security, advocating for an integrated defense-in-depth strategy that fosters resilience against evolving threats.

17.1 INTRODUCTION

The concept of Zero Trust security represents a paradigm shift in the way organizations approach cybersecurity, and its relevance has never been more pronounced than in the dynamic landscape of cloud computing. Historically, traditional network security models relied on perimeter defenses, assuming that threats could be kept at bay by safeguarding the outer boundaries of the network [1]. However, this approach has become increasingly inadequate in the face of evolving cyber threats, the proliferation of remote work, and the widespread adoption of cloud technologies. Zero Trust, in contrast, is a security model predicated on the fundamental assumption that trust should never be granted implicitly based on a user's location or network segment. Instead, it advocates for continuous verification and authentication of every user, device, and application, irrespective of their position within or outside the organizational network [1–4]. In the realm of cloud computing, where data and applications are hosted across diverse and geographically dispersed environments, Zero Trust becomes pivotal.

It recognizes that the cloud's flexibility and scalability, while driving business innovation, also open up new attack vectors. By treating every access request, whether from an employee, customer, or third-party application, as potentially untrusted, Zero Trust security reinforces a proactive and adaptive approach to safeguarding cloud resources. This not only mitigates the risks associated with unauthorized access and data breaches but also aligns seamlessly with regulatory compliance requirements, making it a cornerstone of modern cloud security strategies. Consequently, understanding, and implementing the principles of Zero Trust within cloud computing environments is paramount in ensuring data integrity, confidentiality, and availability while upholding the resilience of digital ecosystems in an era characterized by ever-evolving cyber threats [5].

As organizations continue their rapid migration to cloud environments, the security landscape evolves with both emerging threats and increasingly stringent regulatory requirements. Traditional security models no longer suffice, necessitating a paradigm shift towards Zero Trust security. This research paper explores the evolution of Zero Trust principles in the context of cloud computing and investigates their application as a multi-mission engineering approach to tackle the challenges posed by emerging threats and regulatory compliance. Moreover, it delves into the integration of blockchain consensus algorithms as a strategic enhancement to mitigate data repository breaches within cloud environments. The paper combines theoretical insights with practical examples, demonstrating the synergy between Zero Trust and blockchain in fortifying cloud security [6].

17.1.1 RESEARCH OBJECTIVES AND SIGNIFICANCE OF STUDY

This chapter holds profound significance in the realm of cybersecurity. In an era marked by constantly evolving threats, stringent data protection regulations, and the ubiquitous adoption of cloud services, this study offers a timely and invaluable contribution. It serves as a beacon of innovation, providing a comprehensive framework that bridges the gap between theory and practical implementation. By exploring the evolution of Zero Trust principles, identifying emerging threats, and addressing the complexities of regulatory compliance, this research equips organizations with a blueprint for resilient and adaptable security strategies. Moreover, its pioneering approach of integrating blockchain enhancements into cloud security not only safeguards data repositories but also stimulates future research and innovation in the pursuit of robust, multi-mission cybersecurity solutions. The research objectives are:

1. **Analyze Zero Trust Evolution:** To examine the evolution of Zero Trust principles and their applicability in contemporary cloud computing security.
2. **Identify Emerging Threats:** To identify and categorize emerging cybersecurity threats and vulnerabilities specific to cloud computing environments.
3. **Assess Regulatory Compliance:** To assess the impact of evolving regulations and compliance standards on cloud computing security, including GDPR, HIPAA, and other relevant regulations.
4. **Propose Blockchain Enhancements:** To propose enhancements to blockchain consensus algorithms that can strengthen the security of data repositories in cloud environments.
5. **Develop a Multi-Mission Engineering Model:** To design a comprehensive multi-mission engineering model that integrates Zero Trust principles, cloud security measures, and blockchain enhancements to address emerging threats and regulatory requirements.
6. **Evaluate the Model:** To evaluate the effectiveness of the proposed multi-mission engineering model in mitigating data repository breaches and ensuring compliance with relevant regulations.

As the cybersecurity landscape continually evolves with new threats and regulations, this study addresses the critical need for adaptive security strategies [1, 3, 4]. It offers insights into how Zero Trust principles, cloud security, and blockchain enhancements can be integrated to create a robust defense against emerging threats. With the increasing emphasis on data protection and privacy regulations worldwide, organizations need practical solutions to ensure compliance. This study provides a framework for aligning security practices with regulatory requirements. As organizations increasingly rely on cloud services, safeguarding data stored in the cloud becomes paramount. The study's focus on enhancing blockchain consensus algorithms offers innovative methods for protecting sensitive data within cloud repositories. The concept of multi-mission engineering represents a holistic approach to cybersecurity that can benefit a wide range of industries and organizations. The study's model provides a blueprint for organizations looking to adopt a comprehensive security strategy. By proposing a model and evaluating its effectiveness, the study offers actionable insights for organizations seeking to enhance their security posture. It bridges the gap between theoretical concepts and real-world applications [3, 7]. The study opens avenues for further research in the field of cloud security, Zero Trust, and blockchain technology. It encourages ongoing innovation in addressing emerging threats and regulatory challenges.

This research holds significant relevance in the rapidly changing landscape of cybersecurity. By defining clear research objectives and emphasizing the study's significance, you can effectively communicate the importance of your work in addressing contemporary security challenges and providing practical solutions for organizations and industries facing these issues.

17.1.2 GROWING IMPORTANCE OF CLOUD COMPUTING AND CHALLENGES

Zero Trust and cloud computing have emerged as pivotal concepts in today's dynamic cybersecurity landscape, fundamentally reshaping the way organizations approach security. They are intrinsically linked, addressing the evolving challenges posed by modern threats, changing IT environments, and the need for robust, scalable, and agile security strategies [8].

Zero Trust represents a paradigm shift in cybersecurity, moving away from the traditional perimeter-based security model. In the past, organizations relied on the notion that once inside the network, everything was trusted, leading to vulnerabilities and breaches. Today, this approach is inadequate given the proliferation of sophisticated threats, insider risks, and the expansion of remote work. Zero Trust advocates the principle of "never trust, always verify." It mandates continuous verification of every user, device, or application trying to access network resources, regardless of their location. This approach aligns with the reality of a perimeter-less network, where users may connect from anywhere, and devices are diverse, including personal smartphones and IoT devices [9]. Zero Trust enforces strict access controls, scrutinizes every request, and applies least-privilege principles. It emphasizes continuous monitoring, real-time threat detection, and adaptive security, making it highly relevant in combating the diverse and dynamic threat landscape of today.

Cloud computing has revolutionized the IT landscape by providing scalable, cost-effective, and flexible computing resources [10]. However, this transformation has also introduced new cybersecurity challenges. Cloud environments are complex and dynamic, often involving a mix of public, private, and hybrid clouds. The shift of critical data and applications to the cloud demands a new approach to security. Cloud providers offer robust security controls, but the shared responsibility model means that organizations are still responsible for securing their data and applications in the cloud. Cloud computing's relevance in cybersecurity lies in its ability to provide resilience, scalability, and agility. It allows organizations to implement security measures that can adapt to their needs, rapidly deploy patches and updates, and leverage advanced security services such as threat intelligence and analytics [11].

The convergence of Zero Trust and cloud computing is particularly relevant today. Zero Trust aligns with the cloud's dynamic and distributed nature by providing a framework for continuous verification, access control, and threat detection across cloud resources. Organizations can implement Zero Trust principles, ensuring that only authorized users and devices gain entry. Moreover, cloud environments provide the scalability and resources required to support the computational needs of Zero Trust, including real-time analytics and machine learning for threat detection. Zero Trust and cloud computing are not mere buzzwords but essential components of a modern cybersecurity strategy. They address the challenges posed by a constantly evolving threat landscape, remote work trends, and the migration of critical assets to the cloud [12–14]. Together, they offer a proactive, adaptable, and robust defense against the complex and sophisticated threats that organizations face today, making them indispensable in safeguarding digital assets and ensuring business continuity.

17.2 ZERO TRUST EVOLUTION

Zero Trust challenges the notion that entities within an organization's network can be inherently trusted, acknowledging the relentless evolution of cyber threats and the porous nature of modern networks. At its core, Zero Trust advocates a fundamental shift in mindset: to "never trust, always verify." This means that, regardless of their location or prior authentication, every user, device, or application attempting to access network resources must undergo continuous scrutiny and verification. This introduction marks the inception of a security philosophy that emphasizes the principle of least privilege, strict access controls, real-time threat detection, and adaptive security mechanisms [15]. As organizations grapple with an increasingly complex and dynamic threat landscape, understanding the evolution of Zero Trust becomes paramount for establishing a robust and resilient cybersecurity posture.

17.2.1 FUNDAMENTAL PRINCIPLES OF ZERO TRUST

The fundamental principles of Zero Trust represent a radical departure from traditional cybersecurity models and have become central to modern security strategies. Zero Trust operates on the core tenet of "never trust, always verify."

This principle fundamentally challenges the traditional notion of trusting entities once they are inside the organizational network perimeter [16]. Instead, Zero Trust posits that trust should never be assumed, even for internal users and devices. There are the key principles that underpin the Zero Trust model as:

1. **Verify Identity:** The first pillar of Zero Trust revolves around identity verification. Every user, device, application, or entity attempting to access network resources must undergo rigorous identity verification, often through multi-factor authentication (MFA).
2. **Least Privilege Access:** It imposes the principle of least privilege, granting the minimum level of access necessary for an entity to perform its specific tasks. This principle reduces the potential damage that can occur in the event of a security breach or insider threat.
3. **Micro-Segmentation:** Instead of relying on a monolithic network perimeter, Zero Trust advocates micro-segmentation. Access between these segments is strictly controlled based on need, limiting lateral movement for attackers.
4. **Continuous Monitoring:** Zero Trust doesn't stop at initial authentication and access approval. It emphasizes continuous monitoring of all network traffic and user behavior. Suspicious activities,

anomalies, or deviations from normal behavior are immediately flagged for investigation.

5. **Implicit Zero Trust:** In traditional models, internal traffic is often considered implicitly trusted. Zero Trust challenges this assumption by scrutinizing all traffic, whether internal or external, equally. This ensures that lateral movement by threats is thoroughly monitored and controlled.
6. **Dynamic Risk Assessment:** Zero Trust employs real-time risk assessment to evaluate the security posture of users, devices, and applications. Depending on the perceived risk, access privileges can be dynamically adjusted or revoked, allowing for adaptive security responses.
7. **Encryption and Data Protection:** Data is a primary target for cyberattacks. Zero Trust emphasizes robust encryption and data protection mechanisms to safeguard sensitive information, both in transit and at rest.
8. **Continuous Authentication:** Rather than relying solely on a single authentication event at the point of entry, Zero Trust advocates continuous authentication throughout a user's session. This means users are re-authenticated periodically or when specific activities trigger the need for re- verification.

In essence, the fundamental principles of Zero Trust represent a holistic approach to cybersecurity that assumes that threats can emerge from both external and internal sources. By incorporating rigorous identity verification, reducing access privileges, segmenting the network, and continuously monitoring and adapting security measures, organizations can establish a resilient defense against the ever-evolving threat landscape [17]. Zero Trust is not just a technology but a mindset shift that acknowledges the dynamic nature of modern threats and adapts accordingly to ensure robust security in an increasingly interconnected world.

17.2.2 TRADITIONAL SECURITY MODELS AND THEIR LIMITATIONS: ZERO TRUST EVOLUTION

Traditional security models, which have long been the cornerstone of cybersecurity, are characterized by their reliance on perimeter-based defenses and a fundamental assumption of trust within the network. The most notable of these models is the Castle-and-Moat or Perimeter-based model, which draws an analogy to a medieval castle surrounded by a moat to protect against external threats. In this model, once a user or device gains entry past the perimeter defenses, they are generally considered trusted, leading to several significant limitations [18, 19].

The traditional security models exhibit a fundamental flaw in their assumption of trust within the network. Once an entity, whether it's an internal user, device, or application, breaches the outer defenses, it often enjoys free access to internal resources, potentially wreaking havoc. This assumption of trust has proven to be a glaring weakness, especially in today's threat landscape, where insider threats and sophisticated cyberattacks continuously challenge this trust model. Additionally, the traditional perimeter-centric approach struggles to accommodate the modern IT landscape [20]. With traditional security models being ill-equipped to handle these dynamic and diverse environments, organizations are left vulnerable to attacks from various entry points.

Another limitation lies in the model's reactive nature. Traditional security relies heavily on signature-based approaches, which are effective at detecting known threats but are easily evaded by novel and sophisticated attacks. Moreover, these models often lack the ability to detect lateral movement within the network, allowing attackers to pivot and escalate privileges once they are inside [17].

The Zero Trust Evolution, in contrast, addresses these limitations head-on. It challenges the notion of inherent trust and advocates a "never trust, always verify" approach. It enforces strict identity verification, least-privilege access, continuous monitoring, and micro-segmentation, effectively removing the concept of an implicitly trusted internal network. This model aligns with the dynamic and diverse nature of modern IT environments and provides a proactive, adaptable, and holistic security approach that is better suited to defend against the evolving threat landscape. As organizations increasingly recognize the shortcomings of traditional models, the adoption of Zero Trust is becoming imperative to ensure a robust and resilient cybersecurity posture in today's interconnected world [19].

17.2.3 IMPORTANCE OF ZERO TRUST IN MODERN CYBERSECURITY

The importance of Zero Trust in modern cybersecurity cannot be overstated. In today's digital landscape, characterized by relentless cyber threats, expanding attack surfaces, and a mobile workforce, the traditional security models have proven to be inadequate. Zero Trust emerges as a critical paradigm shift, offering a holistic approach that addresses the evolving challenges of the digital age.

The importance of Zero Trust lies in its fundamental principle of "never trust, always verify." It signifies a shift away from the outdated notion that once inside the network perimeter, entities can be trusted implicitly. Zero Trust recognizes that threats can emerge from both external and internal sources, and it enforces continuous scrutiny and verification of every user, device, and application attempting to access network resources. This approach aligns with the reality of a perimeter-less network, where users connect from anywhere, and devices are diverse, including personal smartphones and IoT devices [21].

Zero Trust is paramount in mitigating insider threats, which have become a significant concern for organizations. Malicious insiders or compromised accounts can be just as dangerous as external attackers. Zero Trust, by enforcing the principle of least privilege and continuous monitoring, significantly reduces the risk associated with insider threats. It ensures that even legitimate users and devices are granted only the minimum access necessary for their specific tasks, minimizing potential damage in the event of a breach. Another crucial aspect of Zero Trust is its adaptability to modern IT environments. Zero Trust, with its emphasis on micro-segmentation and dynamic risk assessment, is well-suited to secure these dynamic and diverse environments, providing granular control and visibility into network activities [22].

In an era of increasing data breaches, data protection is paramount. Zero Trust incorporates robust encryption and data protection mechanisms, safeguarding sensitive information both in transit and at rest. This is especially critical given the stringent data protection regulations, such as GDPR and HIPAA, which carry hefty fines for non-compliance. Zero Trust offers a framework for aligning security practices with these regulatory requirements. The importance of Zero Trust in modern cybersecurity is multi-faceted. It provides a proactive, adaptable, and holistic security approach that acknowledges the dynamic nature of today's threats and IT environments. By enforcing continuous verification, least privilege access, and rigorous monitoring, Zero Trust empowers organizations to strengthen their defenses, mitigate insider threats, protect sensitive data, and ensure compliance with regulations.

17.3 CLOUD COMPUTING SECURITY AS MULTI-MISSION ENGINEERING

Cloud computing security, viewed through the lens of multi-mission engineering, represents a dynamic and adaptive approach to safeguarding data and applications in the cloud. Multi-mission engineering is a concept borrowed from aerospace and defense, where systems must perform reliably in diverse and complex mission scenarios [23]. One of the fundamental principles of multi-mission engineering in cloud computing security is the recognition that there is no one-size-fits-all solution. Cloud environments can vary significantly in terms of architecture, use cases, and threat landscapes. Multi-mission engineering acknowledges this diversity and emphasizes the need for tailored security measures. It involves the careful assessment of an organization's specific needs, the identification of potential threats, and the design of security controls that can adapt to different scenarios. Another critical aspect of multi-mission engineering in cloud security is the integration of various security layers and technologies. Instead of relying solely on a single security solution, multi-mission engineering advocates a defense-in-depth approach. This involves layering security controls at different levels, from network security and access controls to encryption and threat detection. By combining multiple security measures, organizations can create a robust security posture that is resilient to a wide range of threats [24–26].

The scalability is also a key consideration in multi-mission engineering for cloud security. Cloud environments can scale rapidly to accommodate changing workloads and demands. Security measures must be able to scale in tandem. This means that security controls should be designed to handle fluctuations in traffic, data volume, and resource provisioning without compromising effectiveness. Scalability ensures that security remains consistent even as the cloud environment evolves. Moreover, multi-mission engineering in cloud security emphasizes continuous monitoring and adaptation. Threats are not static; they evolve over time. Security measures must be capable of detecting new threats and adapting to changing attack vectors. This involves the use of real-time threat intelligence, behavioral analytics, and automated response mechanisms to identify and mitigate emerging threats promptly.

The cloud computing security as multi-mission engineering represents a strategic approach to securing cloud environments that are dynamic, diverse, and ever-evolving. It recognizes the need for tailored security measures, the integration of multiple security layers, scalability to accommodate changing demands, and a continuous monitoring and adaptation process [27]. By applying multi- mission engineering principles, organizations can achieve a resilient and effective security posture in the cloud, ensuring the protection of data and applications across a spectrum of scenarios and challenges.

17.3.1 SECURITY CHALLENGES ASSOCIATED WITH CLOUD COMPUTING

The cloud computing landscape offers many advantages, including flexibility, scalability, and cost-efficiency. However, it also presents a variety of security challenges that organizations must address. Some of the key security challenges associated with cloud computing include:

1. **Data Breaches:** Data breaches are a significant concern in cloud computing. Storing sensitive data in the cloud means entrusting a third-party provider with its security. If proper security measures are not in place, data can be exposed through vulnerabilities in the cloud infrastructure or misconfigurations.

2. **Data Loss:** Data loss can occur due to various factors, including hardware failures, software bugs, or human error. Cloud providers typically offer data redundancy and backup services, but organizations must ensure they configure and manage these features correctly to prevent data loss.

3. **Identity and Access Management (IAM):** Managing user identities and access controls can be challenging in cloud environments. Improperly configured IAM policies can lead to unauthorized access, data exposure, or service interruptions. It's essential to implement strict access controls and regularly review and update them.

4. **Compliance and Legal Issues:** Many industries have strict compliance requirements regarding data storage, access, and privacy. Ensuring that cloud providers adhere to these regulations and that organizations maintain compliance in the cloud can be complex. Examples include GDPR for data privacy and HIPAA for healthcare data.

5. **Lack of Control:** Cloud providers handle much of the infrastructure management, which can result in a lack of visibility and control over certain aspects of security. Organizations must trust their providers but also take responsibility for securing their applications and data within the cloud.

6. **Cloud Provider Security:** While reputable cloud providers invest heavily in security, the security of the overall infrastructure ultimately depends on the provider. Organizations must evaluate and

trust their chosen providers' security practices and ensure they adhere to industry standards and best practices.

7. **Vendor Lock-In:** Switching between cloud providers can be challenging and costly. This can lead to a dependency on a single vendor, which may limit an organization's flexibility and negotiating power.
8. **Inadequate Security Posture:** Organizations may assume that their cloud providers handle all security aspects, leading to complacency regarding their own security responsibilities.
9. **Security Monitoring and Incident Response:** Detecting and responding to security incidents in the cloud can be more complex than in on-premises environments [28]. Implementing effective monitoring and incident response mechanisms in the cloud is essential to mitigate threats promptly.
10. **Shadow IT:** Employees may use unauthorized cloud services or solutions, commonly referred to as shadow IT, which can introduce unmanaged security risks. Organizations must educate employees about approved cloud services and provide secure alternatives.

So, addressing these cloud computing security challenges requires a proactive and comprehensive approach. Organizations must conduct thorough risk assessments, implement robust security policies and controls, stay informed about emerging threats, and continuously monitor and adapt their security strategies [29].

17.3.2 NEED FOR HOLISTIC APPROACH TO CLOUD SECURITY

A holistic approach to cloud security is indispensable in the modern computing landscape due to the intricate and evolving nature of cloud environments. It recognizes that cloud security cannot be treated as a siloed concern but must be seamlessly integrated into an organization's broader cybersecurity strategy. Several compelling reasons underscore the need for this comprehensive approach [30].

Cloud computing infrastructures encompass a wide array of services, platforms, and deployment models, including public, private, and hybrid clouds. Each of these configurations introduces unique security challenges. A holistic approach acknowledges this diversity and ensures that security measures are not one-size-fits-all but adaptable to the specific nuances of different cloud configurations [27, 31].

Moreover, in an era where hybrid environments are increasingly common, with organizations blending on-premises and cloud-based resources, security must transcend traditional boundaries. A holistic strategy bridges the gap between physical and virtual environments, ensuring that security measures are consistent and seamlessly applied across both realms. This is particularly crucial for maintaining a unified security posture and preventing security gaps or misconfigurations that can arise when managing multiple environments separately.

The ever-evolving threat landscape is another compelling reason for a holistic cloud security approach. Cyberattacks are becoming increasingly sophisticated, and threat actors are constantly devising new tactics to breach security defenses. A holistic approach recognizes that security is not static but requires ongoing adaptation and vigilance. It incorporates proactive threat intelligence, continuous monitoring, and rapid incident response capabilities to detect and mitigate emerging threats effectively [32]. The regulatory and compliance requirements demand a unified approach to security. Many industries face stringent data protection and privacy regulations, such as GDPR or HIPAA. These regulations apply irrespective of whether data resides in on-premises data centers or the cloud. A holistic approach ensures that organizations can maintain compliance across their entire IT landscape, mitigating legal and financial risks associated with non-compliance.

A holistic approach to cloud security also considers the interconnectedness of systems and applications within the cloud environment. Cloud-native architectures often involve numerous interdependencies. An isolated security focus on individual components or services can leave vulnerabilities unaddressed at the system level. A holistic approach accounts for these interdependencies, aiming to strengthen the overall security posture by addressing system-wide vulnerabilities.

The need for a holistic approach to cloud security is rooted in the recognition that cloud environments are multifaceted, interconnected, and subject to an ever-evolving threat landscape. Such an approach transcends the traditional boundaries between physical and virtual environments, adapts to diverse cloud configurations, incorporates proactive threat management, and ensures regulatory compliance. By taking a comprehensive perspective on cloud security, organizations can effectively mitigate risks and safeguard their data, applications, and digital assets in this dynamic and complex computing paradigm [23].

17.4 BLOCKCHAIN CONSENSUS ALGORITHM ENHANCEMENT IN ZERO TRUST EVOLUTION

Zero Trust, a cybersecurity model that has gained significant traction in recent years, challenges the traditional notion of trust within network architectures. It assumes that threats may come from both internal and external sources and, therefore, nothing should be inherently trusted. This concept aligns seamlessly with blockchain technology, which inherently operates on a principle of trustless trust, making it a natural candidate for the Zero Trust evolution [33]. In the ever-evolving landscape of digital technology, the emergence of blockchain technology has brought about a paradigm shift in how we perceive and implement trust in digital interactions. As we embark on a journey towards a more secure and decentralized future, the enhancement of blockchain consensus algorithms plays a pivotal role in realizing the concept of Zero Trust in the digital realm [14]. Any change in the data would require changing the block's hash, which would then necessitate changing the subsequent blocks, a task virtually impossible due to the decentralized nature of the network.

Consensus algorithms, the fundamental building blocks of any blockchain system, serve as the backbone for achieving agreement among nodes in a decentralized network. Traditional consensus mechanisms like Proof of Work (PoW) and Proof of Stake (PoS) have served their purpose but have become inefficient and unsustainable in terms of energy consumption and scalability. Enhancements in consensus algorithms are crucial to addressing these issues. One notable advancement in this regard is the emergence of Proof of Authority (PoA) and Delegated Proof of Stake (DPoS) consensus algorithms. These consensus mechanisms prioritize identity and reputation, ensuring that nodes that participate in block validation are trusted entities. This approach aligns with the Zero Trust model, as it requires a predetermined level of trust in the network participants, making it difficult for malicious actors to gain control [34].

The enhancement of blockchain consensus algorithms in the Zero Trust evolution represents a pivotal moment in the journey towards truly trustless and secure digital ecosystems. Through innovative consensus mechanisms, cryptographic technologies, and decentralized identity solutions, blockchain networks are becoming more resilient to attacks and manipulation while maintaining the core principles of Zero Trust—continuous verification and the elimination of inherent trust. This convergence is paving the way for a future where trust is not an assumption but a result of rigorous validation and verification at every step of the blockchain journey [23].

17.4.1 *Improved Security in Cloud Environments: Enhancements to Blockchain Consensus Algorithms*

The blockchain technology provides multifarious security advantages such as immutability and decentralization. Adapting consensus algorithms to the unique characteristics of cloud computing can further fortify the security posture. One notable enhancement is the integration of cloud-native security mechanisms. Cloud environments offer a range of security features, including IAM, network security groups, and encryption services. By aligning the blockchain's consensus algorithm with these cloud-native security tools, it becomes possible to enhance access controls and protect against unauthorized interactions with the blockchain network. This integration can also streamline key management and encryption processes, further securing the blockchain's data and transactions [25].

Scalability is another vital consideration in cloud-based blockchain applications. To enhance security, consensus algorithms can be modified to accommodate the dynamic nature of cloud resources. Cloud platforms allow for easy scaling of computing power, which should be reflected in the blockchain's consensus mechanism. Implementing sharding or partitioning techniques, which enable parallel processing of transactions, can significantly enhance the network's scalability while maintaining security. This ensures that the blockchain remains resilient and efficient even as the network's demands fluctuate [29, 35, 36].

Additionally, leveraging cloud-based threat intelligence services can bolster security within blockchain consensus algorithms. Cloud providers often offer comprehensive threat detection and monitoring tools that can be integrated into the blockchain network. These services provide real- time insights into potential security threats, enabling rapid response and mitigation. By continuously monitoring the blockchain for suspicious activities or anomalies, cloud-based threat intelligence services contribute to the overall security of the network [30, 37, 38].

The adoption of advanced encryption techniques is another enhancement that can be integrated into blockchain consensus algorithms in cloud environments. Cloud-based blockchain applications can leverage such encryption methods to protect data while it's processed within the network [17]. Finally, considering the compliance and regulatory landscape is crucial when proposing enhancements to blockchain consensus algorithms in cloud environments. Cloud platforms often adhere to various compliance standards, and these can be incorporated into the consensus mechanism. For instance, blockchain networks can be configured to automatically enforce compliance with data protection regulations, ensuring that transactions and data storage comply with legal requirements. Enhancing blockchain consensus algorithms for improved security in cloud environments involves a multifaceted approach. Integrating cloud-native security mechanisms, addressing scalability challenges, leveraging cloud-based threat intelligence, adopting advanced encryption techniques, and aligning with regulatory requirements are all critical steps in fortifying the security of blockchain applications in the cloud. These enhancements not only bolster the integrity and resilience of the blockchain but also pave the way for secure and scalable blockchain deployments in cloud environments, where data protection and reliability are paramount.

17.4.2 BLOCKCHAIN CONSENSUS ALGORITHMS FOR IMPROVED SECURITY IN CLOUD ENVIRONMENTS AND MITIGATE DATA REPOSITORY BREACHES

The enhancement of blockchain consensus algorithms to bolster security in cloud environments and mitigate data repository breaches is a proactive step toward addressing the evolving challenges of data protection and integrity. These enhancements aim to combine the strengths of blockchain technology with the scalability and accessibility of cloud computing while fortifying security measures. One key enhancement lies in the integration of advanced cryptographic techniques [25]. Homomorphic encryption, for example, allows computations to be performed on encrypted data without decrypting it, preserving data confidentiality. By implementing homomorphic encryption within the consensus algorithm, blockchain networks operating in the cloud can ensure that sensitive data remains encrypted during transaction processing and storage, significantly reducing the risk of data exposure. The consensus algorithm can be enhanced to adapt seamlessly to varying workloads and resource demands. The cloud's scalability features can be leveraged to dynamically adjust the consensus mechanism to accommodate an influx of transactions or participants. Techniques like sharding, which partition the blockchain into smaller, manageable segments, can distribute the processing load and improve overall network performance. This ensures that blockchain remains responsive and efficient, even as it scales to meet growing demands, all while maintaining security.

The use of cloud-native security tools and services can significantly enhance the overall security of blockchain systems. Cloud platforms offer a plethora of security features, including IAM, network security groups, and automated threat detection. Integrating these tools within the blockchain consensus mechanism enables robust access controls and protection against unauthorized interactions. IAM policies can be utilized to ensure that only authenticated and authorized entities can participate in the consensus process, reducing the risk of malicious activity [39]. As continuous monitoring and threat detection are pivotal in mitigating data repository breaches, blockchain consensus algorithms can be enhanced to incorporate cloud-based threat intelligence services that provide real-time insights into potential security threats. This integration allows the blockchain network to detect and respond to anomalies promptly. When suspicious activity is detected, automated responses or alerts can be triggered, enabling swift mitigation measures. Such proactive threat monitoring not only safeguards the blockchain's integrity but also strengthens overall cloud security. By leveraging advanced encryption techniques, addressing scalability challenges, integrating cloud-native security measures, incorporating real-time threat intelligence, and automating regulatory compliance checks, blockchain networks operating in the cloud can offer a robust and resilient security framework. These enhancements not only empower organizations to harness the potential of blockchain and cloud computing but also provide confidence in data protection and secure transactions, ultimately contributing to a safer and more trustworthy digital landscape.

17.5 MODELING SOLUTIONS FOR CLOUD SECURITY

A theoretical model for integrating Zero Trust, cloud computing security, and blockchain consensus enhancements represents a forward-thinking approach to building a robust and adaptable security framework that meets the complex demands of modern digital ecosystems.

At the core of this theoretical model is the Zero Trust philosophy, which advocates a "never trust, always verify" mindset. In the context of cloud computing and blockchain, this means that trust should not be assumed for any entity, whether it's a user, device, or application, regardless of its location within the network. The Zero Trust model enforces rigorous identity verification, continuous monitoring, and dynamic access control, ensuring that every interaction is subject to scrutiny.

To integrate Zero Trust principles into cloud computing security, the model incorporates advanced access control mechanisms. IAM plays a pivotal role in ensuring that only authorized users or devices gain access to cloud resources and blockchain networks. MFA is employed to strengthen user verification, while dynamic policies adapt to changing conditions and risks. This seamless integration of Zero Trust into cloud security creates a granular and adaptive access control framework, mitigating the risk of unauthorized access and lateral movement within cloud environments [40].

Blockchain consensus enhancements are integrated to provide an additional layer of security and data integrity. By incorporating advanced consensus algorithms that align with cloud scalability, the model ensures that blockchain networks remain robust, even as they expand to meet increasing transaction volumes. Techniques such as sharding or partitioning can distribute the computational load, facilitating efficient and secure transaction processing within the cloud environment. This enhances the blockchain's responsiveness while maintaining the security features that make it tamper-resistant and trustworthy. The model leverages blockchain's cryptographic capabilities to enhance data security within cloud repositories. Blockchain-based encryption and hashing mechanisms safeguard sensitive data both in transit and at rest. Data stored within the cloud is cryptographically sealed, and any attempt to tamper with it would be immediately detected by the blockchain's immutable ledger [41]. This additional layer of data protection aligns with data privacy regulations and compliance requirements.

The real-time threat intelligence and monitoring services are an integral part of this model. Cloud-native security tools and services continuously analyze network traffic and user behavior, flagging suspicious activities or deviations from established baselines. The blockchain's transparency and immutability further assist in tracking and auditing these security events. Automated responses to threats, such as isolating compromised devices or revoking access credentials, help mitigate risks promptly. By unifying these elements, organizations can establish a robust security posture that ensures the integrity and confidentiality of data, enforces stringent access controls, adapts to evolving threats, and aligns with regulatory compliance requirements. This holistic approach paves the way for secure and resilient cloud-based blockchain applications, offering both trust and efficiency in an increasingly interconnected digital landscape.

17.6 EVALUATION AND DISCUSSION

The effectiveness of the integrated model, which combines Zero Trust principles, cloud computing security measures, and blockchain consensus enhancements, in addressing security threats and regulatory compliance is multifaceted and promising. This holistic approach offers several advantages that contribute to its effectiveness in tackling the complex challenges associated with cybersecurity and regulatory requirements.

The model's implementation of Zero Trust principles ensures that no entity is granted implicit trust within the network. Rigorous identity verification, continuous monitoring, and dynamic access control mechanisms are fundamental to this approach. As a result, the model is highly effective in mitigating security threats, especially those stemming from insider threats or unauthorized access attempts. By consistently verifying the authenticity and integrity of users, devices, and applications, it reduces the attack surface and minimizes the risk of data breaches or malicious activities within the network [40].

The incorporation of cloud-native security tools and services further enhances the model's effectiveness. Cloud platforms offer advanced threat detection, monitoring, and automated response capabilities that can promptly identify and address security threats. This real-time threat intelligence significantly improves the model's ability to detect and respond to emerging threats, reducing the potential impact of security incidents. Additionally, these cloud-native security features align well with regulatory compliance requirements, as they facilitate continuous monitoring and reporting of security events. Blockchain consensus enhancements play a crucial role in ensuring data integrity and tamper resistance. The model's advanced consensus algorithms, capable of scaling to meet the demands of cloud environments, fortify the blockchain's security. By incorporating features such as sharding, the model maintains blockchain efficiency while preserving the trustworthiness of transactions. The blockchain's immutability and cryptographic protections enhance data security and privacy, contributing to its effectiveness in safeguarding sensitive information.

In terms of regulatory compliance, the integrated model offers a comprehensive solution. Zero Trust's strict access controls and identity verification align with compliance requirements that mandate data protection and user privacy [32]. Cloud-native security tools provide the means to monitor and report security events, facilitating compliance auditing. However, it's essential to acknowledge that the effectiveness of this integrated model also depends on its proper configuration, ongoing maintenance, and user training. Ensuring that access policies, threat detection rules, and blockchain consensus algorithms are correctly set up and continuously updated is crucial for maintaining the model's security efficacy. Additionally, user education and awareness are vital to prevent social engineering attacks and ensure that individuals adhere to security policies.

The integrated model effectively addresses security threats and regulatory compliance by combining Zero Trust principles, cloud computing security measures, and blockchain consensus enhancements. Its multifaceted approach strengthens access controls, enhances threat detection and response, ensures data integrity, and aligns with regulatory standards. Nevertheless, successful implementation and ongoing management are essential to realizing the full potential of this model in securing cloud-based blockchain applications and maintaining compliance in a rapidly evolving digital landscape.

17.6.1 ADDRESSING EMERGING THREATS AND REGULATORY REQUIREMENTS: RELEVANCE AND IMPLICATIONS OF FINDINGS

The proposed approach of "Zero Trust Evolution and Cloud Computing Security as Multi-Mission Engineering: Addressing Emerging Threats, Regulations, and Modeling Solutions via Enhancing Blockchain Consensus Algorithm Mitigating Data Repository Breaches" offers a holistic and innovative perspective on addressing contemporary cybersecurity challenges. However, like any comprehensive security framework, it comes with both strengths and weaknesses. The major points of strengths are as:

1. **Holistic Approach:** One of the most significant strengths of this approach is its holistic nature. By integrating Zero Trust principles, cloud security measures, and blockchain enhancements, it provides a comprehensive and unified strategy to tackle a wide range of security challenges. This holistic approach recognizes the interconnectedness of various security components and aims to address them collectively.
2. **Adaptability:** The model's focus on multi-mission engineering and cloud-native security measures allows it to adapt to the dynamic nature of modern IT environments. This adaptability is crucial in an era where cloud adoption, remote work, and evolving cyber threats require flexible and scalable security solutions.
3. **Data Protection:** The incorporation of blockchain technology enhances data protection by providing cryptographic security, immutability, and transparency. This is particularly valuable in ensuring the integrity and confidentiality of sensitive information, which is a top priority in many industries.
4. **Regulatory Compliance:** The model's emphasis on aligning with regulatory standards and requirements is a significant advantage.

Compliance with data protection regulations is a legal imperative for many organizations, and this approach streamlines compliance efforts by embedding them within the security framework.

5. **Real-Time Threat Monitoring:** The integration of cloud-based threat intelligence services and monitoring tools enhances the model's ability to detect and respond to security threats in real time. This proactive threat management contributes to early threat mitigation and reduced potential impact.

There are some aspects of weaknesses as:

1. **Complexity:** The holistic nature of this approach, while a strength, can also be a weakness. Implementing and managing such a comprehensive security framework may be complex and resource-intensive, particularly for smaller organizations with limited IT resources.
2. **Cost:** Achieving the desired level of security through this approach may entail significant costs. Cloud-based security services and blockchain technology can have associated expenses, making it essential for organizations to evaluate the cost-benefit ratio carefully.
3. **Requirements:** Effective implementation of this model requires expertise in multiple domains, including cloud security, blockchain technology, and regulatory compliance. Organizations may need to invest in training and expertise development.
4. **Integration Challenges:** Integrating various security components and ensuring seamless interoperability can be challenging. Misconfigurations or compatibility issues may create security gaps if not addressed correctly.
5. **Scalability:** While the model addresses scalability concerns, implementing, and managing scalability in practice can still be a challenge, especially for organizations with rapidly growing workloads.
6. **Dependence on Cloud Providers:** The model's reliance on cloud providers for certain security features means that the effectiveness of the security framework partially depends on the security practices and infrastructure of the chosen cloud provider.

The approach of combining Zero Trust principles, cloud computing security, and blockchain enhancements represents a forward-thinking and comprehensive strategy to address contemporary cybersecurity challenges. While it offers significant strengths, including holistic coverage, adaptability, data protection, regulatory compliance, and real-time threat monitoring, organizations should be aware of the potential complexities, costs, and expertise requirements associated with its implementation. Careful planning and consideration of the specific needs and constraints of the organization are essential to realizing the full benefits of this approach [28].

17.7 CONCLUSION AND FUTURE SCOPE

This chapter represents a visionary and comprehensive response to the multifaceted challenges that contemporary cybersecurity faces. This holistic framework embodies the evolving nature of technology and security demands, offering a dynamic strategy that brings together cutting-edge principles from Zero Trust, cloud computing security, and blockchain enhancements. One of the standout strengths of this approach lies in its ability to adapt to the complexities of modern IT environments [42]. As organizations increasingly adopt cloud technologies and grapple with diverse cyber threats, a security framework that can scale and adjust in real-time is paramount. By integrating cloud-native security measures and blockchain's data integrity capabilities, this model is equipped to meet these challenges head-on, providing a robust defense against emerging threats.

The emphasis on data protection and regulatory compliance within the framework is a compelling aspect. In an era where data privacy is a fundamental concern, the incorporation of blockchain's cryptographic security and transparency aligns seamlessly with the stringent requirements of regulatory standards like GDPR and HIPAA. This not only mitigates the risks of data breaches but also streamlines compliance efforts, reducing potential legal liabilities.

The proactive approach to threat monitoring and real-time intelligence adds an essential layer of resilience to the model. This capability is particularly crucial in a landscape where cyber threats continue to evolve in sophistication and frequency. It's essential to recognize that implementing this holistic security model is not without its challenges. Complexity, cost considerations, expertise requirements, and potential integration difficulties may pose hurdles for organizations seeking to adopt this approach. Therefore, careful planning, resource allocation, and ongoing maintenance are paramount to realizing the full benefits of this comprehensive security framework. By integrating Zero Trust principles, cloud computing security measures, and blockchain enhancements, it offers a robust and adaptable defense against emerging threats, regulatory scrutiny, and data breaches. While challenges exist, the potential benefits in terms of data protection, regulatory compliance, and real-time threat management make this approach a compelling choice for organizations committed to safeguarding their digital assets in a rapidly evolving and interconnected world.

The future scope of "Zero Trust Evolution and Cloud Computing Security as Multi-Mission Engineering: Addressing Emerging Threats, Regulations, and Modeling Solutions via Enhancing Blockchain Consensus Algorithm Mitigating Data Repository Breaches" is poised to be both promising and evolving, reflecting the dynamic nature of cybersecurity and technological advancements. Here are some key areas where the research and framework can expand as-

17.7.1 ADVANCED BLOCKCHAIN CONSENSUS ALGORITHMS

Future research can focus on the development of even more advanced consensus algorithms tailored specifically for cloud-based blockchain applications. These algorithms could further enhance scalability, efficiency, and security, addressing the evolving demands of distributed ledger technology.

17.7.2 QUANTUM-SAFE SECURITY

As quantum computing technology advances, it poses a potential threat to existing encryption methods. Future work can explore quantum-resistant encryption and cryptographic techniques to ensure the long-term security of cloud-based blockchain systems.

AI and Machine Learning Integration: Leveraging artificial intelligence (AI) and machine learning for threat detection and automated response within the framework can be a significant area of growth. These technologies can enhance the model's ability to identify and mitigate emerging threats in real- time.

17.7.3 INTEROPERABILITY STANDARDS

With the increasing adoption of blockchain technology across various industries, developing interoperability standards for secure data exchange between different blockchain networks and cloud environments will be crucial.

- **Zero Trust for IoT:** As the Internet of Things (IoT) continues to expand, integrating Zero Trust principles with IoT device security within the cloud environment will be a pressing concern. Future research can explore strategies to ensure the security and trustworthiness of IoT data and communications.

17.7.4 REGULATORY EVOLUTION

Regulatory frameworks governing data privacy and cybersecurity are continually evolving. Future research can monitor and adapt the framework to align with emerging regulations, ensuring ongoing compliance.

17.7.5 BLOCKCHAIN AND CLOUD SERVICE PROVIDERS

Collaborations with major cloud service providers (e.g., AWS, Azure, Google Cloud) to develop standardized and integrated security solutions based on this framework can be an area of interest. This could lead to cloud-native security services specifically designed to align with Zero Trust and blockchain enhancements.

Education and skill development: Given the multidisciplinary nature of this approach, future efforts can focus on education and training programs to equip professionals with the expertise needed to implement and manage this framework effectively.

17.7.6 GLOBAL ADOPTION AND CASE STUDIES

Further research can explore the global adoption of this framework across different industries and regions. Real-world case studies and best practices can provide valuable insights into its practical implementation and impact.

17.7.7 ETHICAL CONSIDERATIONS

As blockchain and cloud technologies continue to evolve, ethical considerations related to data ownership, transparency, and privacy may become more prominent. Research can delve into ethical frameworks and guidelines within this context.

So, the future scope of "Zero Trust Evolution and Cloud Computing Security as Multi-Mission Engineering" is expansive and dynamic. It encompasses technological advancements, regulatory changes, interdisciplinary collaborations, and ethical considerations. By staying attuned to emerging trends and adapting to the evolving landscape, this framework can remain at the forefront of cybersecurity solutions, addressing emerging threats and ensuring data security in an ever-changing digital world. "Blockchain Consensus Algorithm Mitigating Data Repository Breaches" is profoundly promising. The framework's scalability and adaptability make it well-suited to accommodate the evolving needs of cloud-based blockchain applications and distributed systems. Collaboration with major cloud service providers to develop standardized security solutions based on this framework could streamline its adoption and implementation. Additionally, the model's alignment with data privacy regulations positions it as a crucial asset for organizations navigating a complex regulatory landscape. Moreover, research into quantum-resistant encryption and evolving ethical considerations surrounding data ownership and transparency can ensure the continued relevance and effectiveness of this approach. Ultimately, the future of this framework lies in its ability to evolve alongside emerging technologies and security challenges, providing organizations with a robust and dynamic defense against evolving threats and regulatory demands in the digital age.

KEYWORDS

- **Blockchain**
- **cloud computing**
- **cybersecurity**
- **data repository**
- **emerging threats**
- **internet of things**
- **zero trust**

REFERENCES

1. Park, J. H., & Park, J. H. (2017). Blockchain security in cloud computing: Use cases, challenges, and solutions. *Symmetry, 9*(8), 164. https://doi.org/10.3390/sym9080164.
2. Carlin, S., & Curran, K. (2013). Cloud computing security. In *Pervasive and Ubiquitous Technology Innovations for Ambient Intelligence Environments* (pp. 12–17). IGI Global.
3. Campbell, M. (2020). Beyond zero trust: Trust is a vulnerability. *Computer, 53*(10), 110–113. https://doi.org/10.1109/MC.2020.2991369.

4. Singh, B. (2023). Blockchain technology in renovating healthcare: Legal and future perspectives. In *Revolutionizing Healthcare Through Artificial Intelligence and Internet of Things Applications* (pp. 177–186). IGI Global.
5. Chen, Y., Paxson, V., & Katz, R. H. (2010). *What's New About Cloud Computing Security*. University of California, Berkeley Report No. UCB/EECS-2010-5, January, 20(2010), 2010-5.
6. Stafford, V. A. (2020). *Zero Trust Architecture* (NIST Special Publication 800-207). National Institute of Standards and Technology.
7. Pavithra, S., Ramya, S., & Prathibha, S. (2019). A survey on cloud security issues and blockchain. In *2019 3rd International Conference on Computing and Communications Technologies (ICCCT)* (pp. 136–140). IEEE. https://doi.org/10.1109/ICCCT.2019.8702843.
8. Almorsy, M., Grundy, J., & Müller, I. (2016). An analysis of the cloud computing security problem. *arXiv preprint arXiv:1609.01107*. https://arxiv.org/abs/1609.01107 (accessed on 11 November 2024).
9. Ghigliotti, C. J., Sprinkle, T. R., & Tesch, K. W. (2022). Artificial intelligence-enabled multi-mission resource allocation tactical decision aid. *Proceedings of the 2022 IEEE International Conference on Systems, Man, and Cybernetics* (SMC), 1247–1252.
10. Velmurugadass, P., Dhanasekaran, S., Anand, S. S., & Vasudevan, V. (2021). Enhancing blockchain security in cloud computing with IoT environment using ECIES and cryptography hash algorithm. *Materials Today: Proceedings, 37*, 2653–2659. https://doi.org/10.1016/j.matpr.2020.09.669.
11. Gai, K., Guo, J., Zhu, L., & Yu, S. (2020). Blockchain meets cloud computing: A survey. *IEEE Communications Surveys & Tutorials, 22*(3), 2009–2030. https://doi.org/10.1109/COMST.2020.2988714.
12. Esposito, C., De Santis, A., Tortora, G., Chang, H., & Choo, K. K. R. (2018). Blockchain: A panacea for healthcare cloud-based data security and privacy? *IEEE Cloud Computing, 5*(1), 31–37. https://doi.org/10.1109/MCC.2018.011771759.
13. Zou, J., He, D., Zeadally, S., Kumar, N., Wang, H., & Choo, K. R. (2021). Integrated blockchain and cloud computing systems: A systematic survey, solutions, and challenges. *ACM Computing Surveys (CSUR), 54*(8), 1–36. https://doi.org/10.1145/3425397.
14. Subramanian, N., & Jeyaraj, A. (2018). Recent security challenges in cloud computing. *Computers & Electrical Engineering, 71*, 28–42. https://doi.org/10.1016/j.compeleceng.2018.03.018.
15. Wylde, A. (2022). Questions of trust in norms of zero trust. In *Science and Information Conference* (pp. 837–846). Cham: Springer International Publishing. https://doi.org/10.1007/978-3-030-86247-9_85.
16. Singh, B. (2022). Understanding legal frameworks concerning transgender healthcare in the age of dynamism. *Electronic Journal of Social and Strategic Studies, 3*, 56–65.
17. Gupta, A., Siddiqui, S. T., Alam, S., & Shuaib, M. (2019). Cloud computing security using blockchain. *Journal of Emerging Technologies and Innovative Research (JETIR), 6*(6), 791–794.
18. Hawkins, M., Wang, A. K., Chang, T., Rowe, D., Martin, B., Enoch, M., & Pandya, J. (2022). Heterogeneous network-driven data fabrics to enable multi-mission autonomy. In *MILCOM 2022–2022 IEEE Military Communications Conference (MILCOM)* (pp. 259–264). IEEE.

19. Sharma, A., & Singh, B. (2022). Measuring impact of e-commerce on small-scale business: A systematic review. *Journal of Corporate Governance and International Business Law, 5*(1), 1–11.
20. Kuyoro, S. O., Ibikunle, F., & Awodele, O. (2011). Cloud computing security issues and challenges. *International Journal of Computer Networks (IJCN), 3*(5), 247–255.
21. Embrey, B. (2020). The top three factors driving zero trust adoption. *Computer Fraud & Security, 2020*(9), 13–15. https://doi.org/10.1016/S1361-3723(20)30093-4.
22. Al-Abdullah, M., Alsmadi, I., AlAbdullah, R., & Farkas, B. (2020). Designing privacy-friendly data repositories: A framework for a blockchain that follows the GDPR. *Digital Policy, Regulation and Governance, 22*(5, 6), 389–411. https://doi.org/10.1108/DPRG-08-2020-0044.
23. Li, S., Iqbal, M., & Saxena, N. (2022). Future industry internet of things with zero-trust security. *Information Systems Frontiers*, 1–14. https://doi.org/10.1007/s10796-022-10295-5.
24. Khalil, I. M., Khreishah, A., & Azeem, M. (2014). Cloud computing security: A survey. *Computers, 3*(1), 1–35. https://doi.org/10.3390/computers3010001.
25. Haddon, D. A. E. (2021). Zero trust networks: The concepts, the strategies, and the reality. In *Strategy, Leadership, and AI in the Cyber Ecosystem* (pp. 195–216). Academic Press.
26. Singh, B. (2022). Relevance of agriculture-nutrition linkage for human healthcare: A conceptual legal framework of implication and pathways. *Justice and Law Bulletin, 1*(1), 44–49.
27. Kumar, N., & LaRoy, N. (2021). Zero trust in the context of the utility industry. In *Proceedings of the Future Technologies Conference (FTC) 2020*, (*Vol. 3*, pp. 947–967). Springer International Publishing. https://doi.org/10.1007/978-3-030-70564-4_90.
28. Chaudhry, U. B., & Hydros, A. K. (2023). Zero-trust-based security model against data breaches in the banking sector: A blockchain consensus algorithm. *IET Blockchain, 3*(2), 98–115. https://doi.org/10.1049/btc2.12034.
29. Zissis, D., & Lekkas, D. (2012). Addressing cloud computing security issues. *Future Generation Computer Systems, 28*(3), 583–592. https://doi.org/10.1016/j.future.2011.07.009.
30. Gong, J., & Navimipour, N. J. (2022). An in-depth and systematic literature review on the blockchain-based approaches for cloud computing. *Cluster Computing, 25*(1), 383–400. https://doi.org/10.1007/s10586-022-03483-0.
31. Singh, B. (2022). COVID-19 pandemic and public healthcare: Endless downward spiral or solution via rapid legal and health services implementation with patient monitoring program. *Justice and Law Bulletin, 1*(1), 1–7.
32. Bertino, E., & Brancik, K. (2021). Services for zero trust architectures—A research roadmap. In *2021 IEEE International Conference on Web Services (ICWS)* (pp. 14–20). IEEE. https://doi.org/10.1109/ICWS52209.2021.00016.
33. Liu, W. (2012). Research on cloud computing security problem and strategy. In *2012 2nd International Conference on Consumer Electronics, Communications and Networks (CECNet)* (pp. 1216–1219). IEEE. https://doi.org/10.1109/CECNet.2012.6202419.
34. Assunção, P. (2019). A zero trust approach to network security. In *Proceedings of the Digital Privacy and Security Conference* (Vol. 2019). Porto, Portugal.
35. Zhang, P., Tian, C., Shang, T., Liu, L., Li, L., Wang, W., & Zhao, Y. (2021). Dynamic access control technology based on zero-trust light verification network model.

In *2021 International Conference on Communications, Information System and Computer Engineering (CISCE)* (pp. 712–715). IEEE. https://doi.org/10.1109/CISCE51857.2021.9378487.

36. Singh, B. (2019). Profiling public healthcare: A comparative analysis based on the multidimensional healthcare management and legal approach. *Indian Journal of Health and Medical Law, 2*(2), 1–5.
37. Tripathi, A., & Mishra, A. (2011). Cloud computing security considerations. In *2011 IEEE International Conference on Signal Processing, Communications and Computing (ICSPCC)* (pp. 1–5). IEEE. https://doi.org/10.1109/ICSPCC.2011.6084182.
38. Muralidhara, P. (2017). The evolution of cloud computing security: Addressing emerging threats. *International Journal of Computer Science and Technology, 1*(4), 1–33.
39. Singh, B. (2020). Global science and jurisprudential approach concerning healthcare and illness. *Indian Journal of Health and Medical Law, 3*(1), 7–13.
40. Barron, C., Yu, H., & Zhan, J. (2013). Cloud computing security case studies and research. In *Proceedings of the World Congress on Engineering* (Vol. 2, No. 2, pp. 1–6).
41. Mughal, M. H., Shaikh, Z. A., Ali, K., Ali, S., & Hassan, S. (2022). IPFS and blockchain-based reliability and availability improvement for integrated rivers' streamflow data. *IEEE Access, 10*, 61101–61123. https://doi.org/10.1109/ACCESS.2022.3183198.
42. Bhattacharjee, A., Badsha, S., Shahid, A. R., Livani, H., & Sengupta, S. (2020). Block-phasor: A decentralized blockchain framework to enhance security of synchrophasor. In *2020 IEEE Kansas Power and Energy Conference (KPEC)* (pp. 1–6). IEEE. https://doi.org/10.1109/KPEC49248.2020.9234045.

CHAPTER 18

Zero-Trust Approach for Secure Healthcare System

RAMANDEEP SANDHU,[1,*] HARPREET KAUR CHANNI,[2,*]
DEEPIKA GHAI,[3,*] MADHURI,[1,*] and MANDEEP KAUR[4,*]

[1]*School of Computer Science and Engineering, Lovely Professional University, Punjab, India*

[2]*Department of Electrical Engineering, Guru Nanak Dev Engineering College, Ludhiana, Punjab, India*

[3]*School of Electronics and Electrical Engineering, Lovely Professional University, Punjab, India*

[4]*Department of Computer Science and Engineering, Chitkara University Institute of Engineering and Technology, Chitkara University, Punjab, India*

*Corresponding author

ABSTRACT

In the dynamic intersection of healthcare and cybersecurity, the Zero Trust approach has emerged as an innovative initiative dedicated to safeguarding the most precious assets: patient data and trust. The proposed approach unveils the latest advancements and visionary strategies that are reshaping the landscape of healthcare security. The healthcare sector, a veritable hub of sensitive patient information, has evolved beyond the confines of traditional security paradigms. This chapter journeys through the core principles of Zero Trust, where the age-old axiom of 'trust but verify' is reinvented into "never trust, always verify." It mainly delves into the practical applications of this philosophy, emphasizing the critical importance of identity verification, data encryption, and adaptive

Zero-Trust Learning: Applications in Modern Network Security. Chetna Kaushal, Amandeep Kaur, Mohit Angurala, Aryan Chaudhary (Eds.)

access controls. Drawing from real-world scenarios, this chapter showcases how healthcare organizations are at the forefront of implementing Zero Trust. From the protection of Electronic Health Records (EHR) to the fortification of telemedicine platforms, these entities are leveraging cutting-edge technologies and strategies to secure healthcare ecosystems against relentless cyber threats. In an era where data breaches and ransomware attacks loom large, the "Zero Trust Approach for Secure Healthcare Systems" offers a strict approach to follow. It explores the symbiotic relationship between trust and verification, creating a paradigm where the sanctity of patient data is upheld, regulatory compliance is seamlessly achieved, and healthcare providers can continue their noble mission with unwavering confidence. For healthcare professionals, IT leaders, and policymakers committed to patient-first security, this chapter provides invaluable insights into the future of healthcare cybersecurity. It illuminates the path to a world where patient data remains inviolable, trust is fortified by a proactive security posture, and healthcare systems stand resilient in the face of emerging threats.

18.1 INTRODUCTION

The traditional approach to network security, which frequently relies on the principle of "trust but verify," is called by the cybersecurity framework and strategy known as "Zero Trust." In a Zero Trust approach, businesses don't presume to trust any user or device, inside or beyond the boundaries of their network [1, 2]. Instead, they put in place stringent access controls, ongoing surveillance, and other security procedures to make sure every user and device trying to access resources is real as shown in Figure 18.1.

The existing perimeter-based security strategy used by the majority of healthcare organizations won't be sufficient in the future, given the interconnected nature of everything from Internet of Medical Things (IoMT) devices to augmented reality to robotics and more. Healthcare firms must continue to invest in the fundamentals while fundamentally changing from the castle-and-moat strategy to a zero-trust model to stay ahead of these developments. Figure 18.2, as depicted below, represents the various entities involved in attaining a Zero Trust approach in existing secure systems. This approach treats every device using an application as a threat vector. According to the basic fundamental of zero trust, if some action is not verified, just block it and do not proceed or allow it in any situation [3, 4]. Regardless of where each is based, whether inside or outside the security perimeter, none of the following should ever be trusted by default: users, workloads, devices, and systems.

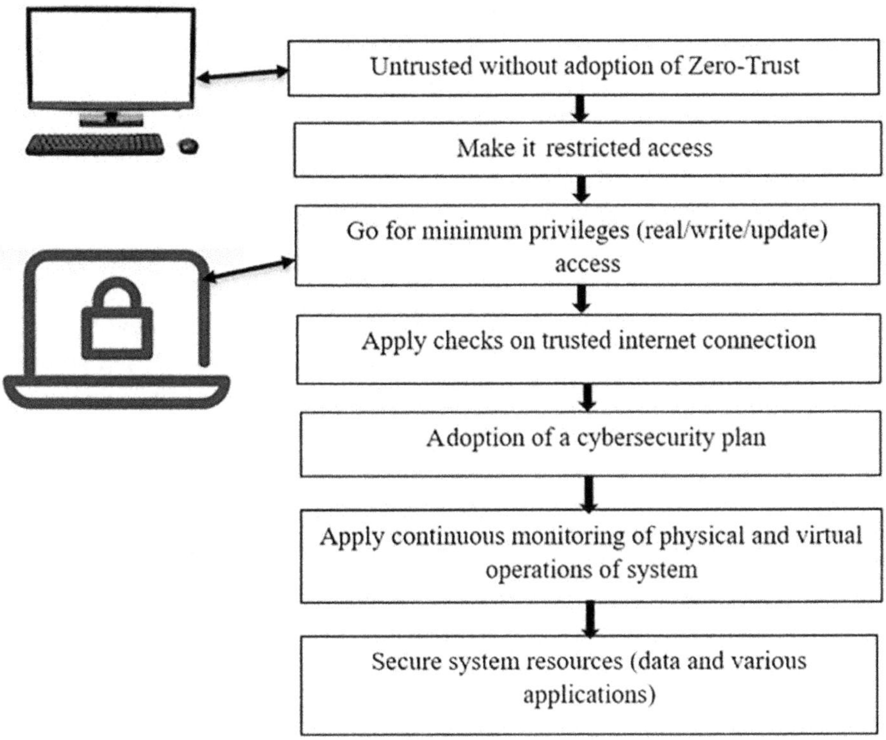

FIGURE 18.1 Abstraction level of zero trust.

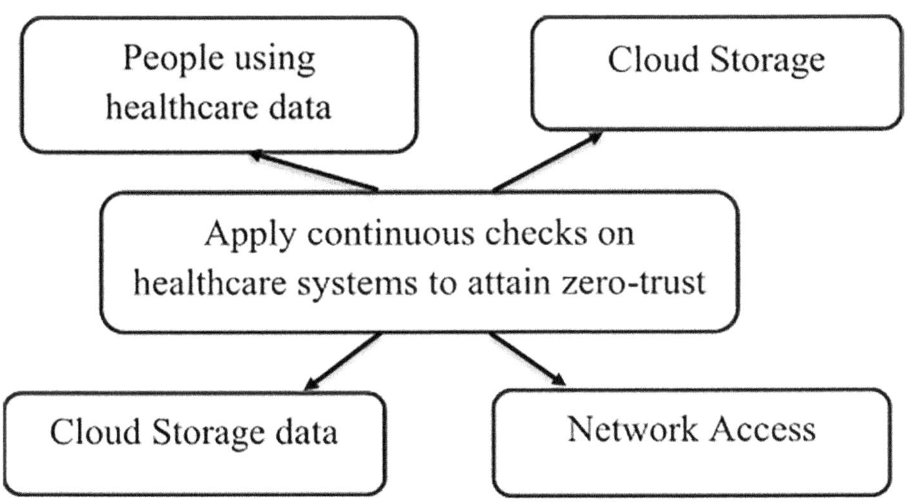

FIGURE 18.2 Entities involved in zero trust.

The demand for reliable and secure healthcare systems is higher than ever in today's age of rapid technological breakthroughs and widespread medical digitalization. The healthcare industry, which is tasked with protecting patients' personal information, confronts several cybersecurity threats that endanger data privacy, information security, and system performance [5, 6]. Cybersecurity concerns have increased as the healthcare sector has become more reliant on EHRs, telemedicine, and connected medical equipment. Once sufficient, traditional security techniques are now inadequate in securing healthcare systems and patient data. In light of these alarming trends, a new cybersecurity paradigm known as the Zero Trust Approach has evolved. By supposing that risks may originate from both external and internal sources and that confidence should never be assumed regardless of the location or context of network access, this novel approach intends to revolutionize the way healthcare organizations think about security [7]. Figure 18.3 shows the adoption of Zero Trust in healthcare. In this chapter, we explore the urgent problem of how to apply the Zero Trust Approach to the healthcare industry. This chapter delves into the reasoning for this framework's widespread adoption, its guiding principles, and the potential gains for healthcare providers, patients, and the industry as a whole [8].

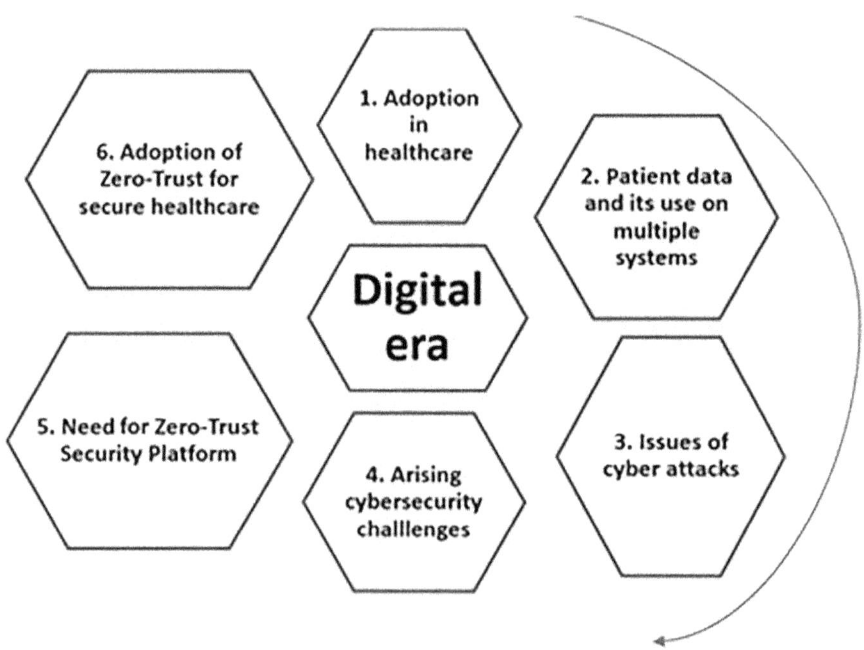

FIGURE 18.3 Adoption of zero trust in healthcare.

It is crucial to have an understanding of the severe cybersecurity issues now confronting the healthcare business before diving into the Zero Trust Approach. Data breaches in the healthcare industry are a major concern, and cyber-attacks are a growing issue. Furthermore, the healthcare industry has been criticized for being unprepared in terms of cybersecurity. The prevalence of security threats like ransomware, data breaches, and phishing in the healthcare industry calls for a preventative and all-encompassing security strategy. Notably, many security breaches in healthcare occur due to internal sources, including human mistakes, ignorance, or deliberate activities by insiders. This necessitates a security policy that views all entry points as suspect and uses them to strictly regulate user access [9, 10].

18.1.1 THE ZERO TRUST APPROACH: A PARADIGM SHIFT IN HEALTHCARE SECURITY

The Zero Trust Approach challenges the conventional "trust but verify" model, which assumes that once inside the network, users, and devices can be trusted. In contrast, Zero Trust operates on the fundamental principle that trust is never implicit and that verification must be continuous. This approach encompasses various aspects, including network security, identity, and access management, data protection, and more, all aimed at minimizing security risks and vulnerabilities [11]. This chapter will explore the core principles of Zero Trust and how they can be applied in the healthcare context to create a secure and resilient healthcare system. It will also address the potential challenges and barriers to implementation and offer insights into best practices and real-world examples of healthcare organizations that have successfully adopted a Zero Trust security model. By adopting the Zero Trust Approach, healthcare organizations can significantly enhance their cybersecurity posture, protect sensitive patient data, and ensure the integrity and availability of critical healthcare services. In doing so, they not only safeguard their patients but also contribute to the overall trust and resilience of the healthcare ecosystem in an increasingly digital world [8].

18.1.2 UNDERSTANDING ZERO TRUST IN HEALTHCARE

Zero Trust is a cybersecurity model that has gained increasing relevance within the healthcare sector as the industry grapples with the challenges of safeguarding sensitive patient data and maintaining the integrity of medical services in an increasingly digital landscape. At its core, Zero Trust challenges the traditional notion of implicit trust once users and devices are inside the network. Instead, it mandates continuous verification and validation of all entities, whether inside or outside the network perimeter. In healthcare, this approach emphasizes robust identity verification, least privilege access, micro-segmentation, continuous monitoring, and data encryption [12]. By implementing Zero Trust, healthcare organizations can enhance their cybersecurity posture, reduce the risk of unauthorized access, and protect patient confidentiality, contributing to the overall trust and resilience of the healthcare ecosystem.

In the evolving landscape of healthcare, where digitization and connectivity have transformed the way patient information is managed and medical services are delivered, the importance of cybersecurity cannot be overstated. Healthcare organizations are entrusted with sensitive patient data, making them prime targets for cyber threats. As a response to the growing complexity of these challenges, the Zero Trust model has gained prominence as an effective approach to securing healthcare systems. This chapter aims to provide a comprehensive understanding of the Zero Trust model and how it can be applied within the healthcare sector. We will explore the core principles of Zero Trust, its relevance in healthcare, and practical steps for implementation [2–6, 13].

18.2 THE FOUNDATION OF ZERO TRUST

The Zero Trust model is built upon a fundamental concept: trust should never be assumed, regardless of the user's location or context within the network. Traditionally, organizations adopted a perimeter-based security approach, where once inside the network, users, and devices were trusted implicitly. However, this approach has proven inadequate in the face of today's sophisticated cyber threats.

Zero Trust challenges this conventional wisdom by requiring continuous verification and validation of users and devices, both inside and outside the network perimeter. Trust is no longer based on location but on a combination of identity, device health, and contextual factors [14, 15].

18.2.1 KEY PRINCIPLES OF ZERO TRUST IN HEALTHCARE

Verify Identity: In healthcare, verifying the identity of users and devices is paramount. Strong authentication mechanisms, such as Multi-Factor

Authentication (MFA), should be in place to ensure that only authorized personnel access patient records and critical systems [16, 17].

1. **Least Privilege Access:** The principle of granting the least privilege access ensures that users have only the permissions necessary for their specific role. In healthcare, this minimizes the risk of unauthorized access to patient data.
2. **Micro-Segmentation:** Micro-segmentation divides the network into smaller, isolated segments, limiting lateral movement for potential attackers. In a healthcare setting, this prevents unauthorized access to different parts of the network, including patient records and medical devices.
3. **Continuous Monitoring:** Real-time monitoring and behavioral analysis of users and devices help detect abnormal activities promptly. This is crucial in healthcare to identify and respond to potential threats quickly.
4. **Data Encryption:** Encrypting data at rest and in transit ensures that even if a breach occurs, the data remains protected. Healthcare organizations must prioritize data encryption to maintain patient confidentiality.

18.2.2 IMPLEMENTING ZERO TRUST IN HEALTHCARE

Implementing a Zero Trust model in healthcare requires a multi-faceted approach:

1. **Assessment and Inventory:** Begin by assessing existing security measures and creating an inventory of devices and systems. Identify vulnerabilities and areas that require improvement.
2. **Access Controls:** Implement robust access controls, including MFA, role-based access control (RBAC), and strict authentication policies. Ensure that users have access only to the data and systems necessary for their responsibilities.
3. **Segmentation:** Divide the network into segments and apply micro-segmentation to restrict lateral movement. This is particularly critical in healthcare to protect sensitive patient data.
4. **Continuous Monitoring:** Deploy advanced monitoring and analytics tools to detect anomalies and potential threats in real time [18, 19].

5. **Education and Training:** Educate healthcare staff on cybersecurity best practices and the importance of adhering to Zero Trust principles. Human error remains a significant threat, and well-informed employees are the first line of defense.
6. **Vendor Assessment:** Assess the cybersecurity posture of third-party vendors and service providers who have access to your healthcare systems.

As the healthcare sector continues to embrace digital transformation, the Zero Trust model emerges as a crucial strategy for safeguarding patient data and maintaining the integrity of medical services. By adopting Zero Trust principles, healthcare organizations can mitigate the risks associated with cyber threats, ensuring the privacy and security of patient information while maintaining the trust of patients and the broader healthcare community.

18.3 CHALLENGES IN HEALTHCARE SECURITY

Healthcare security faces a multitude of challenges that stem from the digitization of medical records, interconnected systems, and the high value placed on patient data. Data breaches and cyberattacks are persistent threats, with ransomware attacks and phishing scams causing disruptions and compromising patient privacy [6–10, 20, 21]. Human error and insider threats compound the risks, as staff may inadvertently expose sensitive information. Legacy systems and outdated technology create vulnerabilities, while the proliferation of interconnected medical devices expands the attack surface. Compliance with regulations like the Health Insurance Portability and Accountability Act of 1996 (HIPAA) and the General Data Protection Regulation (GDPR) is essential but complex, often involving resource constraints and intricate privacy concerns. Supply chain risks, insufficient cybersecurity awareness, and the evolving threat landscape further exacerbate these challenges. Addressing these issues necessitates a comprehensive approach that combines robust cybersecurity measures, staff training, and constant adaptation to emerging threats to ensure the security and privacy of patient data in the healthcare sector.

1. **Patching and Vulnerability Management:** Ensuring that all software and systems are up-to-date with security patches can be challenging in healthcare, where system downtime can impact patient care. Delayed or missed patches can leave systems vulnerable to exploitation.

2. **Data Sharing and Interoperability:** Healthcare institutions often need to share patient data with other providers and organizations. Ensuring secure data exchange and interoperability while maintaining patient privacy is a complex task.
3. **Mobile Device Management:** The use of mobile devices in healthcare, such as smartphones and tablets, introduces new security challenges. These devices can be lost or stolen, potentially exposing sensitive patient information.
4. **Cloud Security:** The adoption of cloud services for storing and processing healthcare data requires robust security measures to protect data in transit and at rest, as well as ensuring the security of cloud-based applications and platforms.
5. **Cybersecurity Talent Shortage:** There is a shortage of skilled cybersecurity professionals in the healthcare industry, making it difficult for organizations to recruit and retain experts who can effectively manage and respond to security threats.
6. **Budget Constraints:** Many healthcare organizations operate on tight budgets, which can limit their ability to invest in advanced cybersecurity solutions and resources needed to adequately protect their systems.
7. **Patient Trust:** Security breaches can erode patient trust in healthcare institutions. Patients may be hesitant to share sensitive information or participate in telehealth services if they are concerned about the security of their data.
8. **Geopolitical Threats:** Healthcare organizations can be targeted by state-sponsored cyberattacks and other geopolitical threats, further elevating the level of sophistication and potential impact of security breaches.
9. **Misalignment of Priorities:** In some cases, the focus on patient care and immediate health concerns may overshadow cybersecurity concerns, leading to potential vulnerabilities that are not adequately addressed.
10. **Healthcare IoT Vulnerabilities:** The increasing use of IoT devices in healthcare, such as remote monitoring devices, can introduce security weaknesses if these devices are not properly secured and monitored.

11. **Legal and Ethical Concerns:** Balancing the legal and ethical obligations of sharing patient information for research or public health purposes with the need to protect patient privacy and security is a complex ethical dilemma in healthcare security.

Addressing these diverse challenges in healthcare security requires a multifaceted approach, combining technology, policy development, staff training, and a strong commitment to data protection. Healthcare organizations must prioritize cybersecurity to maintain patient trust, safeguard sensitive data, and ensure the continued delivery of high-quality care in an increasingly digital healthcare landscape.

18.3 COMPONENTS OF A ZERO TRUST FRAMEWORK

A Zero Trust Framework is a cutting-edge approach to cybersecurity that assumes no inherent trust within a network, challenging the conventional belief that once inside the perimeter, entities can be trusted implicitly. Instead, it mandates rigorous verification and authentication for every user and device, regardless of their location or network connection [22, 23]. The components of a Zero Trust Framework are multifaceted and interconnected, each contributing to the overall security posture.

The key components of a Zero Trust architecture are listed below:

- Identity and Access Management (IAM);
- Least Privilege Access;
- Network Segmentation;
- Continuous Monitoring and Assessment;
- Device Trustworthiness;
- Application Security;
- Encryption;
- Zero Trust Network Access;
- User and Entity Behavior Analytics;
- Incident Response and Remediation;
- Policy Enforcement; and
- Education and Awareness.

Figure 18.4 represents the list of the above-mentioned elements of the zero trust system.

At its core, IAM forms the foundation of Zero Trust. It requires strong authentication methods like MFA to ensure that users and devices are who they claim to be. Least Privilege Access, another fundamental principle, mandates

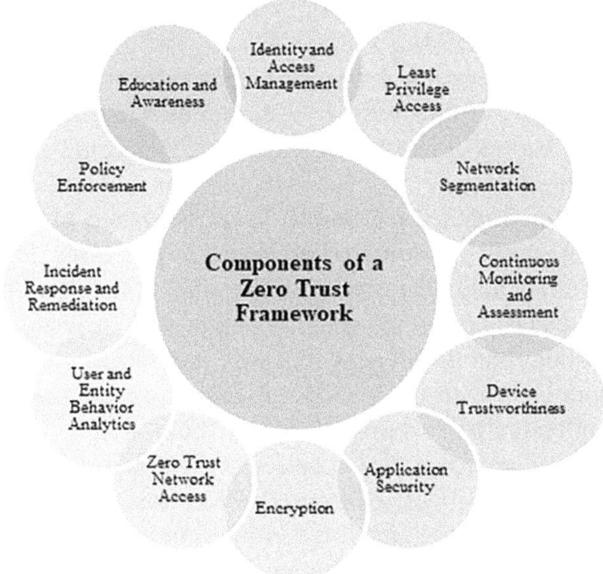

FIGURE 18.4 Key components of a zero trust architecture.

granting users and devices only the minimal access privileges necessary for their specific roles, mitigating the potential damage from credential compromises [8, 11–13]. Network Segmentation further divides the network into smaller, more manageable segments, or micro-segments, where strict access controls and firewall rules govern communication. Continuous Monitoring and Assessment are imperative in Zero Trust, involving the constant scrutiny of user and device behavior for anomalies and potential threats, alongside regular security assessments and vulnerability scans.

Device Trustworthiness focuses on ensuring the integrity of devices seeking access. This includes enforcing security standards, checking for up-to-date security patches, and verifying that devices are free from compromise. Application Security extends the Zero Trust principles to applications, necessitating robust security measures like web application firewalls, secure coding practices, and routine code reviews to uncover and rectify vulnerabilities. Encryption is a cornerstone of Zero Trust, encompassing both data in transit and at rest. It employs strong encryption protocols and meticulous encryption key management practices to protect against eavesdropping and unauthorized access. Zero Trust Network Access (ZTNA) solutions introduce secure, on-demand access to resources for authorized users, irrespective of their location or network, often employing a Software-Defined Perimeter (SDP) approach [15–19, 24].

User and Entity Behavior Analytics (UEBA) leverage machine learning and behavioral analysis to scrutinize user and device actions for deviations from the norm, rapidly identifying anomalies that could indicate security threats. Incident Response and Remediation, an essential aspect, necessitate a well-defined plan for promptly addressing and mitigating security incidents, augmented by automated remediation processes. Policy Enforcement ensures strict adherence to security policies that govern access and data handling, ensuring that all activities align with security requirements. Finally, Education, and Awareness focus on the human element, providing employees and users with the knowledge and tools they need to recognize and respond effectively to security threats. In sum, a Zero Trust Framework demands a holistic approach to cybersecurity, combining various technologies, policies, and best practices to establish a proactive, adaptive security strategy. It signifies a paradigm shift from perimeter-centric thinking to one that recognizes the fluidity of the modern threat landscape. By embracing the principles and components of Zero Trust, organizations can substantially bolster their cybersecurity posture, fortifying themselves against an ever-evolving digital world filled with potential threats [25].

18.4 ZERO TRUST FRAMEWORK IN HEALTHCARE

Zero trust networks assist hospitals in enhancing cybersecurity both inside and outside the network, as they are one of the most frequently attacked and compromised sectors and the ones containing the most sensitive types of data. Achieving a state where healthcare industries are ensuring that every transaction is being carried out by an entity with the authority to do it is the key objective. It works as a patient-centric approach as the main motive is effective patient care. The proposed model of security with zero trust is not only "trust but verify," but it is "always verify but never trust." Therefore, healthcare organizations can monitor any attempts to exploit these online connections' and applications' weaknesses [1–10, 26].

Incorporating a Zero Trust framework in healthcare necessitates a paradigm shift from traditional perimeter-based security models to a holistic and adaptive approach. It starts with identity-centric access control, requiring stringent authentication methods such as MFA for all users, including healthcare professionals and patients. The principle of least privilege ensures that access permissions align with specific roles, limiting the potential damage from credential breaches. It is a balance in security awareness, functionality, and security needs.

Network segmentation is fundamental, isolating critical systems and patient data through rigorous access controls to curtail lateral movement by potential threats. Continuous monitoring and anomaly detection, driven by UEBA, enable swift detection of unusual activities and deviations from normal behavior, enhancing proactive threat response. Endpoint security and data encryption are critical components in safeguarding medical devices and patient data from breaches and unauthorized access. Application security measures, such as application firewalls and vulnerability patching, bolster the resilience of healthcare applications, including Electronic Health Records (EHR) systems. ZTNA solutions ensure secure and flexible access to healthcare resources for authorized users, regardless of their location or network connection, facilitating telehealth and remote healthcare services [27].

A robust incident response plan tailored to healthcare-specific threats, coupled with automated incident response mechanisms, allows for swift containment and mitigation of security incidents. Policy enforcement aligns security practices with healthcare compliance regulations, such as HIPAA, ensuring ongoing adherence to industry standards. Finally, ongoing user education and awareness training empower healthcare staff to recognize and respond effectively to potential threats, including phishing attacks and social engineering attempts. By adopting these Zero Trust principles and practices, healthcare organizations can proactively address the ever-evolving cybersecurity challenges and provide a safer environment for patients and sensitive data. Figure 18.5 shows the framework of Zero Trust architecture (ZTA).

18.5 CASE STUDIES OF ZERO TRUST

The case studies of Zero Trust are enlisted as follows.

18.5.1 CASE STUDY 1: SECURING PATIENT RECORDS

A large urban hospital, XYZ Medical Center, faced challenges in securing patient records within its EHR system. With increasing incidents of data breaches in the healthcare sector, they needed to enhance security without disrupting patient care. XYZ Medical Center adopted a Zero Trust approach. They implemented strict access controls and network segmentation, ensuring that only authorized healthcare providers had access to specific patient

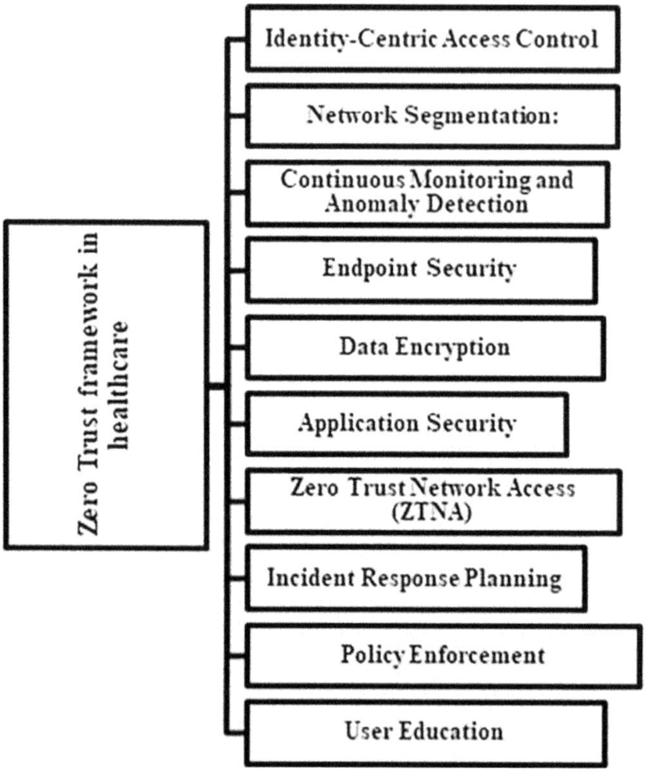

FIGURE 18.5 Methodology of zero trust architecture.

records. MFA became mandatory for all staff, and they employed continuous monitoring to detect unusual activities. The hospital successfully safeguarded patient records. Unauthorized access attempts triggered immediate alerts, allowing for a prompt response. Patient data remained confidential, and the hospital enhanced its reputation for data security [1–10, 28].

18.5.2 CASE STUDY 2: SECURING IOT MEDICAL DEVICES

A regional healthcare network, ABC Health, faced a growing challenge with their Internet of Things (IoT) medical devices. These devices were essential for patient care, but they also introduced potential vulnerabilities. ABC Health implemented a ZTNA solution to secure IoT device access. They segmented their network and deployed micro-segmentation techniques, limiting device communication to necessary resources. Continuous monitoring was used to detect device anomalies, and strong authentication was enforced for device connections. The healthcare network effectively protected IoT devices from cyber threats. Unauthorized access attempts were blocked, and ABC Health ensured uninterrupted patient care while mitigating security risks [8, 11–13, 29].

18.5.3 CASE STUDY 3: TELEMEDICINE SECURITY

Amidst the COVID-19 pandemic, a rural healthcare clinic, RuralCare Clinic, rapidly expanded its telemedicine services to reach patients remotely. However, they needed a secure solution to protect sensitive patient data during virtual consultations. RuralCare Clinic adopted a Zero Trust approach by implementing secure video conferencing platforms with end-to-end encryption. They enforced strict identity verification for both patients and healthcare providers, using video-based MFA for remote consultations. Additionally, they conducted security awareness training for staff and patients. RuralCare Clinic successfully provided secure telemedicine services, ensuring the confidentiality of patient discussions and data. Patient trust in virtual consultations increased, and the clinic expanded its reach to underserved areas.

These hypothetical case studies highlight how a Zero Trust approach can be applied in various healthcare scenarios to enhance security without compromising patient care or data accessibility. Implementing such measures is crucial to safeguard patient information and maintain the integrity of healthcare systems in an increasingly digital age [30].

18.6 BENEFITS AND ROI OF ZERO TRUST

Implementing a Zero Trust security model offers substantial benefits and the potential for a significant return on investment (ROI) for organizations. By adopting a Zero Trust approach, organizations can enhance their security posture through reduced attack surfaces, improved data protection, and reduced risks of data breaches and non-compliance penalties. Zero Trust also enables more efficient incident response, supports remote work, and may lead to cost savings over time, including lower IT expenses and potentially reduced cybersecurity insurance premiums. Additionally, Zero Trust enhances business agility, user experiences, and brand reputation while providing a competitive edge in the market. While there may be upfront costs associated with implementing Zero Trust, the long-term benefits and ROI in terms of security and cost reduction make it a compelling investment for organizations seeking to fortify their cybersecurity defenses.

Moreover, the Zero Trust model supports remote work and facilitates secure access to cloud resources, enhancing business continuity and agility. It also contributes to an improved user experience by enabling seamless, secure access to resources. By preserving brand reputation through robust security practices, organizations can maintain trust with customers and partners. Additionally, Zero Trust can be a market differentiator, showcasing an organization's commitment to cybersecurity and data protection. While there are costs associated with implementing Zero Trust, the long-term benefits and potential cost savings, coupled with reduced risks and improved security, make it a strategic investment with a compelling ROI for organizations concerned about their cybersecurity posture and the safeguarding of critical assets and data. Furthermore, the Zero Trust model has become increasingly relevant in the face of evolving cybersecurity threats, making it a proactive approach to staying ahead of malicious actors. This can translate into savings by avoiding the financial fallout from security incidents, including reputational damage and regulatory fines. Additionally, as organizations continue to adopt cloud technologies and expand their digital presence, Zero Trust provides a scalable and adaptable security framework that aligns with these modernization efforts, potentially reducing the complexity and costs associated with legacy security measures. Overall, the benefits and ROI of Zero Trust extend beyond immediate financial gains to encompass long-term resilience, adaptability, and enhanced trustworthiness in an ever-changing digital landscape [5–9, 18].

18.7 FUTURE TRENDS AND EMERGING TECHNOLOGIES FOR ACHIEVING ZERO TRUST

The future is poised to witness a transformative wave of technological advancements and trends. AI will continue its rapid evolution, driving automation and insights across industries. Edge computing will enable real-time processing for the IoT. The rollout of 5G networks promises high-speed, low-latency connectivity, powering innovations like augmented and virtual reality. Quantum computing holds the potential to revolutionize problem-solving. Biotechnology and genomics will usher in personalized medicine and gene editing breakthroughs. Blockchain will extend beyond cryptocurrencies to reshape various sectors, while cybersecurity will evolve to counter increasingly sophisticated threats. Sustainability will drive green technologies, and autonomous vehicles will reshape transportation. Space exploration is advancing, and augmented reality and virtual reality will find applications beyond entertainment. Biometric technologies will enhance security, and human-machine integration will empower new ways of interacting with technology. These emerging technologies will redefine industries, presenting both opportunities and challenges in the years to come [31].

Additionally, as we look ahead, the convergence of these emerging technologies is expected to create synergies and open new frontiers. For instance, AI-powered healthcare solutions could benefit from genomics data, revolutionizing diagnostics and treatment. Quantum computing may enhance encryption and cybersecurity, while 5G networks enable faster data transmission for IoT devices. Sustainability efforts could incorporate AI and green technologies to optimize resource usage. The future may see the integration of biometrics with blockchain for secure identity verification. Space exploration could leverage AI for autonomous navigation and robotics. Augmented reality and virtual reality might find applications in training AI-driven systems, creating immersive educational experiences. As these technologies interconnect and mature, they have the potential to redefine industries, disrupt traditional business models, and shape our daily lives in ways we are only beginning to imagine.

Furthermore, societal shifts and challenges are likely to influence the trajectory of emerging technologies. For instance, the COVID-19 pandemic accelerated the adoption of remote work and telemedicine, driving innovation in collaboration tools, telehealth solutions, and digital healthcare. Climate change concerns are propelling the development of sustainable energy solutions and eco-friendly transportation options. Ethical considerations surrounding AI, privacy, and data security are sparking discussions on responsible technology use and regulation. As emerging technologies continue to advance, addressing these ethical, legal, and social implications will be essential for ensuring their responsible and equitable deployment. Additionally, global collaboration and partnerships between governments, industries, and research institutions will play a crucial role in accelerating the development and adoption of these technologies while addressing challenges related to cybersecurity, data governance, and ethical guidelines. In the coming years, the intersection of technological innovation, societal needs, and ethical considerations will shape the future landscape of emerging technologies in profound ways [32–35].

The rapid advancement and integration of emerging technologies also bring forth a range of challenges and considerations. Cybersecurity remains a paramount concern as more systems and devices become interconnected, increasing the potential attack surface for cyber threats. Ensuring the privacy and security of data, particularly in the context of AI and IoT, requires robust safeguards and encryption protocols. Ethical dilemmas surrounding AI, including biases in algorithms and the responsible use of AI in decision-making, demand careful consideration. Questions about data ownership, consent, and transparency in data collection and usage continue to be hot topics. The widening digital divide is a concern, as not everyone has equal access to the benefits of emerging technologies, exacerbating societal inequalities. Efforts to bridge this gap and ensure equitable access are vital. Environmental sustainability is another consideration, especially as the energy consumption of data centers and the electronic waste from technology disposal increase. Balancing technological progress with environmental responsibility is a complex challenge [36, 37]. Regulatory and legal frameworks must evolve to keep pace with technological advancements, addressing issues like data protection, liability, and intellectual property rights. International cooperation is crucial, as many technologies operate globally. Lastly, there's the challenge of preparing the workforce for the jobs of the future. Continuous reskilling and upskilling are necessary to adapt to the changing technological landscape and prevent job displacement in some sectors. Navigating these challenges and considerations requires a proactive and collaborative approach involving governments, businesses, academia, and civil society. Striking a balance between innovation and addressing the ethical, social, and environmental implications of emerging technologies is essential for a sustainable and responsible technological future [38].

18.8 BEST PRACTICES AND RECOMMENDATIONS FOR ZERO TRUST

Best practices and recommendations for effectively harnessing emerging technologies include establishing a robust cybersecurity framework that prioritizes data protection, privacy, and threat mitigation. Implementing ethical guidelines in AI and data-driven decision-making is crucial to avoid biases and ensure responsible use. Promoting digital inclusivity through policies and initiatives that bridge the digital divide is essential for equitable access to technology benefits. Sustainability efforts should focus on energy-efficient technology deployment and responsible e-waste management. A proactive approach to regulatory and legal compliance is necessary, with organizations staying informed about evolving laws and standards. Lastly, investing in workforce development and education is vital to empower individuals with the skills needed to thrive in the evolving technological landscape. Collaboration between stakeholders, from governments to industry leaders, is key to addressing these best practices and ensuring that emerging technologies contribute positively to society while minimizing potential risks and challenges [13–17, 37].

Additionally, organizations should prioritize data governance, ensuring transparent data collection, storage, and usage practices. Establishing robust data backup and recovery strategies is crucial to safeguard against data loss and downtime. Embracing a culture of continuous learning and innovation is essential. Organizations should encourage employees to adapt to technological changes, fostering a culture that values experimentation and learning from failures. Engagement in open collaboration and partnerships, both domestically and internationally, can foster innovation and tackle global challenges. Sharing best practices and collaborating on research and development initiatives can accelerate technological progress. To address environmental concerns, organizations should explore eco-friendly technologies and sustainable practices, from energy-efficient data centers to responsible materials sourcing. Lastly, maintaining a strong focus on user experience and customer-centric design principles is critical. Technology should be intuitive, user-friendly, and designed to enhance the lives and experiences of individuals. Incorporating these best practices and recommendations into strategic planning and decision-making can help organizations harness the full potential of emerging technologies while mitigating risks and ensuring responsible and sustainable deployment [21–23, 36].

18.9 CONCLUSION

In this era of rapid technological advancement, embracing emerging technologies while addressing the associated challenges is a dynamic journey. It requires a proactive approach, collaboration, and a commitment to responsible innovation. As we continue to explore the frontiers of AI, IoT, quantum computing, and more, we must remain vigilant about cybersecurity, ethics, and inclusivity. Moreover, the pursuit of sustainability and environmental responsibility in technology is paramount to combat climate change and minimize the ecological footprint of our digital world. Regulatory compliance and legal frameworks must evolve to keep pace with technology, striking a balance between innovation and protection. At the heart of this transformative journey is the workforce of the future. By investing in education, upskilling, and fostering a culture of curiosity and adaptability, we empower individuals to thrive in the rapidly evolving technological landscape. In conclusion, emerging technologies offer a wealth of opportunities to shape a brighter, more connected, and sustainable future. To seize these opportunities and mitigate challenges, we must collectively commit to responsible and ethical technology adoption, ensuring that the benefits are enjoyed by all and that our digital future prioritizes the well-being of individuals and our planet. The world of emerging technologies presents boundless opportunities for innovation, growth, and positive societal impact. However, it also poses complex challenges and ethical considerations that demand careful attention. The successful integration of emerging technologies hinges on a foundation of robust cybersecurity, ethical AI practices, and responsible data governance.

Equally critical is the commitment to digital inclusivity, ensuring that the benefits of these technologies are accessible to all, irrespective of geographic location or socio-economic status. Sustainability should be at the forefront, with a focus on environmentally responsible practices and energy-efficient solutions. To navigate the evolving technological landscape, organizations must remain adaptable, fostering a culture of continuous learning and collaboration. Embracing open partnerships and international cooperation can drive innovation and address global challenges effectively. Ultimately, the future of emerging technologies lies in the hands of those who prioritize responsible, ethical, and sustainable deployment. By adhering to best practices and recommendations, we can harness the potential of these technologies to usher in a brighter and more inclusive future for all.

KEYWORDS

- **cybersecurity**
- **data encryption**
- **healthcare**
- **identity verification**
- **patient data**
- **security paradigms**
- **zero trust**

REFERENCES

1. Wang, L., Ma, H., Li, Z., Pei, J., Hu, T., & Zhang, J. (2022). A data plane security model of SR-BE/TE based on zero-trust architecture. *Scientific Reports, 12*(1), 20612. https://doi.org/10.1038/s41598-022-25499-2.
2. Shore, M., Zeadally, S., & Keshariya, A. (2021). Zero trust: The what, how, why, and when. *Computer, 54*(11), 26–35. https://doi.org/10.1109/MC.2021.3072055.
3. Wylde, A. (2021). Zero trust: Never trust, always verify. In *2021 International Conference on Cyber Situational Awareness, Data Analytics and Assessment (CYBERSA)* (pp. 1–4). IEEE. https://doi.org/10.1109/CYBERSA52776.2021.9554027.
4. Scott, B. (2018). How a zero trust approach can help to secure your AWS environment. *Network Security, 2018*(3), 5–8. https://doi.org/10.1016/S1353-4858(18)30026-6.
5. Ali, B., Gregory, M. A., & Li, S. (2021). Uplifting healthcare cyber resilience with a multi-access edge computing zero-trust security model. In *2021 31st International Telecommunication Networks and Applications Conference (ITNAC)* (pp. 192–197). IEEE. https://doi.org/10.1109/ITNAC52697.2021.9639436.
6. Vukotich, G. (2023). Healthcare and cybersecurity: Taking a zero trust approach. *Health ServicesInsights, 16*, 11786329231187826. https://doi.org/10.1177/11786329231187826.
7. Buck, C., Olenberger, C., Schweizer, A., Völter, F., & Eymann, T. (2021). Never trust, always verify: A multivocal literature review on current knowledge and research gaps of zero-trust. *Computers & Security, 110*, 102436. https://doi.org/10.1016/j.cose.2021.102436.
8. Sarkar, S., Choudhary, G., Shandilya, S. K., Hussain, A., & Kim, H. (2022). Security of zero trust networks in cloud computing: A comparative review. *Sustainability, 14*(18), 11213. https://doi.org/10.3390/su141811213.
9. Hussin, H., Rashid, I. M. A., Samah, I. H. A., Borhanordin, A. H., Roni, M., Shafiai, S., & Elsiefy, E. (2023). The importance of zero trust model among digital employees in migration to healthcare industry 4.0. In *AIP Conference Proceedings* (Vol. 2608, No. 1). AIP Publishing. https://doi.org/10.1063/5.0130854.
10. Vukotich, G. (2023). Healthcare and cybersecurity: Taking a zero trust approach. *Health ServicesInsights, 16*, 11786329231187826. https://doi.org/10.1177/11786329231187826.
11. Sultana, M., Hossain, A., Laila, F., Taher, K. A., & Islam, M. N. (2020). Towards developing a secure medical image sharing system based on zero trust principles and blockchain technology. *BMC Medical Informatics and Decision Making, 20*(1), 1–10. https://doi.org/10.1186/s12911-020-01349-w.
12. Adahman, Z., Malik, A. W., & Anwar, Z. (2022). An analysis of zero-trust architecture and its cost-effectiveness for organizational security. *Computers & Security, 122*, 102911. https://doi.org/10.1016/j.cose.2022.102911.
13. Wang, Z., Yu, X., Xue, P., Qu, Y., & Ju, L. (2023). Research on medical security system based on zero trust. *Sensors, 23*(7), 3774. https://doi.org/10.3390/s23073774.
14. Hong, S., Xu, L., Huang, J., Li, H., Hu, H., & Gu, G. (2023). SysFlow: Toward a programmable zero trust framework for system security. *IEEE Transactions on Information Forensics and Security, 18*, 2794–2809. https://doi.org/10.1109/TIFS.2023.3259234.
15. Ramezanpour, K., & Jagannath, J. (2022). Intelligent zero trust architecture for 5G/6G networks: Principles, challenges, and the role of machine learning in the

context of O-RAN. *Computer Networks, 217*, 109358. https://doi.org/10.1016/j.comnet.2022.109358.
16. Bevish Jinila, Y., Prayla Shyry, S., & Christy, A. (2022). A multi-component-based zero trust model to mitigate the threats in internet of medical things. In *Data Engineering for Smart Systems: Proceedings of SSIC 2021* (pp. 605–613). Springer Singapore. https://doi.org/10.1007/978-981-16-4606-9_57.
17. Johanes, A., Radzuan, N. F. M., & Abdullah, Z. H. Implementation of a zero-trust approach in smart home among the houseowners in Kota Kinabalu, Sabah. *Unpublished.*
18. Jansen, J. N., & Tokerud, S. (2022). Designing the extended zero trust maturity model: A holistic approach to assessing and improving an organization's maturity within the technology, processes, and people domains of information security (Master's thesis, University of Agder). https://hdl.handle.net/11250/3000232 (accessed on 11 November 2024).
19. Manan, A., Min, Z., Mahmoudi, C., & Formicola, V. (2022). Extending 5G services with zero trust security pillars: A modular approach. In *2022 IEEE/ACS 19th International Conference on Computer Systems and Applications (AICCSA)* (pp. 1–6). IEEE. https://doi.org/10.1109/AICCSA55971.2022.9742908.
20. Mehraj, S., & Banday, M. T. (2020). Establishing a zero trust strategy in cloud computing environment. In *2020 International Conference on Computer Communication and Informatics (ICCCI)* (pp. 1–6). IEEE. https://doi.org/10.1109/ICCCI.2020.9188083.
21. Kitchin, R. (2017). Thinking critically about and researching algorithms. *Information, Communication & Society, 20*(1), 14–29. https://doi.org/10.1080/1369118X.2016.1154087.
22. Shin, D. (2023). Embodying algorithms, enactive artificial intelligence and the extended cognition: You can see as much as you know about algorithm. *Journal of Information Science, 49*(1), 18–31. https://doi.org/10.1177/01655515221098824.
23. Assunção, P. (2019). A zero trust approach to network security. In *Proceedings of the Digital Privacy and Security Conference* (Vol. 2019). Porto, Portugal.
24. Sandhu, R., Kumar, P., Bhasin, C., Fatima, N., & Pal, A. (2023). ResNet-34 model for human activity recognition on smartphone sensor data. *International Journal of Intelligent Systems and Applications in Engineering, 11*(9s), 644–652. https://doi.org/10.18280/ijisae.1109s.08.
25. Faiz, M., Sandhu, R., Akbar, M., Shaikh, A. A., Bhasin, C., & Fatima, N. (2023). Machine learning techniques in wireless sensor networks: Algorithms, strategies, and applications. *International Journal of Intelligent Systems and Applications in Engineering, 11*(9s), 685–694. https://doi.org/10.18280/ijisae.1109s.09.
26. Ahmed, I., Nahar, T., Urmi, S. S., & Taher, K. A. (2020). Protection of sensitive data in zero trust model. In *Proceedings of the International Conference on Computing Advancements* (pp. 1–5). IEEE. https://doi.org/10.1109/ICCA51762.2020.9363286.
27. DeCusatis, C., Liengtiraphan, P., Sager, A., & Pinelli, M. (2016). Implementing zero trust cloud networks with transport access control and first packet authentication. In *2016 IEEE International Conference on Smart Cloud (SmartCloud)* (pp. 5–10). IEEE. https://doi.org/10.1109/SmartCloud.2016.35.
28. Ferrario, A., Loi, M., & Viganò, E. (2020). In AI we trust incrementally: A multi-layer model of trust to analyze human-artificial intelligence interactions. *Philosophy & Technology, 33*, 523–539. https://doi.org/10.1007/s13347-019-00379-4.

29. Araujo, T., Helberger, N., Kruikemeier, S., & De Vreese, C. H. (2020). In AI we trust? Perceptions about automated decision-making by artificial intelligence. *AI & Society, 35*, 611–623. https://doi.org/10.1007/s00146-019-00904-7.
30. Sandhu, R., & Lakhwani, K. (2022). Improved scientific workflow scheduling algorithm with distributed HEFT ranking and TBW scheduling method. In *International Conference on Wireless Sensor Networks, Ubiquitous Computing and Applications* (Vol. 244, pp. 1–9). Springer. https://doi.org/10.1007/978-981-16-2919-8_1.
31. Ghai, D., Tripathi, S. L., Saxena, S., Chanda, M., & Alazab, M. (2022). *Machine Learning Algorithms for Signal and Image Processing*. John Wiley & Sons.
32. Shukla, P. K., Sandhu, J. K., Ahirwar, A., Ghai, D., Maheshwary, P., & Shukla, P. K. (2021). Multiobjective genetic algorithm and convolutional neural network based COVID-19 identification in chest X-ray images. *Mathematical Problems in Engineering, 2021*, 1–9. https://doi.org/10.1155/2021/6624085.
33. Ghai, D., Tiwari, S., & Das, N. N. (2021). Bottom-boosting differential evolution based digital image security analysis. *Journal of Information Security and Applications, 61*, 102811. https://doi.org/10.1016/j.jisa.2021.102811.
34. Tripathi, S. L., Dhir, K., Ghai, D., & Patil, S. (2021). *Health Informatics and Technological Solutions for Coronavirus (COVID-19)*. CRC Press.
35. Tripathi, S. L., Mendiratta, N., Ghai, D., Avasthi, S., & Dhir, K. (2022). Coronavirus: Diagnosis, detection, and analysis. In *Biomedical Engineering Applications for People with Disabilities and the Elderly in the COVID-19 Pandemic and Beyond* (pp. 109–117). Academic Press. https://doi.org/10.1016/B978-0-12-822118-5.00011-3.
36. Tavares, J., & Oliveira, T. (2016). Electronic health record patient portal adoption by health care consumers: An acceptance model and survey. *Journal of Medical Internet Research, 18*(3), e5069. https://doi.org/10.2196/jmir.5069.
37. Lakhwani, K., Sharma, G., Sandhu, R., Nagwani, N. K., Bhargava, S., Arya, V., & Almomani, A. (2023). Adaptive and convex optimization-inspired workflow scheduling for cloud environment. *International Journal of Cloud Applications and Computing (IJCAC), 13*(1), 1–25. https://doi.org/10.4018/IJCAC.324809.
38. Gellert, G. A., Kelly, S. P., Wright, E. W., & Keil, L. C. (2023). Zero Trust and the future of cybersecurity in healthcare delivery organizations. *Journal of Hospital Administration, 12*(1). https://doi.org/10.5430/jha.v12n1p14.

CHAPTER 19

Zero-Trust Blockchain Privacy Solution with Homomorphic Encryption

NIRAJ UPADHAYAYA[1,*] and PRAMOD KUMAR[2,*]

[1]*CSE, SRM University, Amaravati, Andhra Pradesh, India*

[2]*Computer Science and Technology, Ganga Institute of Technology and Management, Kablana, Jhajjar, Haryana, India*

*Corresponding author

ABSTRACT

To tackle the privacy protection and efficiency challenges within the blockchain domain, this chapter introduces a privacy-enhancing solution for copyright blockchains, integrating lightweight homomorphic encryption and zero-knowledge proofs. This innovative approach enhances homomorphic encryption algorithms to streamline key generation and encryption processes while incorporating zero-trust security principles to curtail unnecessary homomorphic operations. Following the application of lightweight homomorphic encryption, sensitive data is transformed into ciphertext and securely added to the blockchain ledger by nodes authorized for accounting purposes. This solution not only rectifies the inherent drawbacks of complete data transparency in blockchain networks but also enhances operational efficiency. Security analysis underscores its qualities, including resistance to tampering and data privacy preservation. Through both performance simulations and theoretical deductions, the paper demonstrates that this approach mitigates efficiency challenges related to the distribution, sharing, and computation of private data in ciphertext form. Ultimately, this proposed methodology proves more effective in upholding customer privacy than traditional digital copyright models.

Zero-Trust Learning: Applications in Modern Network Security. Chetna Kaushal, Amandeep Kaur, Mohit Angurala, Aryan Chaudhary (Eds.)

19.1 INTRODUCTION

Blockchain, an essential framework for constructing the future value Internet, extensively incorporates advanced technologies such as point-to-point connectivity, distributed architecture, consensus mechanisms, and cryptography, among others. Consequently, it is positioned at the forefront of technical progress [1]. The functionality of blockchain is predicated upon its inherent resistance to tampering and its capacity for transparency. In recent years, there has been a growing focus on the emergence of globalized and distributed security aspects within the realm of global technological and economic development. Nevertheless, the proliferation of blockchain-driven apps has given rise to certain technological obstacles pertaining to service quality and privacy protection. These issues encompass aspects such as performance, cost, and scalability. The current state of affairs is progressively escalating in severity. The current privacy protection strategy for blockchain, which relies on classical homomorphic encryption, exhibits evident issues in terms of service quality and privacy protection. These issues include a low verification rate, limited scalability, high energy consumption, and vulnerability to data leakage. Consequently, achieving transparency becomes challenging under this scheme. The provision of services that prioritize customer satisfaction can conflict with the preservation of user privacy, diminish justice, credibility, sustainability, and computing efficiency in job execution, and impede the adoption of blockchain technology.

In the application of digital copyright [2, 3] management, the issue of copyright abuse has always attracted much attention, and disputes have continued to emerge. The reasons mainly include the following two aspects: First, compared with solutions based on distributed ledgers, users It is more difficult to obtain detailed information about the author or owner and permission to use through traditional intellectual property management methods (such as copyright marking); secondly, it is difficult for authors to maintain control of their original work and track its usage. Block The introduction of the chain provides new solutions to copyright abuse and privacy issues. Users can download, recreate, and synchronize their work through smart contracts. Every payment generated by Wet's original works will be automatically distributed and distributed through smart contracts. The corresponding amount is sent to the author and copyright holder. On the other hand, the author can have tampered evidence proving ownership and better control and utilize the creative work. At the same time, the traditional digital copyright blockchain network transfer/transaction process the relevant personal information of users such as transfer/transaction amount, transfer/transaction copyright information, and so forth, will be exposed, causing certain security and privacy issues.

This study focuses on the implementation of digital copyright using blockchain technology. The primary objective is to enhance the homomorphic encryption algorithm, thereby reducing the time required for key generation and encryption. Additionally, the paper proposes the utilization of a zero-knowledge proof technique based on probability verification to mitigate the risk of improper synchronization. In the context of lightweight homomorphic encryption, the morphic operation involves the writing of secret data in the form of ciphertext into a block. The secretary node will provide responses based on the random inquiries formulated by the verifier. In the event that the answer is deemed accurate, the secretary node will be granted authorization to transmit the data to the blockchain network. This method addresses the limitations associated with data exposure in the blockchain network while simultaneously enhancing efficiency concerns.

19.2 RELATED WORK

Blockchain has six major characteristics: decentralization, non-tampering, traceability, autonomy, openness, and anonymity [4]. The outstanding performance of blockchain technology in the financial field has also attracted many researchers from different research fields. Experts and scholars have conducted research on blockchain, mainly in the fields of logistics, public services, digital copyright, and insurance, using the characteristics of blockchain technology to accelerate the development of its related applications [5]. Blockchain has also attracted a large number of scholars in the field of crowd sensing who are committed to the joint development of blockchain and crowd sensing, and are constantly researching solutions that can simultaneously ensure system service quality and privacy security. Currently, mainstream public chain platforms such as Bitcoin and Ethereum still upload plain text directly to the chain, and since the transfer/transaction data is fully disclosed, the user's private information is leaked. In order to solve related problems, schemes such as those in the literature [6–11] have appeared. These schemes have solved the problem of anonymity to a certain degree, but they are still unable to adapt to the privacy protection of transfer/transaction data in complex application environments. The individual responsible for establishing Ethereum provided a comprehensive overview of the ongoing efforts in safeguarding privacy within the realm of blockchain technology. This examination also delved into Ethereum's four proposed resolutions, namely channels, mixers, ring signatures, and zero-knowledge proofs. This statement underscores the significance of zero-knowledge proofs in addressing Ethereum's privacy concerns. Despite its inefficiency and implementation challenges, zero-knowledge proofs offer a robust solution by not only safeguarding data privacy but also verifying data validity. The article published in 2019 introduced a decentralized ride-sharing platform that utilized a public blockchain. This platform employed smart contracts and zero-knowledge proofs to ensure the security and anonymity of its users.

The topic of discussion pertains to the safeguarding or preservation of something.

In addition, more, and more blockchain applications now choose to upload only the hash value of data to the chain, such as off-chain isolation channel solutions represented by Lightning Network technology [12] and Raiden Network technology [13], as well as multi-chain channel isolation technology represented by Fabric [14]. Obviously, these solutions do not put the real data on the chain. Although they reduce the storage pressure of the chain, the off-chain data has many problems, such as easy leakage and damage, which leads to an increased possibility of data failure during application.

Homomorphic encryption based on symmetric encryption algorithms [15] is an ideal method to handle long data encryption. Before uploading to the chain, the transfer/transaction data and balance information are lightweight encrypted and stored on the chain. When all users access information on the chain, they only see information that has been homomorphically encrypted, thereby effectively protecting private information such as account funds. However, existing blockchain privacy protection applications that use homomorphic encryption only stay in the encryption stage and do not have the function of zero-knowledge proof. This article will combine the improved homomorphic encryption algorithm and zero-knowledge proof technology to achieve full processing of data in the ciphertext state and achieve controllable access to sensitive data.

19.3 PATTERN OF BLOCKCHAIN TRANSFER/TRANSACTION CONSTRAINTS BASED ON LIGHTWEIGHT HOMOMORPHIC ENCRYPTION

19.3.1 LIGHTWEIGHT HOMOMORPHIC ENCRYPTION SCHEME

19.3.1.1 KEY GENERATION ALGORITHM GEN

The key generation technique Gen(n,N) yields the public key b and the private key b^{-1}, with n being the vector's dimension. As the value of n increases, the computational complexity of the method also increases, hence enhancing its level of security. Furthermore, it should be noted that the maximum range of plaintext is constrained inside the interval $(0,2n)$. When subjected to encryption, any plaintext that falls outside of this specified range is unable to be accurately represented as an n-dimensional vector. The fault point, denoted as N, is determined by the condition $\sqrt{n} \leqslant N \leqslant 2^n$. Operations performed within the range $[0,N]$ can ensure homomorphism, but there is a possibility of failure beyond this range.

19.3.1.2 ENCRYPTION ALGORITHM ENC

The encryption algorithm is a computational procedure that transforms plaintext into ciphertext, ensuring the confidentiality and integrity of sensitive information. The notation $\text{Enc}(b,p)$ denotes the encryption of the plaintext p using the key b. The resulting ciphertext c is represented as an n-dimensional column vector. The procedure of transforming an integer value into a vector representation involves the utilization of an array structure to store its binary digits. Utilize bitwise operations as a means to enhance computational efficiency.

19.3.1.3 DECRYPTION ALGORITHM DEC

The algorithm utilized for decryption The expression $\text{Dec}(b,b^{-1},c)$ denotes the process of utilizing the cryptographic keys b and b^{-1} to decipher the encrypted message c and retrieve the original plaintext p.

19.3.1.4 CALCULATION ALGORITHM CAL

The calculation method $\text{Cal}(c_1, c_2, \text{op})$ is designed to compute the ciphertext directly by utilizing the specified operator op, where op belongs to the set of operators $\{+, \times\}$.

19.4 PROTOCOL DESIGN

By implementing zero-knowledge proof to prevent invalid homomorphic operations and reducing the time required for key generation, encryption, and decryption, this protocol substantially improves efficiency and lays the groundwork for practical applications. Based on zero-knowledge proofs and lightweight homomorphic encryption, this article develops a protocol. Figure 19.1 illustrates the primary user role nodes comprising the knowledge-proof digital copyright blockchain protocol.

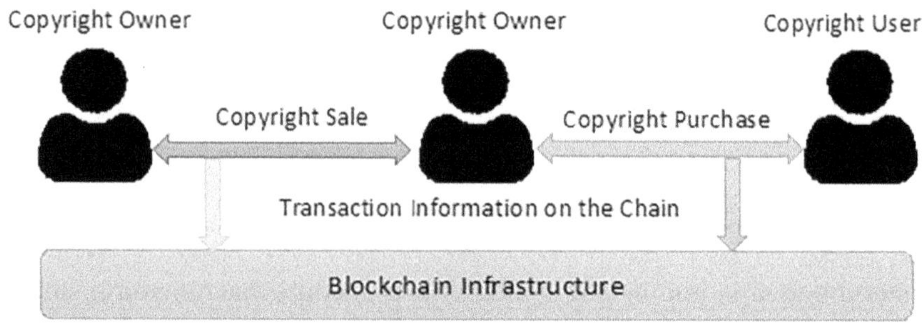

FIGURE 19.1 User role transition diagram.

In the copyright transfer/transaction process, it is divided into three categories: user *node*(U), secretary *node*(S) and verification *node*(Y) . User nodes include copyright owners (C), platform managers (B) and digital copyright users. The person (A) mainly uploads different types of personal identity information and copyright information, and writes it into the copyright blockchain network by the secretary *node*(S) ; the secretary *node*(S) obtains the record through the workload proof-of-work algorithm (POW). Account rights, package relevant data into data blocks and broadcast; verification *node*(Y) is mainly to judge the validity of copyright transfer/transaction.

When the copyright transfer event is triggered, it will involve the transfer/transaction behavior between the copyright user (A) and the copyright platform manager (B). Assume that the fee incurred by the copyright transfer/transaction is T, and the account balance of the copyright user (A). It is T_A. The copyright platform manager (B) needs to judge whether the balance is sufficient and complete the deduction without knowing the clear text information of T and T_A. To achieve this goal, the design protocol is as follows:

- The copyright user (A) uses his own public key to encrypt the fee T_1. The random vector used for encryption is R_{AT}. He divides T_1 into I_1/J_1 and obtains the ciphertext $C_{A,T}=Enc(b_A,I_1)$.
- The copyright user (A) uses the public key of the copyright platform manager (B) to encrypt the fee T_2 . The random vector used for encryption is $R_{B,T}$. The T_2 is divided into I_1/J_1 to obtain the ciphertext $C_{B,T}=\mathrm{Enc}(b_{B,}n\text{-}J_1)$.

- The copyright user (A) needs to prove to the secretary node (S) that the plaintext information of the ciphertext $C_{A,T}$ and $C_{B,T}$ is equal. The entire process can be divided into two stages: zero-knowledge proof generation and transfer/transaction verification.

19.4.1 ZERO-KNOWLEDGE PROOF GENERATION

1. **Step 1:** The copyright user(A) adds the copyright ciphertext $C_{A,T}$ and ciphertext $C_{B,T}$ to the empty zero-knowledge proof π.
2. **Step 2:** Copyright user(A) generates uniformly distributed random numbers $X_i(1 \leqslant i \leqslant k, k$ represents the security parameter, which means generating k random numbers).
3. **Step 3:** The copyright user(A) calculates and generates the ciphertext of these k random numbers. The copyright user(A) first uses his own public key to encrypt the random number X_i as $C_{A,i}$=Enc(b_A ,X_i). Then he uses the copyright The public key encryption random number X_i of the platform manager(B) is $C_{B,i}$=Enc(b_B,X_i), the random vectors used for encryption are $R_{A,i}$ and $C_{B,i}$, and finally $C_{A,i}$ and $C_{B,i}$ are added to zero Knowledge proof π in.
4. **Step 4:** Copyright user(A) calculates the sum of the ciphertext that generates copyright transfer/transaction amount T and the ciphertext of random number X_i. Copyright user (A) first performs homomorphic addition operation $I_{A,i}$=Cal($C_{A,T}$,$C_{A,i}$,+) , and then perform the homomorphic addition operation $I_{B,i}$ =Cal($C_{B,T}$,$C_{B,i}$,+) under the public key of the copyright platform manager(B), and finally add $I_{A,i}$ and $I_{B,i}$ is added to the zero-knowledge proof π.
5. **Step 5:** The copyright user(A) sends the zero-knowledge proof π = $C_{A,T}$,$C_{B,T}$,$C_{A,i}$,$C_{B,i}$,$I_{A,i}$,$I_{B,i}$) to the digital copyright blockchain network.
6. **Step 6:** The secretary node (S) on the digital copyright blockchain issues *a* challenge *h* to the copyright user(A). That is to say, the secretary node(S) sends a string of *k* bits b_i (where b_i is a random number randomly selected from {0,1}.
7. **Step 7:** After receiving *h*, the copyright user(A) gives different answers to the secretary node (S) according to different b_i. That is, if b_i=0, the copyright user(A) sends the copyright of $C_{A,i}$, $C_{B,i}$ The

plaintext information and the random vectors $R_{A,i}$ and $R_{B,i}$ required for encryption are given to the secretary node(S); if $b_i = 1$, the copyright user(A) sends the plaintext information $I_{A,i}$, $I_{B,i}$ and the random vectors required for encryption. Random $R_{A,i}+R_{A,T}$ and $R_{B,i}+R_{B,T}$ are given to the secretary node(S).

19.4.2 TRANSFER/TRANSACTION VERIFICATION

1. **Step 8:** The zero-knowledge proof π generated by the copyright user(A) gives the ciphertext $(R_{A,i}+R_{B,i})$ or $(I_{A,i}+I_{B,i})$, the ciphertext $(R_{A,i}+R_{B,i})$ or $(I_{A,i}+I_{B,i})$ corresponds to the plaintext information X_i or $T+X_i$ respectively, and the randomness used to encrypt X_i or$T+X_i$ into $(R_{A,i}+R_{B,i})$ or $(I_{A,i}+I_{B,i})$ Vector. The secretary node(S) verifies whether the plaintext information given by the copyright user(A), the initiator of the copyright transfer/transaction, can be encrypted into the ciphertext given by the copyright user(A), that is, to verify the consistency.
   - If $b_i = 0$, the verification random number can be encrypted into $(R_{A,i}+R_{B,i})$ using the public key of the copyright user(A) and the public key of the copyright platform manager (B) respectively; if $b_i = 1$, the verification fee T and The sum of random numbers X_i $T+X_i$ can be encrypted into $(I_{A,i}+I_{B,i})$ using the public key of the copyright user(A) and the public key of the copyright platform manager(B) respectively.
2. **Step 9:** The copyright transfer/transaction can be considered as a sufficient transfer/transaction protocol based on lightweight homomorphic encryption if and only if the secretary node (S) verifies that all ciphertexts pass. In the event that the secretary node (S) accepts the proof to be provided, the output will be T. Otherwise, if the secretary node (S) does not accept the proof, the transfer/transaction will be ignored and the output will be F. The copyright transfer/transaction has been finalized.

19.4.3 PRIVACY BLOCKCHAIN

The primary objective of the privacy blockchain is to safeguard the confidentiality of transfer and transaction data within the digital copyright blockchain. Confidential information, such as copyright fees, is encrypted into ciphertext using a lightweight homomorphic encryption technique. Subsequently, the ciphertext information is further converted into ciphertext and stored within the data block. Simultaneously, prior to data storage, it is imperative for the verification node (Y) to authenticate the data using the blockchain transfer/transaction protocol that relies on lightweight homomorphic encryption. Once the verification node (Y) successfully completes the verification process, it will undergo further verification by the secretary. Subsequently, the nodes will be compiled and recorded in the copyright blockchain. Meanwhile, the utilization of homomorphic encryption enables the secretary node (S) to perform immediate updates on the account balance within the ciphertext state, without the need for decryption. Hence, the developed blockchain system, which ensures the safeguarding of data confidentiality, offers the capability to store encrypted private data on the blockchain. Additionally, it enables the verification of data integrity prior to its inclusion on the blockchain, thus optimizing the processing time for erroneous data.

19.4.4 DATA STRUCTURE

Compared with the traditional block data structure of the blockchain, this article will homomorphically encrypt the copyright plaintext DT and then hash it to prevent private data from being directly stored on the blockchain. The process is shown in Figure 19.2.

19.5 BLOCKCHAIN ON-CHAIN

1. **Step 1:** Based on the POW, continuously select the secretary node (S) in the digital copyright blockchain.
2. **Step 2:** The copyright user node (A) initiates a copyright transfer/transaction request and publishes the ciphertext information of the copyright transfer/transaction and the zero-knowledge proof π proving that the transfer/transaction is reasonable on the digital copyright blockchain network. middle.
3. **Step 3:** The verification node (Y) performs calculations based on the zero-knowledge proof provided by the copyright user node (A) to verify whether the copyright transfer/transaction is valid. For the transfer/transaction request that passes the verification, the

verification node (Y) will Put it into the transfer/transaction storage pool and forward it to other nodes.

4. **Step 4:** The secretary node (S) that obtains the accounting rights packages the transfer/transaction together with other transfers/transactions into the copyright block, and joins the copyright block chain after obtaining the consent of the majority of other secretary nodes (S). The process as shown in Figure 19.3.

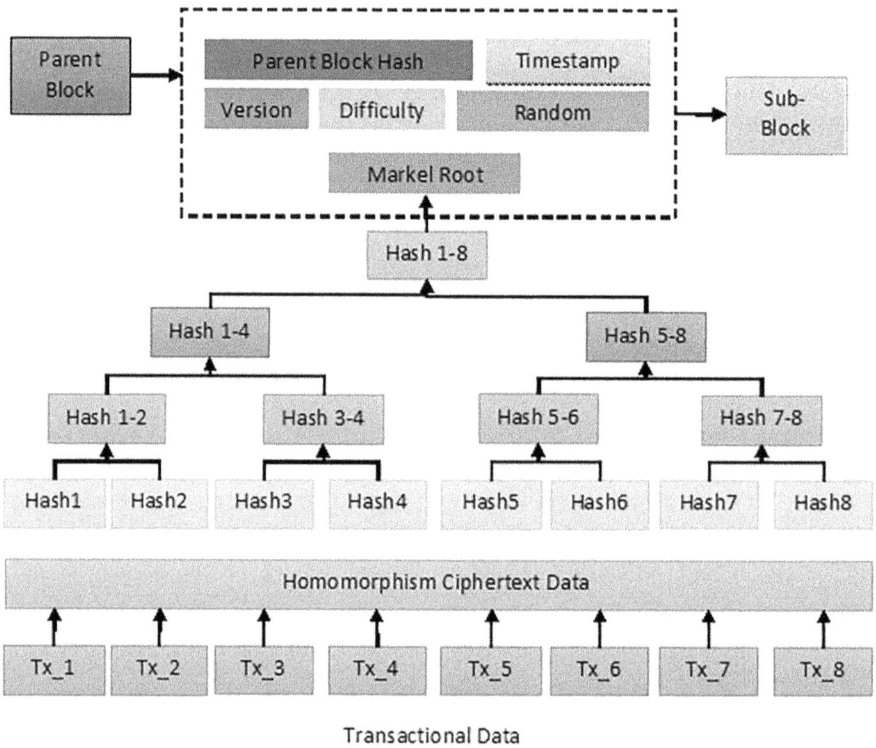

FIGURE 19.2 Block data structure based on homomorphic encryption.

19.6 MODEL WORKING PROCESS

1. **Step 1 User Setting Stage:** For a copyright transfer/transaction, the transfer/transaction entities involved mainly include the copyright user (A) and the copyright platform manager (B). The data privacy protection solution needs to generate the public and private key

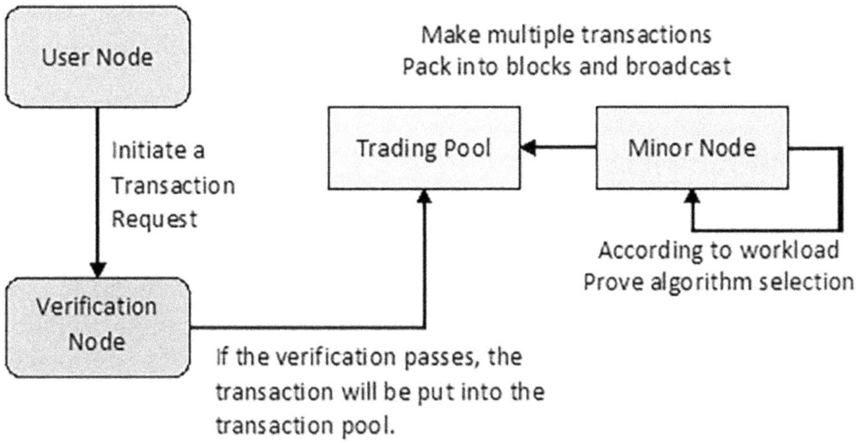

FIGURE 19.3 Blockchain construction process.

information of both at this stage. When the copyright is used When the copyright user (A) initiates a transfer/transaction, the public key b_A, the private key b^{-1} and the ciphertext $C_{A,T}$ of the copyright user (A) need to be generated, and $C_{A,T}$ is written into the data block, where $C_{A,T}=Enc(C_{A,}T)$. At this stage, T, T_A need to be encrypted with the public key of the copyright user (A) to generate ciphertext $C_{A,T}$, $C_{A,TA}$, and at the same time, the copyright platform manager (B). The public key encrypted transfer/transaction amount t generates ciphertext $C_{B,T}$ to achieve zero-knowledge proof verification.

2. **Step 2 Execution Phase:** The account balance of the copyright user (A) is T_A, and the amount T needs to be transferred to the copyright platform manager (B). Basis for the transfer/transaction stage 2. The transfer/transaction protocol described in Section 2 allows the copyright user (A) to first generate a zero-knowledge proof π $(T, b_A, b_B, C_{AT}, C_{BT})$ proving that the transfer/transaction amount is sufficient. Verification node (Y) will verify the legality of this transfer/transaction in the ciphertext state and determine the relationship between the account balance of the copyright user (A) and the transfer/transaction amount. After passing the verification, the transfer contract is executed, and the transfer/transaction is packaged into the transfer/transaction pool. If the verification fails, the result is fed back to the relevant user, and the transfer/transaction process ends.

The secretary node (S) selected by the POW packages the transfer/transaction data. After the transfer/transaction is confirmed, the new block is written to the copyright blockchain.

19.7 MODEL IMPLEMENTATION AND EVALUATION

The implementation of the lightweight homomorphic encryption algorithm used in this article is based on the homomorphic encryption library TFHE. This system is deployed on a Linux system (Centos 6.5). In order to run 3 types of nodes and implement transfer/transaction functions, some operations are required Environment support, so you need to configure the environment and install dependent libraries in advance [16–18] (Table 19.1).

TABLE 19.1 The Environment of Configuration

Python 3.6	Pip3	Anaconda	Requests
Hashlib	Math	Keras	Numpy
Urllib	Json	Time	Uuid
Threading	Flask	Textwrap	socket

19.8 NODE CONFIGURATION

A total of 6 nodes were set up in this experiment, as listed in Table 19.2.

TABLE 19.2 The Setting of Node

Node	CPU Core Number	Memory/GB	Remarks
1	1	1.4	User Node
2	2	2.0	User Node
3	1	2.2	Secretary node (S)
4	1	2.0	Secretary node (S)
5	2	2.0	Secretary node (S)
6	2	2.2	Verification node (Y)

Build a copyright blockchain database: Nodes 1–5 are heavyweight nodes, which are data storage nodes in the copyright blockchain database, and are also users of the copyright blockchain database; node 6 is a light-weight node, which is the copyright area Users of blockchain databases do not have data storage functions.

19.9 ATTACK MODEL

As a blockchain model based on a proof-of-work algorithm and a consensus algorithm [19], it focuses on simulating the impact of a fork attack on the system. Assume that multiple verification nodes (Y) are attacked and try to create a new blockchain branch, set two variables: z, the blockchain length gap between the attacker's control of the forked chain and the main chain, and the probability of the attacker's success in a round of accounting rights competition, q. Fork attacks can occur at different blockchain lengths and attackers. The probability of success under the conditions of probability of success in a round of competition for accounting rights is as listed in Table 19.3.

TABLE 19.3 The Result of the Experiments

	q=0.1	q=0.3	q=0.4
z=1	0.204578	0.724759	0.828920
z=2	0.050872	0.455805	0.736508
z=3	0.013204	0.324610	0.664285
z=4	0.003442	0.239178	0.603398
z=5	0.000896	0.177266	0.550638
z=10	1.242302X10-6	0.041672	0.360259

The simulation results of the double-fork attack are shown in Figure 19.4.

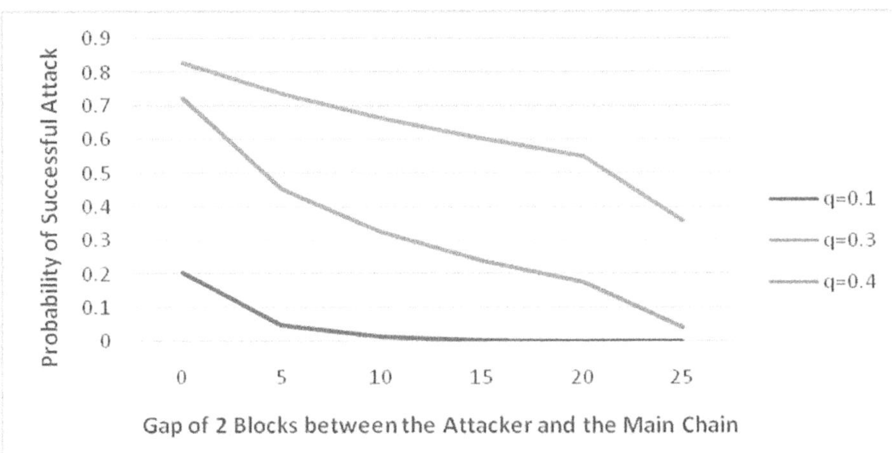

FIGURE 19.4 Double prong attack success probability chart.

It can be seen that the success rate of a fork attack decreases exponentially as z increases. Unless the attacker has close to 50% of the computing power of the entire network, it is almost impossible for a fork attack to succeed in terms of probability.

19.10 PERFORMANCE ANALYSIS

19.10.1 ENCRYPTION TIME ANALYSIS

Upon examination of the data pertaining to key generation time and encryption time [17, 18] presented in Figures 19.5 and 19.6, it is evident that the lightweight homomorphic encryption method exhibits a notable reduction in both key generation time and encryption time when compared to the fully homomorphic encryption technique. Simultaneously, the utilization of zero-knowledge proof [19, 20] enables the preclusion of homomorphic operations that fail to satisfy transfer or transaction prerequisites, hence potentially lowering the duration of such procedures. Additionally, the overall duration of the homomorphic encryption procedure is influenced by the vector's dimension. In scenarios where the dimension is limited, the overall execution time will be significantly reduced, thus satisfying the requirements of realistic implementations of digital copyright blockchain [21].

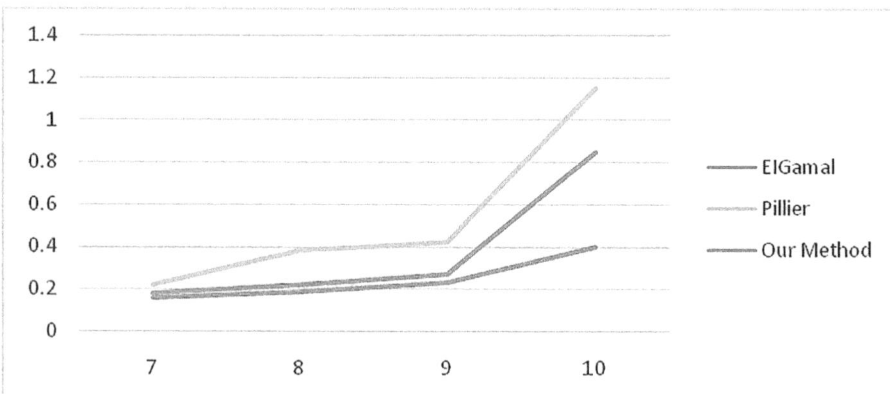

FIGURE 19.5 Encryption time graph.

19.10.2 PROTOCOL CORRECTNESS ANALYSIS

From the properties of the homomorphic encryption scheme, it can be calculated:

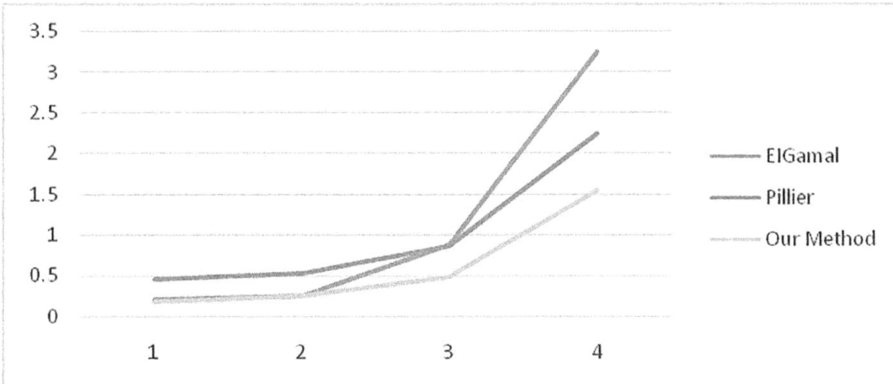

FIGURE 19.6 Decryption time chart.

$$T = T_1 T_2 = E(I_1)^{J2} E(n\text{-}J_1)^{I2} = E\,(I_1J_2)\,E\,(nI_2\text{-}J_1I_2) = E\,(I_1J_2) + (nI_2\text{-}J_1I_2)$$

Decrypting the above results can yield:

$$D\,(T) = I_1J_2 + nI_2 - J_1I_2$$

Because:

$$D\,(T) mod n = (I_1J_2 + nI_2 - J_1I_2) = (I_1J_2 - I_1J_2)\; mod n$$

The above result modulo n can be obtained by taking the remainder:

$$m = I_1J_2 - I_1J_2$$

Also, because:

$$X_A - X_B = I_1/J_1 - I_2/J_2 = (I_1J_2 - I_2J_1) / I_1J_2 = m = J_1J_2)$$

Since $J_1J_2 > 0$, if $m>0$, then $X_A > X_B$; if $m >0$, then $X_A < X_B$; if $m>0$, then $X_A = X_B$. Therefore, this agreement is correct.

19.10.3 PROTOCOL SECURITY ANALYSIS

Since the copyright user (A) and the copyright platform manager (B) process their respective data proportionally in a safe state, both parties cannot know the other party's split value. The copyright user (A) sends the homomorphic ciphertext to the copyright platform manager (B), and the copyright platform manager (B) cannot open the ciphertext without the decryption key, thereby protecting the privacy information of the copyright user (A). The copyright user (A) obtains the result after the decryption is completed: $D(T) = I_1J_2 + nI_2 - J_1I_2$. It can be seen that this is an equation with infinite solutions. The copyright user (A) cannot infer the values of I_1, I_2, and it is also impossible to infer the secret data value of the copyright platform manager (B). Therefore, the designed protocol has good security [22, 23] and robustness [24–26].

19.10.4 DATA PRIVACY ANALYSIS

The secretary node (S) uses zero-knowledge proof to verify whether the digital copyright transfer/transaction meets the conditions for the transfer/transaction account amount to be sufficient. In this process, the transfer/transaction data is completely confidential, and the private data is subjected to lightweight homomorphic encryption [27–29]. Afterwards, it is copied, shared, calculated, and distributed in a ciphertext state, which will not cause the leakage of private information and ensure information security [30].

19.11 CONCLUSION

The present study introduces a novel blockchain privacy protection strategy that leverages lightweight homomorphic encryption and zero-knowledge proof technology. The scheme is further evaluated through experimentation with digital copyright. The enhanced homomorphic encryption technology effectively mitigates the time overhead associated with the encryption and decryption procedures while concurrently introducing negligible computational burden. The implementation of knowledge proof techniques prevents the execution of erroneous homomorphic operations, thereby guaranteeing the confidentiality of user data related to digital copyright within a transparent blockchain sharing ecosystem, all the while enhancing operational efficiency. The present approach offers a resolution to a range of issues prevalent in contemporary digital copyright data sharing practices. An empirically validated solution.

KEYWORDS

- **blockchain privacy**
- **copyright protection**
- **decryption**
- **homomorphic encryption**
- **information security**
- **secretary node**
- **zero trust security**
- **zero-knowledge proofs**

REFERENCES

1. Chen, P. C., Kuo, T. H., & Wu, J. L. (2021). A study of the applicability of ideal lattice-based fully homomorphic encryption scheme to Ethereum blockchain. *IEEE Systems Journal, 15*(2), 1528–1539. https://doi.org/10.1109/JSYST.2021.3064053.
2. Ma, Z., Wang, L., & Zhao, W. (2021). Blockchain-driven trusted data sharing with privacy protection in IoT sensor network. *IEEE Sensors Journal, 21*(22), 25472–25479. https://doi.org/10.1109/JSEN.2020.3046752.
3. Miao, Y., Liu, Z., Li, H., Choo, K. K. R., & Deng, R. H. (2022). Privacy-preserving Byzantine-robust federated learning via blockchain systems. *IEEE Transactions on Information Forensics and Security, 17*, 2848–2861. https://doi.org/10.1109/TIFS.2022.3196274.
4. Costantino, G., De Vincenzi, M., Martinelli, F., & Matteucci, I. (2023). A privacy-preserving solution for intelligent transportation systems: Private driver DNA. *IEEE Transactions on Intelligent Transportation Systems, 24*(1), 258–273. https://doi.org/10.1109/TITS.2022.3217358.
5. Sun, J., Wu, Y., Wang, S., Fu, Y., & Chang, X. (2022). Permissioned blockchain frame for secure federated learning. *IEEE Communications Letters, 26*(1), 13–17. https://doi.org/10.1109/LCOMM.2021.3121297.
6. Mendis, G. J., Wu, Y., Wei, J., Sabounchi, M., & Roche, R. (2021). A blockchain-powered decentralized and secure computing paradigm. *IEEE Transactions on Emerging Topics in Computing, 9*(4), 2201–2222. https://doi.org/10.1109/TETC.2020.2983007.
7. Sen, S., Aydogan, E., & Aysan, A. I. (2018). Coevolution of mobile malware and anti-malware. *IEEE Transactions on Information Forensics and Security, 13*(10), 2563–2574. https://doi.org/10.1109/TIFS.2018.2824250.
8. Peng, W., Li, F., Zou, X., & Wu, J. (2014). Behavioral malware detection in delay tolerant networks. *IEEE Transactions on Parallel and Distributed Systems, 25*(1), 53–63. https://doi.org/10.1109/TPDS.2013.27.

9. Kim, J., Paik, J. Y., & Cho, E. S. (2023). Attention-based cross-modal CNN using non-disassembled files for malware classification. *IEEE Access, 11*, 22889–22903. https://doi.org/10.1109/ACCESS.2023.3253770.
10. Jin, B., Choi, J., Hong, J. B., & Kim, H. (2023). On the effectiveness of perturbations in generating evasive malware variants. *IEEE Access, 11*, 31062–31074. https://doi.org/10.1109/ACCESS.2023.3262265.
11. Das, S., Liu, Y., Zhang, W., & Chandramohan, M. (2016). Semantics-based online malware detection: Towards efficient real-time protection against malware. *IEEE Transactions on Information Forensics and Security, 11*(2), 289–302. https://doi.org/10.1109/TIFS.2015.2491300.
12. Feng, J., Yang, L. T., Zhang, R., & Gavuna, B. S. (2021). Privacy-preserving Tucker train decomposition over blockchain-based encrypted industrial IoT data. *IEEE Transactions on Industrial Informatics, 17*(7), 4904–4913. https://doi.org/10.1109/TII.2020.2968923.
13. Wang, Z., et al. (2023). Towards privacy-driven truthful incentives for mobile crowdsensing under untrusted platform. *IEEE Transactions on Mobile Computing, 22*(2), 1198–1212. https://doi.org/10.1109/TMC.2021.3093552.
14. Wang, Q., Huang, L., Chen, S., & Xiang, Y. (2022). Blockchain enables your bill safer. *IEEE Internet of Things Journal, 9*(16), 14162–14171. https://doi.org/10.1109/JIOT.2020.3016721.
15. Luo, X., Xue, K., Xu, J., Sun, Q., & Zhang, Y. (2021). Blockchain-based secure data aggregation and distributed power dispatching for microgrids. *IEEE Transactions on Smart Grid, 12*(6), 5268–5279. https://doi.org/10.1109/TSG.2021.3099347.
16. Zhao, B., Liu, X., Chen, W. N., & Deng, R. H. (2023). CrowdFL: Privacy-preserving mobile crowdsensing system via federated learning. *IEEE Transactions on Mobile Computing, 22*(8), 4607–4619. https://doi.org/10.1109/TMC.2022.3157603.
17. Rezaeibagha, F., Mu, Y., Huang, K., Chen, L., & Zhang, L. (2023). Toward secure data computation and outsource for multi-user cloud-based IoT. *IEEE Transactions on Cloud Computing, 11*(1), 217–228. https://doi.org/10.1109/TCC.2021.3087614.
18. Chaudhry, J. A., Saleem, K., Alazab, M., Zeeshan, H. M. A., Al-Muhtadi, J., & Rodrigues, J. J. P. C. (2021). Data security through zero-knowledge proof and statistical fingerprinting in vehicle-to-healthcare everything (V2HX) communications. *IEEE Transactions on Intelligent Transportation Systems, 22*(6), 3869–3879. https://doi.org/10.1109/TITS.2021.3066487.
19. Yin, J., et al. (2023). SmartDID: A novel privacy-preserving identity based on blockchain for IoT. *IEEE Internet of Things Journal, 10*(8), 6718–6732. https://doi.org/10.1109/JIOT.2022.3145089.
20. Dieye, M., et al. (2023). A self-sovereign identity based on zero-knowledge proof and blockchain. *IEEE Access, 11*, 49445–49455. https://doi.org/10.1109/ACCESS.2023.3268768.
21. Li, W., Meese, C., Guo, H., & Nejad, M. (2023). Aggregated zero-knowledge proof and blockchain-empowered authentication for autonomous truck platooning. *IEEE Transactions on Intelligent Transportation Systems, 24*(9), 9309–9323. https://doi.org/10.1109/TITS.2023.3271436.
22. Mouris, D., & Tsoutsos, N. G. (2021). Zilch: A framework for deploying transparent zero-knowledge proofs. *IEEE Transactions on Information Forensics and Security, 16*, 3269–3284. https://doi.org/10.1109/TIFS.2021.3074869.

23. Rasheed, A. A., Mahapatra, R. N., & Hamza-Lup, F. G. (2020). Adaptive group-based zero-knowledge proof-authentication protocol in vehicular ad hoc networks. *IEEE Transactions on Intelligent Transportation Systems, 21*(2), 867–881. https://doi.org/10.1109/TITS.2019.2899321.
24. Lu, L., et al. (2008). Pseudo trust: Zero-knowledge authentication in anonymous P2Ps. *IEEE Transactions on Parallel and Distributed Systems, 19*(10), 1325–1337. https://doi.org/10.1109/TPDS.2008.15.
25. Partala, J., Nguyen, T. H., & Pirttikangas, S. (2020). Non-interactive zero-knowledge for blockchain: A survey. *IEEE Access, 8*, 227945–227961. https://doi.org/10.1109/ACCESS.2020.3046025.
26. Chen, X., Yang, A., Weng, J., Tong, Y., Huang, C., & Li, T. (2023). A blockchain-based copyright protection scheme with proactive defense. *IEEE Transactions on Services Computing, 16*(4), 2316–2329. https://doi.org/10.1109/TSC.2023.3246476.
27. Lin, S. D., & Chen, C. F. (2000). A robust DCT-based watermarking for copyright protection. *IEEE Transactions on Consumer Electronics, 46*(3), 415–421. https://doi.org/10.1109/30.883387.
28. Islam, M. M., & In, H. P. (2023). Decentralized global copyright system based on consortium blockchain with proof of authority. *IEEE Access, 11*, 43101–43115. https://doi.org/10.1109/ACCESS.2023.3270627.
29. Xiao, L., Huang, W., Xie, Y., Xiao, W., & Li, K. C. (2020). A blockchain-based traceable IP copyright protection algorithm. *IEEE Access, 8*, 49532–49542. https://doi.org/10.1109/ACCESS.2020.2969990.
30. Chung, T. Y., Hong, M. S., Oh, Y. N., Shin, D. H., & Park, S. H. (1998). Digital watermarking for copyright protection of MPEG2 compressed video. *IEEE Transactions on Consumer Electronics, 44*(3), 895–901. https://doi.org/10.1109/30.713211.

CHAPTER 20

AI-Powered Security in Zero-Trust Learning: Enhancing Safety and Adaptability

MOHIT ANGURALA,[1,*] SHUBHANGI MISHRA,[2] NEHA SHARMA,[2] and YASHIKA SHARMA[2]

[1]*Department of Computer Science, Guru Nanak Dev University College, Pathankot (Constituent College of Guru Nanak Dev University, Amritsar), Punjab, India*

[2]*Apex Institute of Technology (CSE), Chandigarh University, Mohali, Punjab, India*

*Corresponding author

ABSTRACT

This chapter investigates the interaction of AI and security in the context of Zero Trust Learning (ZTL). It begins with a review of the key principles of zero-trust security and the growing role of artificial intelligence (AI) in current security procedures. The chapter then goes into AI-powered Intelligent Tutoring Systems, AI-enhanced user authentication, continuous monitoring, and threat detection in ZTL. Real-world examples and case studies demonstrate the use of AI-powered security solutions in educational settings. Furthermore, under the Zero Trust paradigm, the chapter explores the importance of AI in cloud security. It emphasizes AI's ability to improve authentication, access management, and threat detection, resulting in a more secure and adaptable learning environment.

Zero-Trust Learning: Applications in Modern Network Security. Chetna Kaushal, Amandeep Kaur, Mohit Angurala, Aryan Chaudhary (Eds.)

20.1 INTRODUCTION

The first part presents a brief explanation of the core ideas behind Zero Trust security, highlighting the importance of strong authentication, access control, and constant monitoring. Furthermore, it emphasizes the rising role of AI in modern security measures, using its ability to handle massive volumes of data and draw actionable insights. This section digs into Artificial Intelligence (AI)-powered authentication procedures, illuminating how AI technologies might improve the accuracy and speed of user authentication processes. The talk includes the use of AI-driven user behavior analytics (UBA) to detect aberrant patterns and deviations, allowing for real-time risk assessment based on user behavior and environmental factors.

This section focuses on AI's predictive analytics and machine learning capabilities, with an emphasis on its function in increasing threat detection efforts. Real-world examples and case studies highlight the usefulness of AI-driven threat detection systems. The following article looks at the use of AI in contextual access control, demonstrating how real-time assessments of access requests may be tailored to specific user conditions. The ability of AI to process and analyze large datasets is emphasized, indicating its critical role in enhancing security incident response tactics [1]. The AI in Data Loss Prevention (DLP) section investigates the relationship between AI and data loss prevention initiatives, demonstrating how AI can dynamically categorize and secure sensitive data in real time. Concrete instances of AI-driven DLP systems demonstrate their effectiveness in protecting important information. The transformational potential of AI in developing adaptive access policy is investigated here. The chapter emphasizes the dynamic nature of access control rules impacted by real-time AI-driven risk assessments, assuring compliance and security efficiently. This section focuses on the role of AI in automating labor-intensive security jobs. It discusses how AI-driven procedures might enable faster responses to security issues while decreasing manual intervention, hence increasing overall efficiency. The convergence of AI and cloud security is examined, demonstrating how AI's real-time monitoring and anomaly detection capabilities strengthen cloud security.

20.1.1 BRIEF OVERVIEW OF ZERO TRUST SECURITY PRINCIPLES

Zero Trust is a security concept that fundamentally challenges the traditional perimeter-based cybersecurity strategy. In a conventional security approach, businesses normally trust people and devices after they enter their network perimeter and deploy security measures to keep dangers out [2]. However, in today's extremely dynamic and linked digital environment, this technique has proven insufficient. Zero Trust, on the other hand, is based on the premise that no entity, whether a person, device, or program, should be intrinsically trusted, regardless of its location. The following are the main principles of Zero Trust as shown in Figure 20.1.

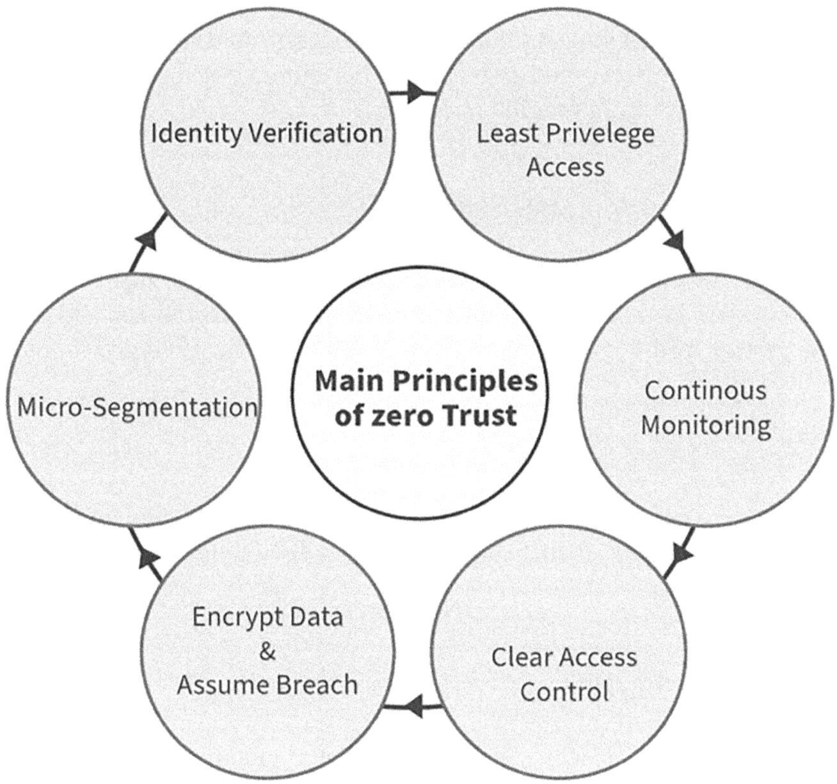

FIGURE 20.1 Main principles of zero trust.

20.1.1.1 IDENTITY VERIFICATION

Zero Trust highlights the importance of continually verifying the identity of every person and device attempting to access resources, regardless of location. As a result, strong authentication mechanisms like multi-factor authentication (MFA) are critical components of the system.

20.1.1.2 LEAST PRIVILEGE ACCESS

Users and devices should only be provided the bare minimum of access required to complete their responsibilities. This approach restricts the possible damage that a security compromise might produce.

20.1.1.3 MICRO-SEGMENTATION

Networks are divided into smaller, isolated zones, with traffic between these zones strictly monitored and regulated. This minimizes the attack surface and stops attackers from moving laterally.

20.1.1.4 CONTINUOUS MONITORING

It is critical to continuously monitor user and device activity as well as network traffic in order to spot abnormalities and potential security risks. Any departure from regular conduct may result in notifications and inquiries.

20.1.1.5 CLEAR ACCESS CONTROL

Resource access is granted based on clear regulations and context. Policies consider not just user identification, but also device health, location, and behavior.

20.1.1.6 ENCRYPT DATA

To prevent unwanted access, data should be encrypted both in transit and at rest.

20.1.1.7 ASSUME BREACH

Zero Trust is built on the premise that breaches may and will occur. As a result, the emphasis moves from simply preventing breaches to swiftly recognizing and responding to them. Implementing Zero Trust principles can be difficult since it necessitates a confluence of technology, policy, and cultural adjustments within an organization. It is a comprehensive security strategy that matches with the increasing threat landscape and the requirement for increased protection against both internal and external threats.

20.2 THE INCREASING IMPORTANCE OF AI IN MODERN SECURITY

The growing significance of AI in modern security reflects the increasing complexity and size of cybersecurity concerns [3, 4]. For numerous reasons, AI technologies have become crucial to security strategies (*Refer* Figure 20.2).

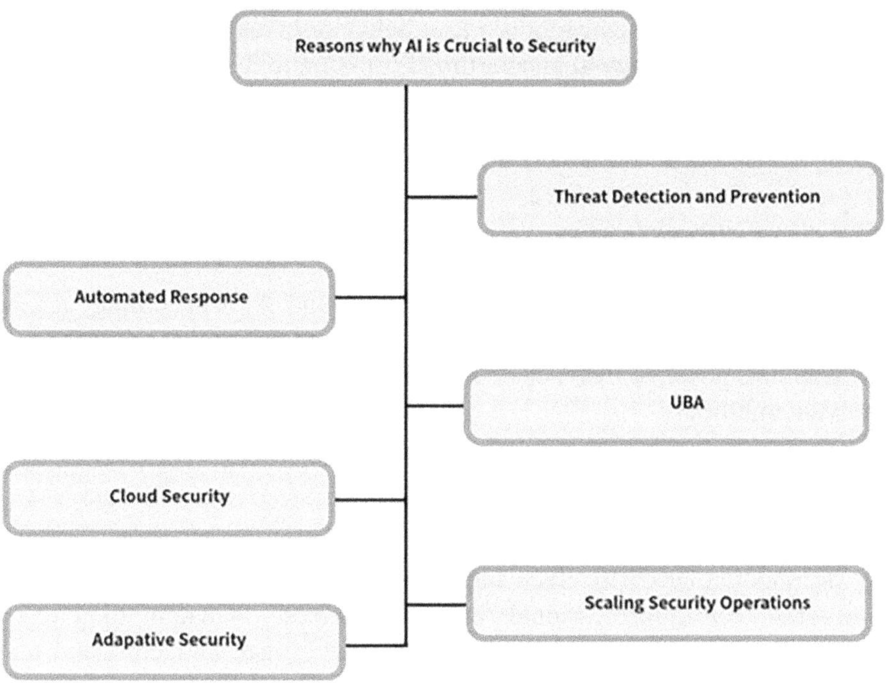

FIGURE 20.2 Reasons why AI is crucial to security.

20.2.1 THREAT DETECTION AND PREVENTION

Because AI excels at evaluating large datasets in real-time, it is extremely successful at finding patterns suggestive of cyber risks. Machine learning algorithms can detect abnormalities and deviations from typical behavior, assisting in the detection of complex threats that traditional signature-based systems may miss.

20.2.2 AUTOMATED RESPONSE

AI provides automated responses to security problems. When a danger is recognized, AI-powered systems may perform specified actions, such as isolating compromised devices or blocking harmful communications, without the need for human participation. This quick response can assist to lessen the consequences of security breaches.

20.2.3 UBA

AI-powered UBA systems monitor user behavior in real time and can detect aberrations that may signal insider threats or compromised accounts. This is especially useful in the context of Zero Trust, where user identification and conduct must be verified continuously.

20.2.4 CLOUD SECURITY

As businesses shift their data and apps to the cloud, AI is becoming increasingly important in monitoring and safeguarding cloud settings. Cloud security solutions powered by AI can detect and respond to cloud-specific threats and vulnerabilities.

20.2.5 SCALING SECURITY OPERATIONS

As the amount of security alerts and events grows, AI-powered technologies may assist security professionals in more efficiently managing their workloads. AI can help analysts prioritize warnings, triage incidents, and give important insights.

20.2.6 ADAPTIVE SECURITY

AI can adjust security measures in real-time in response to changing threat landscapes. This flexibility is particularly important in the context of Zero Trust, where security policies and access control rules must be responsive and adaptable.

"A proactive and dynamic approach to cybersecurity is exemplified by the integration of AI with current security methods, such as Zero Trust. It helps enterprises to effectively fight against a wide range of threats while also simplifying security processes and decreasing incident response time. As AI technologies advance, they will almost certainly play a larger role in safeguarding digital assets and data in the future.

20.3 APPLICATIONS OF AI IN ZERO TRUST LEARNING

The introduction of the Zero Trust Learning (ZTL) paradigm symbolizes a paradigm shift in education, defined by individualized and adaptable learning experiences. The integration of AI, which revolutionizes traditional teaching approaches, is at the heart of this transformation [4, 5]. This section delves into various AI applications within the framework of ZTL, offering insight on how AI-driven technologies are transforming the learning environment.

20.3.1 AI-POWERED INTELLIGENT TUTORING SYSTEMS (ITS)

The development of ITS is a major application of AI in ZTL. These systems use machine learning algorithms to provide students with dynamic and individualized learning experiences. ITS, unlike traditional educational systems, does not rely on a one-size-fits-all content delivery model. Instead, it tailors the learning process to each student's unique requirements, preferences, and progress.

20.3.2 AI-ENHANCED USER AUTHENTICATION

In the world of ZTL, strong user authentication is critical for protecting important educational data and resources. AI plays a crucial role in enhancing user authentication procedures on educational platforms. Traditional authentication techniques, such as passwords or PINs, have flaws, while AI-powered authentication strengthens security by introducing dynamic and MFA procedures.

20.3.3 CONTINUOUS MONITORING AND AI-DRIVEN UBA

Continuous monitoring is a core principle of ZTL, and AI-driven UBA plays an important role in this regard. User behaviors, interactions, and activities inside educational platforms are constantly monitored by UBA systems.

These AI systems assess user activity in real time, indicating deviations or aberrant trends that may signal security issues or academic dishonesty.

20.3.4 AI IN PERSONALIZED CONTENT DELIVERY

One of ZTL's main objectives is to provide students with tailored learning experiences. Individual students' learning habits, skills, and shortcomings are examined by AI algorithms. Using this data, educational platforms may provide customized content, activities, and evaluations to students, ensuring that they receive the most relevant and effective learning resources. Furthermore, based on real-time performance data, AI may dynamically adjust the complexity of assignments, offering challenges that fit the student's current level of comprehension.

20.3.5 AI-ENHANCED STUDENT SUPPORT

AI-powered chatbots and virtual assistants are becoming more common in the world of ZTL. These technologies provide students with rapid help and advice, answering questions, resolving concerns, and explaining concepts. They can even provide comments on assignments and exams in real time. As a result, students may get help whenever they need it, encouraging self-directed learning and easing the strain on instructors.

20.3.6 AI-DRIVEN LEARNING ANALYTICS

AI-powered learning analytics enable schools to acquire insights about their students' performance, engagement, and areas of difficulty. These analytics make use of AI algorithms to process massive volumes of data generated by students' interactions with educational platforms. Educators may utilize this information to make data-driven choices, identify students who may need extra help, and improve course materials to enhance learning outcomes. The incorporation of AI into the ZTL paradigm is altering the educational environment. Intelligent Tutoring Systems deliver adaptive learning experiences, AI-powered authentication improves security, and constant monitoring using UBA maintains educational systems' integrity [6]. AI also enables individualized material distribution, provides virtual assistants to students, and offers useful learning metrics. These applications herald the start of a new age in education, one in which personalized learning, adaptive assessment, and robust security coexist, ultimately benefiting both educators and learners. As ZTL evolves, AI will play an increasingly important role in determining the future of education.

Robust security measures are required in the context of ZTL to secure educational resources and sensitive student data. This section investigates how AI might improve user identification, access management, and continuous monitoring, hence increasing educational platforms' security posture.

20.3.7 AI-POWERED USER AUTHENTICATION

In order to protect educational resources, user authentication is the first line of security. By implementing dynamic and MFA systems, AI may greatly improve user authentication such as:

1. **Behavioral Biometrics:** AI can develop unique biometric profiles by analyzing human behavior patterns such as typing speed, keystroke dynamics, and mouse movements. By continually confirming user identities based on their behaviors, behavioral biometrics give an extra degree of protection.
2. **Facial Recognition:** AI-powered facial recognition may be used for user verification. AI algorithms may validate a user's identification in real-time by studying facial characteristics and expressions, making it impossible for unauthorized users to acquire access.

AI-powered voice recognition systems can authenticate users based on their voiceprints. These systems evaluate and match voice features.

20.3.8 AI IN ACCESS CONTROL

Access control is a key component of ZTL, with the goal of restricting user access based on the concept of least privilege. AI improves access control by adding context-aware policies and real-time evaluations include:

1. **Contextual Access Policies:** When analyzing access requests, AI algorithms consider numerous contextual aspects such as user location, device health, and time of access. This dynamic strategy guarantees that access is provided or refused based on the circumstances, which reduces the danger of unwanted access.

AI continually monitors user behavior and environmental conditions in real time. Any divergence from usual behavior patterns causes real-time risk evaluations to be performed. If AI identifies questionable actions or access attempts, it can urge extra authentication procedures or restrict access until the user's identity is validated.

20.3.9 CONTINUOUS MONITORING AND AI-DRIVEN UBA

Continuous monitoring is a basic ZTL concept, and AI-driven UBA plays an important role in this regard [7]. AI continually analyzes user activity, network traffic, and system behavior in real time. This proactive method enables AI to spot abnormalities and possible security problems as they arise.

1. **Aberrant Behavior Detection:** AI-powered UBA systems examine user behavior patterns and departures from the norm. If a person shows strange behavior, such as accessing resources from an unusual place or time, AI can flag these behaviors for further examination.
2. **Threat Identification:** AI-powered UBA can detect abnormal user activity to identify possible dangers such as insider threats or compromised accounts [8]. This proactive threat detection enables educational institutions to respond quickly to security threats.

The use of AI into user authentication, access control, and continuous monitoring dramatically improves educational platforms' security posture in the context of ZTL. Behavioral biometrics and face recognition, for example, are AI-powered authentication technologies that enable strong identity verification. Access is allowed depending on current conditions thanks to context-aware access control policies and dynamic rule changes. Continuous monitoring with AI-powered UBA allows for real-time threat identification and preemptive security solutions.

As ZTL evolves, AI will become increasingly important in safeguarding educational resources and promoting a secure and conducive learning environment. The marriage of AI with Zero Trust principles creates a potent synergy that enables educational institutions to embrace the future of safe and adaptable learning.

20.4 INTELLIGENT TUTORING SYSTEMS

Education is experiencing a paradigm transition, fueled by advances in AI and machine learning technology. One of the most promising uses of AI in education is the creation of ITS, which provide students with tailored learning experiences. These systems use AI algorithms to adjust to individual student requirements, preferences, and progress, ultimately transforming education. We will look at how AI-driven Intelligent Tutoring Systems may deliver individualized learning experiences, the benefits they provide, and the obstacles they may encounter in this study.

20.4.1 PERSONALIZATION IN EDUCATION

Personalized learning acknowledges that each student is unique and learns at their own rate, rendering the one-size-fits-all approach to education ineffective. Traditional classroom environments frequently struggle to meet students' diverse needs, which can lead to disengagement and decreased academic success. AI-powered Intelligent Tutoring Systems overcome this issue by personalizing the learning experience for each learner.

20.4.2 ADAPTIVE CONTENT DELIVERY

One of the most important characteristics of AI-driven ITS is the capacity to customize content delivery depending on individual student performance and preferences. These systems constantly analyze a student's knowledge, skills, limitations, and learning style, and then adjust the curriculum as needed. For example, if a student excels in one subject, the system might assign more difficult material to keep them engaged and motivated. In contrast, if a learner is having difficulty with a given topic, the system can provide additional materials and practice tasks until mastery is reached.

20.4.3 REAL-TIME FEEDBACK AND ASSESSMENT

AI-powered ITS give students rapid and constructive feedback, which is critical for their learning process. In contrast to typical classroom settings, where feedback may be delayed, these systems can identify errors or misunderstandings as they occur, allowing students to fix their faults in real-time. This immediate feedback increases not just learning outcomes but also student motivation by offering a sense of success.

20.4.4 ADAPTIVE LEARNING PATHS

AI-powered ITS provide adaptable learning routes, ensuring that students follow a course that is tailored to their specific learning needs. These systems can determine whether a learner is ready to move on to more sophisticated ideas or whether they need more practice. This adaptability ensures that students are not overwhelmed or stifled, providing a sense of autonomy and control over their learning experience.

20.4.5 DATA-DRIVEN INSIGHTS

Intelligent Tutoring Systems collect massive volumes of information on student performance. This information can be evaluated to understand more about individual and group learning tendencies. Teachers and educators can utilize this data to make informed curriculum changes, identify areas where students may need extra help, and even forecast learning issues before they become major impediments.

20.4.6 OVERCOMING CHALLENGES

While AI-powered Intelligent Tutoring Systems have enormous promise, they also confront a number of obstacles [9, 10]. Concerns regarding data privacy, fairness in technological access, and the necessity for experienced instructors to properly incorporate these tools into the classroom are among these. Addressing these issues is critical to ensuring that all students can reap the benefits of individualized learning.

20.5 AI-POWERED AUTHENTICATION IN EDUCATION

AI-powered authentication is becoming increasingly important in improving user security and convenience on educational platforms. Traditional authentication techniques, such as passwords and PINs, are frequently vulnerable to breaches, and the stakes are especially high in the education sector, where sensitive student and teacher data is at stake. AI provides a complex solution to these problems.

To begin, AI-driven authentication analyzes a user's activity, such as typing patterns, device recognition, and biometrics, to generate a unique user profile. This multifactor authentication not only improves security but also simplifies the user experience by eliminating the need for long passwords and frequent password changes. AI can identify abnormalities and respond quickly to possible security risks by continually learning and adapting to the user's behavior.

Additionally, AI-driven UBA is useful in real-time risk assessment. UBA tracks and analyzes user activity on the educational platform, looking for odd trends or departures from expected behavior [11]. This real-time analysis allows for quick reactions to possible problems like illegal access or suspicious activities. AI can analyze the risk level of each user interaction and dynamically alter authentication requirements by integrating historical data and real-time monitoring. This strategy not only improves security but also provides a consistent user experience by subjecting authorized users to fewer authentication hurdles, whereas suspicious actions trigger additional verification processes.

Finally, AI-powered authentication and UBA are changing educational platform security. They offer a strong defense against possible dangers while also simplifying the user experience. As educational institutions continue to implement AI-driven solutions, they will be able to protect sensitive data while still maintaining the integrity of their online learning environments.

20.6 AI FOR THREAT DETECTION IN EDUCATIONAL ENVIRONMENTS

In recent years, the education sector has seen an increase in the demand for powerful threat detection techniques to safeguard students, faculty, and sensitive data from a variety of threats, including cyberattacks and physical security breaches. AI has emerged as a valuable ally in enhancing educational institution security measures. AI-powered threat detection systems provide a holistic approach to recognizing and mitigating possible dangers, making them a useful tool in today's increasingly complex educational contexts.

The analysis of massive volumes of data is one of the key ways AI helps to improve danger detection in educational institutions [12–14]. In real time, machine learning algorithms may scan and interpret data from a variety of sources, including security cameras, access control systems, network traffic records, and even social media. AI can spot unexpected patterns or activities that may suggest a possible threat by continually examining this data. AI may, for example, detect unauthorized users seeking to gain access to restricted areas, detect strange network traffic patterns indicative of a cyberattack, and even monitor social media posts for terms connected with violence or threats. Real-world examples show how AI-driven threat detection works in educational contexts. Consider the following scenario: a huge institution with many campuses. They put in place an AI-powered video surveillance system that employs face recognition to identify people on campus [15]. This technology can identify known dangers, such as expelled students or people with restraining orders, and promptly notify campus security. Furthermore, the system can identify and respond to suspicious conduct, such as lingering in prohibited areas or attempting to gain illegal entry to buildings, and send out real-time notifications. This proactive strategy has dramatically improved campus security and response times.

AI-driven threat detection is equally important in the field of cybersecurity. Large volumes of sensitive data, such as student records, financial information, and research data, are frequently stored in educational institutions. Cyberattacks on sensitive data are on the rise, necessitating the implementation of strong protection methods. AI-powered cybersecurity systems may continually monitor network traffic, detect unexpected patterns or abnormalities, and respond to possible attacks quickly [16]. For example, an AI-driven intrusion detection system (IDS) at a university can identify a suspicious surge in network traffic caused by an illegal device and instantly isolate that device to prevent a possible breach.

AI may also evaluate user activity to discover potential insider risks. AI-driven UBA revealed an employee who was stealing important research data in a case study of a famous academic organization. The technology detected odd data access patterns and marked the behavior for additional examination, resulting in the discovery and avoidance of a major data breach.

Finally, AI-powered threat detection systems have become critical tools for educational institutions in protecting their campuses, data, and communities. These systems use AI to handle and analyze massive volumes of data, allowing for real-time danger detection and response. Real-world examples and case studies demonstrate the real benefits of applying AI in educational settings, ranging from improved physical security to strengthened cybersecurity measures.

20.7 AI AND CLOUD SECURITY

Security has become a top priority in the shifting educational scene, as remote and online learning are becoming increasingly popular. In this context, the ZTL idea, inspired by the Zero Trust security model, highlights the necessity to continually check and authenticate people and devices before giving access to educational materials, independent of their location or network. Within the ZTL architecture, AI plays a critical role in enhancing cloud security [17–19]. AI's capacity to provide continuous and adaptive authentication is one of its most important contributions to ZTL. Traditional security solutions sometimes rely on static credentials, such as usernames and passwords, which may be compromised. AI-driven authentication solutions, on the other hand, assess a variety of characteristics, such as user behavior, location, device, and biometrics, to develop a dynamic user profile. This enables real-time risk assessment and detection of abnormalities that might indicate unwanted access. For example, if a student regularly logs in from a given place and then tries to access resources from a different location, AI can identify this as a possible security issue and request for extra verification.

Furthermore, AI-powered threat detection solutions are critical for protecting cloud resources. Because cloud systems are very dynamic and complex, they are appealing targets for cyberattacks. AI-powered security technologies can continually monitor network traffic, recognize unusual trends, and respond quickly to possible attacks [20–23]. For example, if an AI system identifies an abnormally large volume of data being transmitted from an educational cloud server, it may generate an alert and take preventive measures to reduce the danger of data exfiltration. AI is also important in preventing phishing and social engineering assaults. These attacks frequently target people via fraudulent emails or texts. AI-powered email security systems can scan incoming communications in real time, detect phishing efforts, and prevent harmful information.

In conclusion, the incorporation of AI into cloud security within the scope of ZTL is critical to maintaining the protection of educational resources and data [24]. It enables organizations to transcend beyond old security paradigms and adopt a proactive, constantly adaptable security approach. Educational institutions may build a safer online learning environment by employing AI for continuous authentication, threat detection, and phishing avoidance. Trust is created based on real-time, data-driven evaluations rather than static credentials [25]. As the educational environment embraces technology, the role of AI in cloud security remains critical for protecting important educational assets as well as the privacy of students and instructors.

20.8 CONCLUSION

The chapter emphasizes the crucial role of AI in increasing security within educational institutions in an era of fast technological breakthroughs and a trend toward remote and online learning. Educational platforms may apply AI-driven solutions to improve user authentication, access management, and real-time threat detection by embracing the concepts of ZTL, which stress continuous verification and authentication. Intelligent Tutoring Systems driven by AI transform individualized learning experiences, while behavioral biometrics, facial recognition, and contextual access control enhance security. Continuous monitoring and user behavior analytics provide proactive threat detection, allowing educational institutions to respond quickly to possible threats. Furthermore, the integration of AI with cloud security improves data protection and access management, supporting the credibility of online learning environments.

KEYWORDS

- **artificial intelligence**
- **cloud security**
- **data loss prevention**
- **intelligent tutoring systems**
- **machine learning algorithms**
- **security problems**
- **zero trust learning**
- **zero trust principles**

REFERENCES

1. Sousa, M. J., & Rocha, A. (2019). Digital learning: Developing skills for digital transformation of organizations. *Future Generation Computer Systems, 91*, 1–12. https://doi.org/10.1016/j.future.2018.08.035.
2. Buck, C., Olenberger, C., Schweizer, A., Völter, F., & Eymann, T. (2021). Never trust always verify: A multivocal literature review on current knowledge and research gaps of zero-trust. *Computers and Security, 110*, 102368. https://doi.org/10.1016/j.cose.2021.102368.

3. Wiafe, I., Koranteng, F. N., Obeng, E. N., Assyne, N., Wiafe, A., & Gulliver, S. R. (2020). Artificial intelligence for cybersecurity: A systematic mapping of literature. *IEEE Access, 8*, 25409–25432. https://doi.org/10.1109/ACCESS.2020.2974239.
4. Yan, X., & Wang, H. (2020). Survey on zero-trust network security. *Communications in Computer and Information Science, 1252*, 45–57. https://doi.org/10.1007/978-3-030-43117-0_6.
5. Xin, Y., et al. (2018). Machine learning and deep learning methods for cybersecurity. *IEEE Access, 6*, 22413–22434. https://doi.org/10.1109/ACCESS.2018.2837045.
6. Daniloff, N. (2013). Suzanne Massie trust but verify: Reagan Russia and me. *Journal of Cold War Studies, 18*(4), 380–382. https://doi.org/10.1162/JCWS_a_00408.
7. Liu, Q., Li, P., Zhao, W., Cai, W., Yu, S., & Leung, V. C. M. (2018). A survey on security threats and defensive techniques of machine learning: A data-driven view. *IEEE Access, 6*, 61353–61369. https://doi.org/10.1109/ACCESS.2018.2879894.
8. Musumeci, F., Fidanci, A. C., Paolucci, F., Cugini, F., & Tornatore, M. (2022). Machine-learning-enabled DDoS attacks detection in P4 programmable networks. *Journal of Network and Systems Management, 30*(1), 1–20. https://doi.org/10.1007/s10922-022-09571-3.
9. Ferretti, L., Magnanini, F., Andreolini, M., & Colajanni, M. (2021). Survivable zero trust for cloud computing environments. *Computers and Security, 110*, 102375. https://doi.org/10.1016/j.cose.2021.102375.
10. Noor, U., Anwar, Z., Amjad, T., & Choo, K. K. R. (2019). A machine learning-based FinTech cyber threat attribution framework using high-level indicators of compromise. *Future Generation Computer Systems, 96*, 129–140. https://doi.org/10.1016/j.future.2019.01.018.
11. Li, J. H. (2018). Cybersecurity meets artificial intelligence: A survey. *Frontiers of Information Technology and Electronic Engineering, 19*(12), 1559–1574. https://doi.org/10.1631/FITEE.1800623.
12. Embrey, B. (2020). The top three factors driving zero trust adoption. *Computer Fraud and Security, 2020*(9), 8–11. https://doi.org/10.1016/S1361-3723(20)30097-9.
13. Benzaid, C., & Taleb, T. (2020). AI-driven zero-touch network and service management in 5G and beyond: Challenges and research directions. *IEEE Network, 34*(2), 73–80. https://doi.org/10.1109/MNET.001.1900424.
14. Frost & Sullivan. (2017). *The 2017 Global Information Security Workforce Study: Benchmarking Workforce Capacity and Response to Cyber Risk.* Retrieved from: https://www.frost.com (accessed on 11 November 2024).
15. Taddeo, M., McCutcheon, T., & Floridi, L. (2019). Trusting artificial intelligence in cybersecurity is a double-edged sword. *Nature Machine Intelligence, 1*(12), 585–588. https://doi.org/10.1038/s41563-019-0461-6.
16. Kinyua, J., & Awuah, L. (2021). AI/ML in security orchestration automation and response: Future research directions. *Intelligent Automation and Soft Computing, 28*(2), 379–388. https://doi.org/10.32604/iasc.2021.014237.
17. Alevizos, L., Ta, V. T., & Eiza, M. H. (2022). Augmenting zero trust architecture to endpoints using blockchain: A state-of-the-art review. *Security and Privacy, 5*(1), e269. https://doi.org/10.1002/spy2.269.
18. Mita, R. (2020). Improving trust in a zero trust architecture (ZTA). *Getting it Right–Collaborating for Mission Success, 10*(4), 2. https://doi.org/10.1145/3443527.

19. National Institute of Standards and Technology. (2020). *Zero Trust Architecture (SP 800–207)*. Retrieved from: https://doi.org/10.6028/NIST.SP.800-207 (accessed on 11 November 2024).
20. Decusatis, C., Liengtiraphan, P., Sager, A., & Pinelli, M. (2016). Implementing zero trust cloud networks with transport access control and first packet authentication. *Proceedings of the IEEE International Conference on Cloud Computing, 1*, 15–22. https://doi.org/10.1109/CloudComp.2016.21.
21. Mehraj, S., & Banday, M. T. (2020). Establishing a zero trust strategy in cloud computing environment. *International Journal of Computer Applications, 178*(7), 23–28. https://doi.org/10.5120/ijca2020919426.
22. Cunningham, C. (2018). *The Zero Trust Extended (ZTX) Ecosystem. Cambridge, MA: Forrester.* Retrieved from: https://go.forrester.com/research/ (accessed on 11 November 2024).
23. Keeriyattil, S. (2019). Microsegmentation and zero trust: Introduction. *Zero Trust Networks with VMware NSX*, 1–8. https://www.vmware.com (accessed on 11 November 2024).
24. Hierons, R. (1999). Machine learning. *Tom M. Mitchell. Software Testing Verification and Reliability, 9*(3), 230–234. https://doi.org/10.1002/(SICI)1099-1689(199909)9:3<230::AID-STVR227>3.0.CO;2-T.
25. Ward, R., & Beyer, B. (2014). BeyondCorp: A new approach to enterprise security. *Login: The Magazine of USENIX & SAGE, 39*(6), 23–30. https://www.usenix.org/ (accessed on 11 November 2024).

CHAPTER 21

The Future of AI in Zero-Trust Learning

MAITRI DULLA and NEHA JAIN*

Department of Computer Engineering, MMCOE, Pune, Maharashtra, India

*Corresponding author

ABSTRACT

As education evolves to integrate advanced technologies, the convergence of Artificial Intelligence (AI) and Zero Trust Learning (ZTL) represents a transformative shift. This chapter explores how AI can amplify the principles of ZTL to create personalized, secure, and adaptive learning environments. By leveraging AI's capacity for real-time data analysis, anomaly detection, and customization, ZTL emphasizes the importance of continuous verification in learning interactions, mirroring its security roots. This fusion enables the development of AI-driven solutions such as identity authentication, adaptive content delivery, and predictive learning analytics. The chapter examines how these technologies can address challenges like algorithmic biases and over-reliance on automated systems while ensuring fairness and data privacy. It also highlights the enduring role of human educators, whose empathy and expertise remain irreplaceable in fostering meaningful learning experiences.

21.1 INTRODUCTION

In a future where education and technology are quickly combining, an exciting synergy is forming at the intersection of both domains: a combination of artificial intelligence (AI) and the notion of zero trust. We are heading into unknown areas as we investigate "The Future of AI in Zero Trust Learning

Zero-Trust Learning: Applications in Modern Network Security. Chetna Kaushal, Amandeep Kaur, Mohit Angurala, Aryan Chaudhary (Eds.)

(ZTL)," with the possibility of dramatic improvements in how we approach education, security, and customization.

Historically, AI has been associated with innovation in a variety of fields, transforming businesses with its capacity to mimic human cognition, identify patterns, and assist well-informed decision-making. But why bring together these seemingly disparate forces? The answer rests in the need for an educational paradigm that fosters resilience, adaptation, and critical thinking, as well as knowledge. Enter ZTL, an approach that extends verification procedures to knowledge acquisition, mirroring the security idea. Learners must be prepared with the tools to investigate, assess, and authenticate knowledge sources in a world overloaded with information and ever-changing risks.

This chapter's fundamental goal is to reveal the subtle marriage of AI and zero trust in the sphere of education. We will look at how AI may be used to strengthen the concepts of zero trust, providing an environment where learning is not just individualized but also safe. The possible applications range from AI-driven authentication methods that confirm identities to anomaly detection algorithms that monitor user activity.

21.2 THE CONVERGENCE OF AI AND ZERO TRUST LEARNING (ZTL): SHAPING THE FUTURE OF EDUCATION AND SECURITY

As we embark on this intellectual journey, each chapter will serve as a doorway into a different aspect of this confluence. We will trace the beginnings of ZTL and understand its key ideas. Moving forward, we will delve into the world of artificial intelligence (AI), investigating its potential to improve education through customization, adaptive learning, and other means.

Throughout our journey, we will envision AI-augmented learning environments that are geared to meet the different requirements of learners by bridging educational gaps and constructing individualized experiences. Nonetheless, with each step AI makes, new obstacles and limits emerge. We will explore these barriers, from algorithmic biases to the possible hazards of over-reliance on AI.

As we near the end of our journey, we will look to the future, speculating on the progress of AI and ZTL. Our vista will be a mix of excitement and reflection, from the integration of developing technology to the potential alterations in educational institutions. The journey ahead promises to combine new thinking with practical implementations.

As we move forward in the chapters, we invite you to join us in this examination of AI and ZTL, establishing links between technological advancement and educational enrichment. Our goal with this synthesis is to create a colorful image of a future in which learning is not only dynamic and individualized but also robust and safe (Figure 21.1).

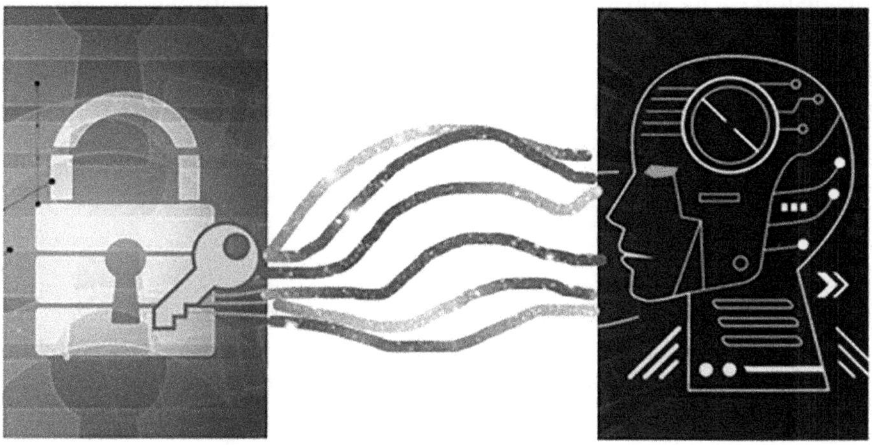

FIGURE 21.1 AI threats.

Source: ScienceSoftCybersecurity. https://www.scnsoft.com/security/ai-threats-cybersecurity

21.3 THE FUTURE OF AI-POWERED ZERO TRUST LEARNING

21.3.1 ENVISIONING THE EVOLUTION OF EDUCATION

As we traverse the terrain of AI integration in education, it's essential to peer into the horizon and anticipate the future trajectory. This chapter examines the future of AI-powered ZTL, exploring emerging technologies, potential shifts in educational paradigms, and the role of human educators in this dynamic landscape.

21.3.2 THE ACCELERATION OF TECHNOLOGICAL CONVERGENCE

The coming years promise an acceleration of technological convergence, with AI at the forefront. Emerging technologies such as virtual reality (VR), augmented reality (AR), and mixed reality (MR) hold the potential to transform education by immersing learners in dynamic and interactive environments. AI's ability to personalize these experiences further enhances the learning journey, as educators craft immersive lessons tailored to individual needs.

21.3.3 AI-DRIVEN LEARNING ANALYTICS

The fusion of AI and learning analytics heralds a future where educators gain unprecedented insights into learner behavior, engagement patterns, and performance trends. These insights enable educators to make informed decisions about instructional strategies, content adjustments, and personalized interventions. With AI-powered predictive analytics, educators can identify potential challenges before they arise, proactively addressing learning roadblocks.

21.3.4 LIFELONG LEARNING AND CONTINUOUS ADAPTATION

In the era of rapid technological advancement, the notion of education as a static endeavor is fading. AI-powered zero-trust learning paves the way for lifelong learning and continuous adaptation. Learners of all ages can access resources and engage in learning experiences that evolve alongside their interests and goals. This adaptive model aligns with the principles of zero trust, as learners continuously validate and update their knowledge.

21.3.5 THE ROLE OF HUMAN EDUCATORS

While AI has the potential to revolutionize education, the role of human educators remains indispensable. AI enhances efficiency, personalization, and data analysis, but human educators bring empathy, inspiration, and nuanced understanding to the educational process. The future envisions a collaborative partnership between AI-driven tools and human educators, where technology augments, rather than replaces, the human touch.

21.3.6 EXPANDING ACCESS TO EDUCATION

AI-driven ZTL has the potential to democratize education, bridging geographical, economic, and social gaps. Virtual classrooms, digital resources, and personalized learning experiences can reach learners in remote areas, underserved communities, and those with unique learning needs [1–28]. As AI-enhanced education becomes more accessible, the future of learning becomes more inclusive.

21.4 THE CHALLENGE OF ETHICAL AI

As AI becomes increasingly integrated into education, ethical considerations become more complex. Ensuring algorithmic fairness, safeguarding data privacy, and mitigating bias become critical priorities. Educators, policy-makers, and developers must collaborate to establish ethical guidelines that align with the principles of zero trust, ensuring that AI deployment is responsible and equitable.

21.5 PIONEERING THE FUTURE OF EDUCATION

The marriage of AI and zero trust principles illuminates a path toward a future where education is not only knowledge transfer but also a journey of continuous validation, personalization, and empowerment. This future envisions an educational ecosystem where AI-powered tools enhance the learning experience while upholding security, fairness, and privacy. The journey through the exploration of AI in ZTL has led us to a promising horizon—a horizon where education and technology converge to create a dynamic and secure ecosystem. The future of AI-powered ZTL envisions immersive experiences, personalized pathways, and an ongoing partnership between AI tools and human educators.

21.6 CONCLUSION AND REFLECTIONS

21.6.1 A JOURNEY THROUGH INNOVATION AND INTEGRITY

As we near the end of our exploration into the realm of AI and ZTL, it's an opportune moment to reflect on the insights gained, the promises unveiled, and the responsibilities that lie ahead. This chapter encapsulates the journey we've embarked upon, highlighting the key takeaways and encouraging contemplation on the future of education in a digital age.

21.6.2 A FUSION OF INNOVATION AND SECURITY

The fusion of AI and zero-trust principles has presented a unique opportunity to reimagine education in a way that marries innovation with security. Throughout this book, we've delved into the myriad ways AI can enhance learning environments, empower educators, and create personalized experiences. At the same time, we've explored how zero-trust principles, rooted in continuous validation and verification, can ensure the integrity of educational ecosystems.

21.6.3 NAVIGATING ETHICAL CONSIDERATIONS

The integration of AI into education is not without its ethical considerations. From algorithmic bias to data privacy and transparency, the responsible deployment of AI requires careful navigation of these concerns. Our exploration has emphasized the importance of aligning AI integration with ethical principles, ensuring that the benefits of technology are harnessed while respecting learner rights and values.

21.6.4 EMPOWERING LEARNERS AND EDUCATORS

At the heart of AI-powered zero-trust learning lies a shared goal: to empower both learners and educators. AI's ability to personalize learning experiences, provide real-time feedback, and enhance engagement is a testament to its potential to transform education. Equally important is the role of educators who bring their expertise, empathy, and mentorship to the learning journey.

21.6.5 A CONTINUOUS JOURNEY OF VERIFICATION

The concept of ZTL offers a departure from traditional paradigms of implicit trust. Instead, it introduces a continuous journey of verification and validation, mirroring the ever-evolving nature of the digital landscape. Just as learners continuously assess information and engage in critical thinking, educational institutions and AI-driven systems must maintain vigilance to ensure the trustworthiness of resources and interactions.

21.6.6 EMBRACING CHANGE AND INNOVATION

The future of education is one where change and innovation are embraced with open arms. The integration of AI-driven technologies invites us to envision a landscape where learning is not confined to traditional boundaries, but rather a dynamic and adaptive experience. This future hinges on our ability to adapt to technological advancements while upholding the core values of education.

21.6.6 CULTIVATING LIFELONG LEARNING

As we conclude our journey, let us recognize that education is a lifelong endeavor. The integration of AI and zero trust principles invites us to cultivate a mindset of continuous learning, where verification and validation extend beyond the realm of technology to encompass the evolution of knowledge and skills.

21.6.7 A CALL TO ACTION

The intersection of AI and ZTL is not just a theoretical exploration—it's a call to action. It invites educators, policymakers, developers, and learners to collaborate in shaping an educational landscape that is innovative, secure, and ethical. By leveraging the potential of AI while embracing the principles of verification and validation, we have the opportunity to create an educational ecosystem that empowers individuals to thrive in a rapidly changing world.

21.7 CONCLUSION

As we near the end of our intellectual journey through the realms of AI and ZTL, we find ourselves at a crossroads, reflecting on the insights learned and the potential shown. This trip has shown a way that combines innovation and security, providing a chance to reinvent education in the digital era.

The combination of AI with zero trust principles has opened up countless possibilities for improving learning environments, empowering instructors, and customizing experiences. We have seen the potential for AI to improve education by offering individualized learning experiences, real-time feedback, and enhanced engagement through these investigations. Parallel to this, the zero trust principles, which are distinguished by continual validation and verification, serve as guardians of educational integrity.

KEYWORDS

- **AI-augmented learning**
- **augmented reality**
- **data privacy**
- **mixed reality**
- **virtual reality**
- **zero trust learning**
- **zero-trust principles**

REFERENCES

1. Singh, R., Srivastav, G., Kashyap, R., & Vats, S. (2023). Study on zero-trust architecture, application areas & challenges of 6G technology in the future. In *Proceedings of ICDT 2023.*
2. Sedjelmaci, H., Senouci, S. M., Ansari, N., & Boualouache, A. (2022). A trusted hybrid learning approach to secure edge computing. *IEEE, 11*(3), 1–14. https://doi.org/10.1109/ACCESS.2022.3109832.
3. Wylde, A. (2023). Intelligent virtual assistants (IVAs): Trust and zero trust. In *Proceedings of the [Conference Name], 8*, 1–13. https://doi.org/10.1007/978-3-031-37963-5_8.
4. O'Neil, C. (2016). *Weapons of Math Destruction: How Big Data Increases Inequality and Threatens Democracy*. Broadway Books.
5. Müller, V. C. (2020). The ethics of artificial intelligence. *Stanford Encyclopedia of Philosophy*. https://plato.stanford.edu/entries/ethics-ai/ (accessed on 11 November 2024).
6. Doshi, N., Cebrian, M., & Larson, K. (2017). The spread of true and false news online. *Science, 359*(6380), 1146–1151. https://doi.org/10.1126/science.aap9559.
7. Ramezanpour, K., & Jagannath, J. (2022). Intelligent zero trust architecture for 5G/6G networks: Principles, challenges, and the role of machine learning in the context of O-RAN. *Computer Networks, 217*, 109358. https://doi.org/10.1016/j.comnet.2022.109358.
8. Brundage, M., Avin, S., Clark, J., Toner, H., Eckersley, P., Garfinkel, B., & Zeitzoff, T. (2018). *The Malicious Use of Artificial Intelligence: Forecasting, Prevention, and Mitigation. arXiv preprint arXiv:1802.07228. https://arxiv.org/abs/1802.07228.*
9. van der Walt, S., & Bernard, E. (2018). Effective cybersecurity using artificial intelligence and machine learning. *IT Professional, 20*(3), 35–41. https://doi.org/10.1109/MITP.2018.031021106.
10. Cebrian, M., & Larson, K. (2017). The spread of true and false news online. *Science, 359*(6380), 1146–1151. https://doi.org/10.1126/science.aap9559.

11. Qazi, F. A. (2022). Study of zero trust architecture for applications and network security. In *Proceedings of the 2022 IEEE 19th International Conference on Smart Communities: Improving Quality of Life Using ICT, IoT and AI (HONET)* (pp. 111–116). Marietta, GA, USA. https://doi.org/10.1109/HONET56683.2022.10019186.
12. *Ethics of Artificial Intelligence and Robotics*. (2020). *Stanford Encyclopedia of Philosophy*. https://plato.stanford.edu/entries/ethics-ai/ (accessed on 11 November 2024).
13. *Toward trustworthy AI development: Mechanisms for supporting verifiable Claims*. (2021). *National Academies of Sciences, Engineering, and Medicine*. https://doi.org/10.17226/25759.
14. *Digital learning*. (n.d.). *University of Leeds*. https://digitallearning.leeds.ac.uk (accessed on 11 November 2024).
15. Kindervag, J. (2010). *No More Chewy Centers: Introducing the Zero Trust Model of Information Security. Forrester Research*. https://www.forrester.com (accessed on 11 November 2024).
16. *International Society for Technology in Education (ISTE)*. (n.d.). *Organization*. https://www.iste.org (accessed on 11 November 2024).
17. *Journal of Artificial Intelligence in Education*. (n.d.). *Organization*. https://www.springer.com/journal/40593 (accessed on 11 November 2024).
18. *EdSurge*. (n.d.). *Organization*. https://www.edsurge.com (accessed on 11 November 2024).
19. *Educational Technology and Mobile Learning*. (n.d.). *Organization*. https://www.educatorstechnology.com (accessed on 11 November 2024).
20. Kurzweil, R. (2005). *The Singularity is Near: When Humans Transcend Biology*. Penguin Books.
21. O'Neil, C. (2016). *Weapons of Math Destruction: How Big Data Increases Inequality and Threatens Democracy*. Broadway Books.
22. Isaacson, W. (2014). *The Innovators: How a Group of Hackers, Geniuses, and Geeks Created the Digital Revolution*. Simon & Schuster.
23. Horn, M., & Staker, H. (2014). *Teaching and Learning in the Digital Age: Making the Case for a Strong Pedagogical Framework*. Harvard Education Press.
24. Egenrieder, J., O'Connor, K., & Koriak, L. (2015). E-learning, online education, and the use of social media in the classroom. *International Journal of Social Media and Interactive Learning Environments, 3*(1), 22–39. https://doi.org/10.1504/IJSMILE.2015.069051.
25. Kelly, K. (2017). *How AI Can Bring on a Second Industrial Revolution. TED Talks*. https://www.ted.com/talks/kevin_kelly_how_ai_can_bring_on_a_second_industrial_revolution (accessed on 11 November 2024).
26. Brown, B. (2010). *The Power of Vulnerability. TED Talks*. https://www.ted.com/talks/brene_brown_the_power_of_vulnerability (accessed on 11 November 2024).
27. Robinson, K. (2006). *Do SCHOOLS KILL CREATIVITY? TED Talks*. https://www.ted.com/talks/ken_robinson_do_schools_kill_creativity (accessed on 11 November 2024).
28. *Introduction to Machine Learning*. (n.d.). *Stanford University, Online Course*. https://online.stanford.edu/courses/cs229-machine-learning (accessed on 11 November 2024).

Index

B

D

E

Q

R

S

www.ingramcontent.com/pod-product-compliance
Ingram Content Group UK Ltd.
Pitfield, Milton Keynes, MK11 3LW, UK
UKHW022212190926
475702UK00003B/324
9781779643568